Topics in Galois Theory

Gail Gallitano and Shiv K. Gupta

Department of Mathematics

West Chester University

Kendall Hunt
publishing company

Kendall Hunt
publishing company

www.kendallhunt.com
Send all inquiries to:
4050 Westmark Drive
Dubuque, IA 52004-1840

ISBN 978-1-5249-6024-7

Published in the United States of America

Contents

Introduction and Acknowledgments

The main purpose of this book is to provide the readers with a basic understanding of Galois Theory especially over the field of characteristic zero. We shall follow a constructive approach and avoid excessive formalism. Since much of what we discuss is motivated by the corresponding results in the subfields of complex numbers, we often take the examples from the same. So unless otherwise explicitly stated all fields will be assumed to be of characteristic zero and mostly be subfields of the field of complex numbers. We shall assume that the reader is familiar with introductory abstract algebra, in particular with the basic elementary properties of groups, rings, integral domains, fields, vector spaces, and so forth.

This book is rich in illustrative examples and some, such as the one concerning the Galois correspondence of the polynomial $x^{16}-2$, spans over several pages which we have not seen in print. It took many weeks to write this example and having done so we wanted to share it with everyone interested in Galois groups.

Topics in Galois Theory has a concrete approach to problem-solving which, in our experience, is very much appreciated by our students. We have included over 100 problems at various levels of difficulty related to the topics discussed in this book. We have provided detailed solutions to some and hints for most of the others.

The first author has been teaching mathematics, mostly at the university level, for over 45 years. She taught at various institutions including Fairleigh Dickinson University, Teaneck, New Jersey, Montclair University, Montclair, New Jersey, Shippensburg University, Shippensburg, Pennsylvania, The University of Nevada, Las Vegas, Nevada, and West Chester University, West Chester, Pennsylvania. The second author has taught courses on Algebra (Group Theory, Galois Theory, Number Theory) and related topics at various institutions in India, Nigeria, and United States during his long teaching career.

In the bibliography, we have given a list of several books on Galois Theory and related topics which we have consulted, at one time or another, either while teaching or while preparing *Topics in Galois Theory.* There may be some omissions, misprints, and even mistakes in this book. We would of course appreciate learning about them from the readers and will acknowledge them promptly. It is hoped that the readers will find something of interest in our book.

We would also like to thank Michael Fisher, Clayton Garthwait, Peter Glidden, Richard Halpern, Clifford Johnston, James McLaughlin, Viorel Nitica, Lin Tan, Lawrence Washington, and Paul Wolfson who helped us in different ways that they may or may not remember.

—Gail Gallitano and Shiv Gupta

Notation

We shall use the following standard notations in these notes:

$\mathbb{N}$ = Set of natural numbers.

$\mathbb{Z}$ = Set of all integers.

$\mathbb{Q}$ = Set of all rational numbers.

$\mathbb{R}$ = Set of all real numbers.

$\mathbb{C}$ = Set of all complex numbers.

$\mathbb{Z}_n = \mathbb{Z}/n\mathbb{Z}$ (n a positive integer) is the ring of congruence classes modulo n. If n is a prime number p then $\mathbb{Z}_p = \mathbb{Z}/p\mathbb{Z}$ is a field.

$\mathbb{Z}_n^*$ = The multiplicative group of units in $\mathbb{Z}_n$.

C_n = Cyclic group of order n.

$\mathbb{F}[x]$ = Polynomial ring over a ring or a field $\mathbb{F}$.

$\mathbb{F}(x)$ = The field of rational function over a field $\mathbb{F}$.

$\mathbb{F}_q$ = Finite field of order q, where q is a power of a prime.

$\mathbb{F}_p$ or $\mathbb{Z}_p$ = Finite field of prime order p.

$\operatorname{Irr}(\alpha, \mathbb{F})$ = Irreducible polynomial of the algebraic number α over the field $\mathbb{F}$.

$\mathbb{L}/\mathbb{K}$ = The extension of the field $\mathbb{K}$ by the field $\mathbb{L}$.

$[\mathbb{L} : \mathbb{K}]$ = The dimension of the (field) $\mathbb{L}$ considered as a vector space over the field $\mathbb{K}$.

$\overline{\mathbb{Q}}$ = The algebraic closure of the field $\mathbb{Q}$ of rational numbers. Same as the field of all algebraic numbers.

${}_nC_r$ or $\binom{n}{r}$ = Binomial coefficient (Combinations of n taken r at a time).

S_n = Symmetric group on n letters.

A_n = Alternating group on n letters.

$\phi(n)$ = Euler's totient function.

$\Phi_n(x)$ = n-th cyclotomic polynomials.

CHAPTER 1

Preliminaries

Galois theory is certainly not the first course in algebra to be taught to the students. Therefore we shall assume some knowledge of introductory abstract algebra on the part of the readers of these notes. Specifically we shall assume that the reader is familiar with the following: The definitions of group, ring, integral domain, Euclidean domain, principal ideal domain, unique factorization, field, vector space, the dimension of a vector space, concepts of eigenvalues, eigen vectors and eigen space, polynomial ring over a field, the fact that if $\mathbb{F}$ is a field then $\mathbb{F}[x]$ the polynomial ring over $\mathbb{F}$ is an Euclidean domain, elementary number theory, properties of congruence, Fundamental Theorem of Algebra, the notions of homomorphism, isomorphism, division algorithm, Euclidean algorithm, and so on. Mostly we shall be working within the subfields of complex numbers.

1. Algebraic Numbers

Definition 1.1. A complex number α is said to be an algebraic number if there exists a nonconstant polynomial $f(x) \in \mathbb{Q}[x]$ such that $f(\alpha) = 0$.

Let α be an algebraic number. Then there exists a unique nonconstant polynomial $f_\alpha = f(x) \in \mathbb{Q}[x]$ satisfying the following conditions:

(i) α is a zero of $f(x)$.

(ii) $f(x)$ is irreducible over the field $\mathbb{Q}$ (that it cannot be expressed as a product $f(x) = f_1(x) \times f_2(x)$ of two smaller degree polynomials $f_1(x)$ and $f_2(x)$ in $\mathbb{Q}[x]$).

(iii) $f(x)$ is monic (that is its leading coefficient is unity).

This uniquely determined polynomial f_α is called the **minimal polynomial** of the algebraic number α and often also denoted by $Irr(\alpha, \mathbb{Q})$.

Note. If $f(x)$ is a monic irreducible polynomial in $\mathbb{Q}[x]$, then it is the minimal polynomial of each of its zeros. Minimal polynomial of an algebraic number α is also referred to in the literature as the irreducible polynomial of the algebraic number α.

It is easy to see that any complex number which can be expressed in terms of radicals (involving rational numbers only) is an algebraic number. For example, a number like $\alpha = \sqrt{2}+\sqrt{3}$ is an algebraic number as it is the zero of the polynomial $f(x) = x^4 - 10x^2 + 1$. There are algebraic numbers which cannot be expressed in terms of radicals.

If the degree of the minimal polynomial of an algebraic number α is k then we say that α is of degree k. We may be tempted to believe that all algebraic numbers can be expressed in terms of radicals. As we just mentioned this is *not* true. In fact using the so called *Fundamental Theorem of Galois Theory*, one can argue that the zeros of the polynomial $x^5 - 16x + 2$ (which are algebraic number by definition) cannot be written in terms of radicals.

One may not be easily able to find the degree of an algebraic number (given in the form of a radical expression) by simply looking at the radical expression for it. In fact, as the following table shows the appearance can often be rather deceptive.

Algebraic Number	Minimal Polynomial
$\sqrt{2}$	$x^2 - 2$
$\frac{1}{\sqrt{2}}$	$x^2 - \frac{1}{2}$
$\sqrt{2}+\sqrt{3}$	$x^4 - 10x^2 + 1$
$2\cos\frac{2\pi}{7}$	$x^3 + x^2 - 2x - 1$
$2\sin\frac{2\pi}{7}$	$x^6 - 7x^4 + 14x^2 - 7$
$\cos\frac{2\pi}{7}$	$x^3 + \frac{1}{2}x^2 - \frac{1}{2}x - \frac{1}{8}$
$2\cos\frac{2\pi}{5}$	$x^2 + x - 1$
$2\sin\frac{2\pi}{5}$	$x^4 - 5x^2 + 5$
$2^{\frac{1}{4}} + i$	$x^8 + 4x^6 + 2x^4 + 28x^2 + 1$
$2^{\frac{1}{3}} + \omega$	$x^6 + 3x^5 + 6x^4 + 3x^3 + 9x + 9$
$\sqrt[3]{2+\frac{10}{9}\sqrt{3}}+\sqrt[3]{2-\frac{10}{9}\sqrt{3}}$	$x - 2$
$\sqrt{2}+\sqrt{3}+\sqrt{5}$	$x^8 - 40x^6 + 352x^4 - 960x^2 + 576$

In the above table $\omega = e^{\frac{2\pi i}{3}} = \cos\frac{2\pi}{3} + i\sin\frac{2\pi}{3} = \dfrac{-1+\sqrt{-3}}{2}$ is a primitive cube root of unity.

Remark. What we have called an *algebraic number* are complex numbers which are algebraic over $\mathbb{Q}$. More generally if $\mathbb{F} \subseteq \mathbb{K}$ are fields, $\alpha \in \mathbb{K}$ is called algebraic over $\mathbb{F}$, if

α is a zero of a nonconstant polynomial in $\mathbb{F}[x]$. In such a case the uniquely determined monic irreducible polynomial $f(x) \in \mathbb{F}[x]$ having α as one of its roots is called the minimal polynomial of α over the field $\mathbb{F}$ and is also denoted as $\mathrm{Irr}(\alpha, \mathbb{F})$. In these notes we shall follow the convention that when we simply say that α is an algebraic number then we shall mean that α is algebraic over $\mathbb{Q}$. We shall now prove that the set of algebraic numbers form a field. The following lemma will be needed in the proof.

Lemma 1.1. *Eigenvalues of a (square) matrix over the field of rational numbers are algebraic numbers.*

Proof. Immediate consequence of the definition of an eigenvalue. □

Theorem 1.2. *Sum and product of algebraic numbers are algebraic numbers. If $\alpha \neq 0$ is an algebraic number, then so is $\dfrac{1}{\alpha}$. So the set of algebraic numbers form a field.*

Proof. Let α and β be algebraic numbers of degree m and n respectively and let $f(x)$ and $g(x)$ (respectively) be their minimal polynomials, where

$$f(x) = x^m + a_1 x^{m-1} + \cdots + a_m$$
$$g(x) = x^n + b_1 x^{n-1} + \cdots + b_n.$$

Since $f(\alpha) = g(\beta) = 0$ we can write

$$\alpha^m = -(a_1\alpha^{m-1} + \cdots + a_m)$$
$$\beta^n = -(b_1\beta^{n-1} + \cdots + b_n).$$

It follows that the powers of α higher than $(m-1)$ can be expressed as a linear combination with rational coefficients of $\{1, \alpha, \alpha^2, \cdots, \alpha^{m-1}\}$. Likewise the powers of β higher than $(n-1)$ can be expressed as a linear combination with rational coefficients of $\{1, \beta, \beta^2, \cdots, \beta^{n-1}\}$. Consider the set

$$S = \{\alpha^i\beta^j : 0 \leq i \leq m-1, \quad 0 \leq j \leq n-1\}$$

of $N = mn$ numbers. In view of the preceding remark we can express $\alpha^i\beta^j(\alpha+\beta)$ as a linear combination of members of the set S with rational coefficients. We order the members of S in any specific way and write them as $\{\gamma_k : 1 \leq k \leq N\}$. As stated above, we can write

$$\gamma_k(\alpha+\beta) = \sum_{l=1}^{N} a_{kl}\gamma_l, \qquad 1 \leq k \leq N$$

where a_{kl} are rational numbers. This can be written in matrix form as follows:

$$\begin{bmatrix} a_{11} & a_{12} & a_{13} & \cdots & \cdots & a_{1N} \\ a_{21} & a_{22} & a_{23} & \cdots & \cdots & a_{2N} \\ \vdots & \vdots & \vdots & \cdots & \cdots & \vdots \\ a_{N1} & a_{N2} & a_{N3} & \cdots & \cdots & a_{NN} \end{bmatrix} \cdot \begin{bmatrix} \gamma_1 \\ \gamma_2 \\ \vdots \\ \gamma_N \end{bmatrix} = (\alpha+\beta) \cdot \begin{bmatrix} \gamma_1 \\ \gamma_2 \\ \vdots \\ \gamma_N \end{bmatrix}.$$

From this it follows that $(\alpha+\beta)$ is an eigenvalue of the $N \times N$ rational matrix $[a_{kl}]$. Likewise by multiplying $\gamma_k' s$ by $\alpha\beta$ we observe that $\alpha\beta$ is also an eigenvalue of some $N \times N$ rational matrix. This shows that if α and β are algebraic numbers then so are $\alpha+\beta$ and $\alpha\beta$. Finally if $\alpha \neq 0$ is a zero of the polynomial

$$x^m + a_1 x^{m-1} + \cdots + a_m$$

then α^{-1} is a zero of the polynomial

$$a_m x^m + \cdots + a_1 x + 1.$$

This completes the proof of the theorem. □

Notes.

(i) The set of all algebraic numbers will be denoted by $\overline{\mathbb{Q}}$. This is called the algebraic closure of the field of rational numbers $\mathbb{Q}$. $\overline{\mathbb{Q}}$ is an algebraically closed field—which means that any polynomial which has coefficients in $\overline{\mathbb{Q}}$ has all of its zeros in $\overline{\mathbb{Q}}$. It can be easily shown that $\overline{\mathbb{Q}}$ is a countable set. As the set $\mathbb{C}$ of complex numbers is uncountable, it follows that there are numbers (real or complex) which are not algebraic numbers—namely the transcendental numbers.

(ii) If α is any complex number then the field generated by α is denoted by $\mathbb{Q}(\alpha)$. It is nothing but the quotient field of the integral domain $\mathbb{Q}[\alpha]$ which consists of all polynomials in α with rational coefficients. However, if α is an algebraic number then $\mathbb{Q}(\alpha) = \mathbb{Q}[\alpha]$ as a nonzero polynomial in α has an inverse which is also a polynomial in α. This can be shown as follows:

Let $f(x)$ be the minimal polynomial of the algebraic number α. Assume that $f(x)$ has degree m. Then

$$\mathbb{Q}[\alpha] = \{a_0 + a_1\alpha + \cdots + a_{m-1}\alpha^{m-1} : a_i \in \mathbb{Q}\}$$

as any power of α higher than $(m-1)$ can be written as a linear combination of $\{1, \alpha, \alpha^2, \cdots, \alpha^{m-1}\}$ with rational coefficients—since α is a zero of an m-th degree polynomial $f(x) \in \mathbb{Q}[x]$. Let $g(\alpha)$ be any nonzero element of $\mathbb{Q}[\alpha]$. Since $f(x)$ is minimal polynomial of α, it is irreducible, therefore $GCD(f(x), g(x)) = 1$. Since $\mathbb{Q}[x]$ is an Euclidean domain (and therefore a principal ideal domain), we can find polynomials $a(x)$ and $b(x)$ in $\mathbb{Q}[x]$ such that

$$a(x)f(x) + b(x)g(x) = 1.$$

Since $f(\alpha) = 0$ this gives, $b(\alpha)g(\alpha) = 1$, and so $b(\alpha) \in Q[\alpha]$ is the inverse of $g(\alpha)$.

(iii) If α is an algebraic number of degree m, then the set $\{1, \alpha, \alpha^2, \cdots, \alpha^{m-1}\}$ is linearly independent over $\mathbb{Q}$. This is so because the minimal polynomial of α has degree m. It follows that as a vector space over $\mathbb{Q}$, $\mathbb{Q}[\alpha] = \mathbb{Q}(\alpha)$ has dimension m.

(iv) Using the preceding remarks one can give another proof of the fact that the sum and product of algebraic numbers are algebraic numbers. First, if $\mathbb{K}$ and $\mathbb{L}$ are fields such that $\mathbb{K} \subseteq \mathbb{L}$, one can define (as in the case of $\mathbb{Q}$ and $\mathbb{C}$) the meaning of $\alpha \in \mathbb{L}$ being algebraic over $\mathbb{K}$ and the concept of minimal polynomial of α over $\mathbb{K}$. If $[\mathbb{L} : \mathbb{K}] = n$, and $\alpha \in \mathbb{L}$, then the set of $n+1$ elements $\{\alpha^i, 0 \le i \le n\}$ in $\mathbb{L}$ is linearly dependent over $\mathbb{K}$. This shows that if $\mathbb{K} \subseteq \mathbb{L}$ is a finite extension, then every element of $\mathbb{L}$, is algebraic over $\mathbb{K}$ and therefore a finite extension is an algebraic extension. Now suppose that α and β are algebraic numbers of degree m and n, respectively, over $\mathbb{Q}$. So $[\mathbb{Q}(\alpha) : \mathbb{Q}] = m$ and $[\mathbb{Q}(\beta) : \mathbb{Q}] = n$. Then $[\mathbb{Q}(\alpha, \beta) : \mathbb{Q}(\alpha)] \le n$—as the minimal polynomial of β over $\mathbb{Q}$ may factor over $\mathbb{Q}(\alpha)$. Therefore $[\mathbb{Q}(\alpha, \beta) : \mathbb{Q}] \le mn$. Since every element of a finite extension is algebraic over the base field (in this case $\mathbb{Q}$) it follows that $\alpha + \beta$ and $\alpha\beta$ are both algebraic numbers.

2. Algebraic Integers

Definition 2.1. A complex number α is said to be an algebraic integer if it is an algebraic number and its minimal polynomial f_α has integer coefficients.

We shall denote the set of all algebraic integers by $\mathbb{A}$. As the minimal polynomial of a rational number α is $x - \alpha$, it follows that only algebraic integers among the rational numbers are integers. In other words, $\mathbb{Q} \cap \mathbb{A} = \mathbb{Z}$.
The sum and product of two algebraic integers are algebraic integers. The proof of this statement is similar to the proof of the fact that sum and product of two algebraic numbers is an algebraic number. So the set $\mathbb{A}$ of all algebraic integers is a ring. However the set of algebraic integers do not form a field. The reciprocal of an algebraic integer many not be an algebraic integer. For instance $\frac{1}{2}$ is not an algebraic integer.

3. Finite Fields

Let $\mathbb{K}$ be a field . Let e denote its multiplicative identity. Consider the mapping

$$\eta : \mathbb{Z} \longrightarrow \mathbb{K}, \qquad \eta(n) = n \cdot e.$$

Evidently, η is a ring homomorphism. If kernel $\eta = \{0\}$, then η is one-to-one and the field $\mathbb{K}$ contains an integral domain isomorphic to $\mathbb{Z}$. The field of fractions of this integral domain is a field isomorphic to the field $\mathbb{Q}$ of rational numbers. If such is the case then we say that the field $\mathbb{K}$ has characteristic zero. In other words, a field $\mathbb{K}$ is said to have characteristic zero if and only if it contains a subfield isomorphic to the field $\mathbb{Q}$ of rational numbers.
If kernel $\eta \neq 0$, then the least positive integer n such that $\eta(n) = 0$ must be a prime number and if p is this prime number then kernel $\eta = p\mathbb{Z}$. Then the field $\mathbb{K}$ contains a subfield isomorphic to the finite filed $\mathbb{Z}_p = \mathbb{Z}/p\mathbb{Z}$, which we shall also denote by $\mathbb{F}_p$. In the field $\mathbb{Z}_p$ the two binary operations are addition and multiplication modulo the prime p. If such is the case we say that the field $\mathbb{K}$ has characteristic p. So the characteristic of a field is either zero or a prime number. In the former case the field contains as a subfield isomorphic to the field $\mathbb{Q}$ of rational numbers and in the latter case the field $\mathbb{K}$ contains an isomorphic copy of the finite field $\mathbb{Z}_p$. In what follows we shall ignore the technicality and assume that a field of characteristic zero contains $\mathbb{Q}$ as a subfield and a field of characteristic p contains the finite field $\mathbb{Z}_p$ as a subfield.

Let $\mathbb{K}$ and $\mathbb{L}$ be fields such that $\mathbb{K} \subseteq \mathbb{L}$. Then $\mathbb{L}$ can be regarded as a vector space over K, and we shall say that $\mathbb{L}/\mathbb{K}$ is a field extension. We denote the dimension of $\mathbb{L}$ as a vector space over $\mathbb{K}$ by $[\mathbb{L} : \mathbb{K}]$. If $[\mathbb{L} : \mathbb{K}]$ is finite then we shall say that $\mathbb{L}/\mathbb{K}$ is a finite extension. A sequence of fields such as

$$\mathbb{K}_1 \subseteq \mathbb{K}_2 \subseteq \cdots \mathbb{K}_{r-1} \subseteq \mathbb{K}_r$$

will be referred to as the **Tower of Fields**. The aforementioned observation that a finite field necessarily contains $\mathbb{Z}_p$ as a subfield leads to the following important theorem.

Proposition 3.1. *The number of elements in a finite field is a power of a prime number.*

Proof. Let $\mathbb{K}$ be a finite field. Then there is a prime p such that $\mathbb{K}$ contains $\mathbb{F}_p$ as a subfield. Since $\mathbb{K}$ has only finitely many elements the dimension of $\mathbb{K}$ as a vector space over $\mathbb{F}_p$ is finite. Let $[\mathbb{K} : \mathbb{F}_p] = n$. Let $u_1, u_2, \cdots, u_n$ be a basis of $\mathbb{K}$ over $\mathbb{F}_p$. Then every element of $\mathbb{K}$ can be uniquely written as

$$a_1u_1 + a_2u_2 + \cdots + a_nu_n$$

where $a_i \in \mathbb{F}_p$, $1 \leq i \leq n$. Since each a_i can take p possible values there are p^n such possible (different) expressions. Therefore $\mathbb{K}$ has precisely p^n elements. □

Note. For a prime p and a positive integer n, there is a finite field of order p^n and any two finite fields of order p^n are isomorphic.

Proposition 3.2. *The multiplicative group of a finite field is cyclic.*

Proof. Let $\mathbb{F}_q$ be a finite field of order q, so q is a power of a prime number. Let $G = \mathbb{F}_q^*$ be the multiplicative group of nonzero elements of the field $\mathbb{F}_q$. So G has order $q - 1$. We shall show that for every prime p dividing $q - 1$ the Sylow p-subgroup of G is cyclic. Let $q - 1 = p^k m$ where p does not divide m and P be a subgroup of G of order p^k. If P is not cyclic then every element of P has order dividing p^{k-1} and therefore every element of P is a root of the polynomial $f(x) = x^{p^{k-1}} - 1 \in \mathbb{F}_q[x]$. This would imply that the polynomial $f(x) \in \mathbb{F}_q[x]$ of degree p^{k-1} has at least p^k roots. This is not possible as the number of roots of any nonzero polynomial over a field cannot exceed its degree. Therefore P must be a cyclic group. G is an Abelian group whose every Sylow p-subgroup is cyclic and so G itself must be cyclic. □

CHAPTER 2

Field Embedding and Extensions

Let $\mathbb{L}/\mathbb{K}$ be a field extension. As mentioned earlier, $\mathbb{L}$ can be regarded as a vector space over $\mathbb{K}$. We recall that $\mathbb{L}$ is a finite extension of $\mathbb{K}$ if as a vector space $\mathbb{L}$ is of finite dimension over $\mathbb{K}$. If $\alpha_1, \alpha_2, \cdots, \alpha_n \in \mathbb{C}$ and $\mathbb{K}$ is a subfield of $\mathbb{C}$ then $\mathbb{K}(\alpha_1, \alpha_2, \cdots, \alpha_n)$ denotes the smallest subfield of $\mathbb{C}$ containing the subfield $\mathbb{K}$ and the elements $\alpha_1, \alpha_2, \cdots, \alpha_n$. An extension $\mathbb{L}/\mathbb{K}$ is said to be an algebraic extension if every $\alpha \in \mathbb{L}$ is a root of some nonzero polynomial in $\mathbb{K}[x]$. An extension $\mathbb{L}/\mathbb{K}$ is said to be finitely generated if there exist $\alpha_1, \alpha_2, \cdots, \alpha_n \in \mathbb{L}$ such that

$$\mathbb{L} = \mathbb{K}(\alpha_1, \alpha_2, \cdots, \alpha_n).$$

A finitely generated extension need not be finite. For instance, $\mathbb{K} = \mathbb{Q}(\pi)$ is finitely generated over $\mathbb{Q}$ but is not a finite extension. The field extension $\mathbb{R}/\mathbb{Q}$, where $\mathbb{R}$ is the field of real numbers, is not even finitely generated. However, as we have seen, a finite extension is necessarily algebraic. The extension $\overline{\mathbb{Q}}/\mathbb{Q}$ is not finite but algebraic. A field extension generated by a single element is called a *simple extension*. For subfields of complex numbers it can be shown (as we shall do shortly) that a finitely generated algebraic extension is in fact a simple extension. This is an important and useful result. For example, it can be easily seen that $\mathbb{Q}(\sqrt{2}, \sqrt{3}) = \mathbb{Q}(\sqrt{2} + \sqrt{3})$ and $\mathbb{Q}(2^{\frac{1}{4}}, i) = \mathbb{Q}(2^{\frac{1}{4}} + i)$.

1. Field Embedding

Let $\mathbb{K}, \mathbb{L}$ be fields. A one-to-one mapping $\sigma : \mathbb{K} \longrightarrow \mathbb{L}$ satisfying

$$\sigma(x + y) = \sigma(x) + \sigma(y), \qquad \sigma(xy) = \sigma(x)\sigma(y)$$

is called an embedding of the field $\mathbb{K}$ into the field $\mathbb{L}$. In particular, if $\mathbb{K}$ is a subfield of the filed $\mathbb{L}$ then the identity mapping of $\mathbb{K}$ into $\mathbb{L}$ is an embedding. The only embedding of the field $\mathbb{Q}$ into $\mathbb{C}$ (or $\mathbb{R}$) is the identity embedding. There are two embedding of $\mathbb{Q}(\sqrt{2})$ into $\mathbb{C}$. Each of these embedding is specified by its effect on $\sqrt{2}$. One sends $\sqrt{2}$ to itself (and is therefore the identity embedding) and the other sends $\sqrt{2}$ to $-\sqrt{2}$. If α is an algebraic number of degree n then the field $\mathbb{Q}(\alpha)$ has n embedding into $\mathbb{C}$. Each of these n embedding is characterized by the image of α under it—which can be any of the n zeros of its minimal polynomial. For instance, if $\alpha = 2^{\frac{1}{4}}$, its minimal polynomial over $\mathbb{Q}$, namely $x^4 - 2$ has four zeros, namely $\pm 2^{\frac{1}{4}}$ and $\pm i2^{\frac{1}{4}}$. The field $\mathbb{Q}(\alpha)$ has four embedding into $\mathbb{C}$, corresponding to the four different ways of mapping $\alpha = 2^{\frac{1}{4}}$ onto the zeros of $x^4 - 2$. We shall now prove the following theorem.

Theorem 1.1. *Let $\mathbb{K}$ be a subfield of complex numbers and α be algebraic over $\mathbb{K}$ of degree n. Then an embedding σ of $\mathbb{K}$ into $\mathbb{C}$ has exactly n extensions to embedding of $\mathbb{K}(\alpha)$ into $\mathbb{C}$.*

Proof. Let $f(x)$ be the minimal polynomial of α over $\mathbb{K}$. Since α has degree n over $\mathbb{K}$, $f(x)$ has degree n. Let the zeros of $f(x)$ (in $\mathbb{C}$) be $\alpha = \alpha_1, \alpha_2, \cdots, \alpha_n$. The n mappings

$$\sigma_i : \mathbb{K}(\alpha) \longrightarrow \mathbb{C}$$

$$\sigma_i(\alpha) = \alpha_i, \qquad \sigma_i|\mathbb{K} = \sigma, \qquad 1 \leq i \leq n$$

(where each σ_i restricted to $\mathbb{K}$ coincides with σ) constitute n extensions of σ. These embedding are all distinct as they map α to different elements. Since α must be mapped to one of the zeros of $f(x)$, these are the only extensions of σ to $\mathbb{K}(\alpha)$. □

The following theorem known as *Primitive Element Theorem* essentially says that a finitely generated algebraic extension is a simple extension.

Theorem 1.2. *Let $\alpha, \beta \in \mathbb{C}$ be two algebraic numbers. Then there exists an algebraic number $\gamma \in \mathbb{C}$ such that $\mathbb{Q}(\alpha, \beta) = \mathbb{Q}(\gamma) = \mathbb{Q}[\gamma]$.*

Proof. Let α be algebraic of degree $m > 1$ and β be algebraic of degree $n > 1$ and $f_1(x), f_2(x) \in \mathbb{Q}[x]$ be the minimal polynomials of α and β respectively. Let $\alpha = \alpha_1, \alpha_2, \cdots, \alpha_m \in \mathbb{C}$, be the roots of $f_1(x)$ and $\beta = \beta_1, \beta_2, \cdots, \beta_n \in \mathbb{C}$ be the roots of $f_2(x)$. Consider the set S

$$S = \left\{ \frac{\alpha_i - \alpha}{\beta - \beta_j} : \; 1 \leq i \leq m, 2 \leq j \leq n \right\}.$$

So the set S has at most $m(n-1)$ elements. Let $q \in \mathbb{Q}$ be such that $q \notin S$. Let $\gamma = \alpha + q\beta$. We claim that $\mathbb{Q}(\alpha, \beta) = \mathbb{Q}(\gamma)$. Obviously $\mathbb{Q}(\gamma) \subseteq \mathbb{Q}(\alpha, \beta)$. So we only need to show that

$\mathbb{Q}(\alpha, \beta) \subseteq \mathbb{Q}(\gamma)$. We note that $\gamma \neq \alpha_i + q\beta_j$, $1 \leq i \leq m$, $2 \leq j \leq n$. For

$$\gamma = \alpha + q\beta = \alpha_i + q\beta_j \Longrightarrow q = \frac{\alpha_i - \alpha}{\beta - \beta_j}$$

which contradicts our choice of q. Let $g(x) = f_1(\gamma - qx) = f_1(\alpha + q\beta - qx) \in \mathbb{Q}(\gamma)[x]$. We note that $g(\beta) = f_1(\alpha + q\beta - q\beta) = f_1(\alpha) = 0$. So β is a zero of the polynomial $g(x)$ as well as of the polynomial $f_2(x)$. Therefore $(x - \beta)$ divides both $g(x)$ and $f_2(x)$. The two polynomials $g(x)$ and $f_2(x)$ do not have any other common root. For if β_j for some j, $2 \leq j \leq n$ is another common zero of these two polynomials then

$$0 = g(\beta_j) = f_1(\gamma - q\beta_j) = f_1(\alpha + q\beta - q\beta_j) \Longrightarrow \alpha + q\beta - q\beta_j = \alpha_i, \quad \text{for some } i, \quad 1 \leq i \leq m$$

This would imply that $q = \dfrac{\alpha_i - \alpha}{\beta - \beta_j}$ which is not true. Now $gcd(g(x), f_1(x)) = x - \beta \in \mathbb{Q}(\gamma)[x]$. So $\beta \in \mathbb{Q}(\gamma)$, and so $\alpha = \gamma - q\beta \in \mathbb{Q}(\gamma)$. Hence $\mathbb{Q}(\alpha, \beta) \subseteq \mathbb{Q}(\gamma)$. This completes the proof of the *Primitive Element Theorem.* □

Note. The above proof is valid if the field $\mathbb{Q}$ is replaced by any field $\mathbb{K}$ of characteristic zero and α, β are algebraic over $\mathbb{K}$.

The fact that for algebraic numbers α and β we can find an algebraic number γ such that $\mathbb{Q}(\alpha, \beta) = Q(\gamma)$, implies that α and β can be linearly expressed in terms powers of γ (that is a polynomial in γ). For instance, $\mathbb{Q}(\sqrt{2}, \sqrt{3}) = \mathbb{Q}(\sqrt{2} + \sqrt{3})$ and we have the following expressions for $\sqrt{2}$ and $\sqrt{3}$ in terms of powers of $(\sqrt{2} + \sqrt{3})$.

$$\sqrt{2} = \frac{1}{2}\left(\sqrt{2} + \sqrt{3}\right)^3 - \frac{9}{2}\left(\sqrt{2} + \sqrt{3}\right)$$

$$\sqrt{3} = -\frac{1}{2}\left(\sqrt{2} + \sqrt{3}\right)^3 + \frac{11}{2}\left(\sqrt{2} + \sqrt{3}\right).$$

Let $\alpha = 2^{\frac{1}{4}} + \sqrt{-1} = 2^{\frac{1}{4}} + i$. Then

$$2^{\frac{1}{4}} = \frac{1}{24}(5\alpha^7 + 19\alpha^5 + 5\alpha^3 + 151\alpha)$$

and

$$i = -\frac{1}{24}(5\alpha^7 + 19\alpha^5 + 5\alpha^3 + 127\alpha)$$

2. Some General Comments about Algebraic Numbers

It should be clear that any number (real or complex) which can be expressed (starting with integers) by using $+, -, \times, \div$, and radicals (taking square roots, cube roots etc.) is an algebraic number. Sometimes a rather complicated looking expression of an algebraic number involving radicals turns out to be equal to some rational number. For instance, it is not too hard to see that the radical number

$$\sqrt[3]{2+\frac{10}{9}\sqrt{3}}+\sqrt[3]{2-\frac{10}{9}\sqrt{3}}$$

is equal to 2. On the other hand, it is rather easy to "give" examples of algebraic numbers which cannot be expressed in terms of radicals. The reason that we have put quotes around the word "give" in the previous sentence is that the only means that we have available to us for expressing a number is the use of algebraic operations ($+, -, \times, \div$, and radicals), and if some algebraic numbers happen to be such that they are not expressible using these algebraic operations then how we are to provide (i.e., communicate) examples of such algebraic numbers? The answer lies in the fact that we can give a mathematical proof that roots of certain polynomial equations (which certainly exist) with integral coefficients cannot be expressed by means of algebraic operations only. Attempts to find algebraic expressions for the roots of the equation (with integral coefficients) of degree five and higher resulted in failure and this in a way paved the way for the development of Galois Theory.

In any course on Galois Theory (or Abstract Algebra) there is some discussion of the so called constructible numbers. These are essentially the numbers which can be expressed (starting with integers) by means of algebraic operations but limiting the use of radicals to square root radicals only—that is not requiring cube roots, fifth roots, and so on. It should be noted that a fourth root is nothing but taking two square roots in succession. In view of this we can say that Gauss's proof that a regular 17-gon can be constructed was simply the demonstration that the length of the side of a regular 17-gon inscribed in a unit circle (equivalently the value of $\cos\frac{2\pi}{17}$) can be expressed by standard algebraic operations ($+, -, \times, \div$) and by using only the square root radicals—we repeat to emphasize that this means avoiding the use of cube roots and higher radicals other than the n-th roots where n is a power of two. The inclusion of the square root radicals in standard construction techniques (involving only straight edge and compass) is a consequence of the fact that with the help of a compass it is possible to draw a circular arc which represents a curve of degree two. If we were to allow the use of instruments which can draw curve of degree three or higher it may be possible to extend the class of constructible numbers. In particular, a constructible number is necessarily an algebraic number whose minimal polynomial has degree which is a power of two. The converse of this statement is not true as it is easy to find examples of algebraic numbers of degree four which are not constructible. In other words there are polynomials of degree four with integral coefficients (such as $x^4 - x - 1$) whose roots cannot be expressed using standard arithmetical operations and square root radicals only.

The impossibility of trisecting an arbitrary angle is an immediate consequence of the fact that for certain values of θ (such as $\theta = 20^o$), $\cos 3\theta$ is a constructible number (or simply a rational number) and $\cos\theta$ is a zero of a third degree irreducible polynomial. Since $\cos 3\theta = 4\cos^3\theta - 3\cos\theta$, if we let $\theta = 20^o$, and $\cos\theta = x$, then $8x^3 - 6x - 1 = 0$. As can be easily checked $8x^3 - 6x - 1$ is irreducible over $\mathbb{Q}$, its zeros (namely $\cos 20^o$, $\cos 140^o$, $\cos 260^o$) are not constructible, that is, not expressible in terms of square root radicals. In fact we shall later see that the roots of above equation (although all real) require taking the cube roots of complex numbers in their expression in terms of radicals.

The impossibility of "Doubling a Cube" can be similarly easily explained by the fact that, the length of the side of a cube whose volume is twice that of a unit cube will be a zero of the polynomial $x^3 - 2$, which is irreducible over $\mathbb{Q}$. As is easily seen the zeros of the polynomial $x^3 - 2$ cannot be expressed in terms of square root radicals.

Finally the impossibility of squaring a circle is equivalent to the fact that π (which is equal to the area of a unit circle) does not have any algebraic expression. In fact much more is true, as we shall see later, that π like e is a transcendental number—that is, it is *not* a root of any polynomial with rational coefficients.

The **Fundamental Theorem of Algebra** is the statement that a polynomial of degree $n > 0$, with complex coefficients has n zeros counted according to the multiplicity of their occurrences—so a zero appearing k times is counted as k zeros. Another way to interpret the statement of the Fundamental Theorem of Algebra is that if

$$f(x) = x^n + a_1 x^{n-1} + \cdots + a_n \in \mathbb{C}[x]$$

then there exist complex numbers $\alpha_1, \alpha_2, \cdots, a_n$, such that $f(x)$ can be written as

$$f(x) = (x - \alpha_1)(x - \alpha_2)\cdots(x - \alpha_n)$$

This leads to the fundamental relationship (known as Vieta's formulas) between the roots and coefficients of the equation

$$x^n + a_1x^{n-1} + \cdots + a_n = 0$$

namely,

$$\begin{array}{lcr} \alpha_1 + \alpha_2 + \cdots + \alpha_n & = & -a_1 \\ \alpha_1\alpha_2 + \alpha_1\alpha_3 + \cdots + \alpha_{n-1}\alpha_n & = & a_2 \\ \vdots & & \vdots \\ \alpha_1\alpha_2\cdots\alpha_i + \cdots + \alpha_{n-i+1}\alpha_{n-i+2}\cdots\alpha_n & = & (-1)^i a_i \\ \vdots & & \vdots \\ \alpha_1\alpha_2\cdots\alpha_n & = & (-1)^n a_n \end{array}$$

This in turn suggests another way of stating the Fundamental Theorem of Algebra. The Fundamental Theorem of Algebra is equivalent to the statement that the mapping

$$\begin{array}{ccc} \mathbb{C}^n & \longrightarrow & \mathbb{C}^n \\ (\alpha_1, \alpha_2, \cdots, \alpha_{n-1}, \alpha_n) & \longrightarrow & (a_1, a_2, \cdots, a_{n-1}, a_n) \end{array}$$

where

$$\begin{array}{lcl} a_1 & = & -(\alpha_1 + \alpha_2 + \cdots + \alpha_n) \\ a_2 & = & (\alpha_1\alpha_2 + \alpha_1\alpha_3 + \cdots + \alpha_{n-1}\alpha_n) \\ \vdots & & \vdots \\ a_i & = & (-1)^i \left(\sum \alpha_1\alpha_2\cdots\alpha_i\right) \\ a_n & = & (-1)^n\alpha_1\alpha_2\cdots\alpha_n \end{array}$$

is onto. In above where we write $\sum \alpha_1\alpha_2\cdots\alpha_i$, the summation is taken over all choices of i indices from the set $\{1, 2, \cdots, n\}$, so there are ${}_nC_i$ summands in this sum.

A note on notation

If $\alpha_1, \alpha_2, \cdots, \alpha_n$ are n variables then the elementary symmetric functions of these are defined as follows:

$$\begin{aligned}
\sigma_1 &= \alpha_1 + \alpha_2 + \cdots + \alpha_n \\
\sigma_2 &= \alpha_1\alpha_2 + \alpha_1\alpha_3 + \cdots + \alpha_{n-1}\alpha_n \\
\vdots &\ \ \vdots\ \vdots \\
\sigma_i &= \sum \alpha_1\alpha_2\cdots\alpha_i \\
\vdots &\ \ \vdots\ \vdots \\
\sigma_n &= \alpha_1\alpha_2\cdots\alpha_n
\end{aligned}$$

whereas before in

$$\sigma_i = \sum \alpha_1\alpha_2\cdots\alpha_i$$

the sum is taken over all ${}_nC_i$, i-tuples of $\{1, 2, \cdots, n\}$. It is obvious that each of these elementary symmetric functions is invariant under the action of the symmetric group S_n— on the indices $\{1, 2, \cdots, n\}$. In what follows we shall assume without proof the following theorem about symmetric function. Its proof can be found in several of the texts mentioned in the bibliography.

Theorem 2.1. (Fundamental Theorem of Symmetric Functions)
Let $\mathbb{K}$ be a field and $f(\alpha_1, \alpha_2, \cdots, \alpha_n)$ be any polynomial with coefficients in $\mathbb{K}$ which is invariant under the action of the symmetric group S_n. $f(\alpha_1, \alpha_2, \cdots, \alpha_n)$ can be uniquely expressed as a polynomial in the elementary symmetric functions $\sigma_1, \sigma_2, \cdots, \sigma_n$ of $\alpha_1, \alpha_2, \cdots, \alpha_n$ with coefficients in the field $\mathbb{K}$.

As can be easily checked we have

$$\sum \alpha_i^2 = \left(\sum \sigma_i\right)^2 - 2\sum \alpha_i\alpha_j$$

or

$$\sum \alpha_i^2 = \sigma_1^2 - 2\sigma_2.$$

Likewise

$$\sum \sigma_i^3 = \sigma_1^3 - 3\sigma_1\sigma_2 + 3\sigma_3.$$

These are special cases of what are known as *Newton's Identities* which we now proceed to discuss.

3. Newton's Identities

Theorem 3.1. *(Newton's Identities) Let $\alpha_1, \alpha_2, \cdots, \alpha_n$ be the roots of the following polynomial equation with coefficients in a field $\mathbb{K}$*

$$f(x) = x^n + a_1x^{n-1} + \cdots + a_{n-1}x + a_n = 0$$

Let $\sigma_1, \sigma_2, \cdots, \sigma_n$ denote the elementary symmetric functions of these roots and let

$$s_k = \alpha_1^k + \alpha_2^k + \cdots + \alpha_n^k.$$

Then $s_0 = n$ and for $m \geq n$ we have

$$s_m + a_1s_{m-1} + \cdots + a_{m-1}s_1 + a_ns_{m-n} = 0$$

and for $1 \leq m < n$

$$s_m + a_1s_{m-1} + a_2s_{m-2} + \cdots + a_{m-1}s_1 + ma_m = 0.$$

Proof. Let $m \geq n$. If we multiply the equation

$$x^n + a_1x^{n-1} + \cdots + a_{n-1}x + a_n = 0$$

by x^{m-n} we get,

$$x^m + a_1x^{m-1} + \cdots + a_{n-1}x^{m-n+1} + a_nx^{m-n} = 0$$

Substituting $x = \alpha_1, \alpha_2, \cdots, \alpha_n$ and adding we get

$$s_m + a_1s_{m-1} + \cdots + a_{n-1}s_{m-n+1} + a_ns_{m-n} = 0$$

This proves the Newton's identity for $m \geq n$. For $1 \leq m < n$ we proceed as follows. First we write

$$f(x) = (x - \alpha_1)(x - \alpha_2)\cdots(x - \alpha_n)$$

By logarithmic differentiating and multiplying both sides by $f(x)$ we get

$$f^{'}(x) = \frac{f(x)}{x - \alpha_1} + \frac{f(x)}{x - \alpha_2} + \cdots + \frac{f(x)}{x - \alpha_n}$$

Dividing

$$f(x) = x^n + a_1x^{n-1} + \cdots + a_{n-1}x + a_n$$

by $(x-\alpha_i)$ by the process of synthetic division we get

$$\frac{f(x)}{x-\alpha_i}=x^{n-1}+(\alpha_i+a_1)x^{n-2}+(\alpha_i^2+a_1\alpha_i+a_2)x^{n-3}+\cdots+(\alpha_i^{n-1}+a_1\alpha_i^{n-2}+\cdots+a_{n-2}\alpha_i+a_{n-1}).$$

Summing the above relation from $i=1$ to $i=n$ we get

$$f^{'}(x)=nx^{n-1}+(\sigma_1+na_1)x^{n-2}+(s_2+a_1\sigma_1+na_2)x^{n-3}+\cdots+(s_{n-1}+a_1s_{n-2}+\cdots+a_{n-2}s_1+na_{n-1}).$$

Equating the coefficients of various powers of x in above and in

$$f^{'}(x)=nx^{n-1}+(n-1)a_1x^{n-2}+\cdots+a_{n-1}$$

we get (noting that $\sigma_1=s_1$)

$$\begin{aligned}
s_1+na_1 &= (n-1)a_1\\
s_2+a_1\sigma_1+na_2 &= (n-2)a_2\\
\vdots \quad &\vdots \qquad \vdots\\
s_{n-1}+a_1s_{n-2}+\cdots+a_{n-2}s_1+na_{n-1} &= a_{n-1}
\end{aligned}$$

which can also be written as

$$\begin{aligned}
s_1+a_1 &= 0\\
s_2+a_1\sigma_1+2a_2 &= 0\\
\vdots \quad &\vdots \quad \vdots\\
s_{n-1}+a_1s_{n-2}+\cdots+a_{n-2}s_1+(n-1)a_{n-1} &= 0
\end{aligned}$$

So it follows that the recurrence

$$s_m+a_1s_{m-1}+\cdots+a_{m-1}s_1+ma_m=0$$

holds for $1\leq m\leq n-1$. This completes the proof of the Newton's Identities. □

Note. With the help of Newton's Identities, we can easily calculate the values of the symmetric polynomials s_k for $k\geq 1$. These are in turn helpful in calculating certain other symmetric functions such as the discriminant, which in turn is helpful in characterizing (or identifying) Galois group in some situations as we shall see later. We presently show the usefulness of Newton's Identities in the calculation of the discriminant of a polynomial.

For the monic polynomial

$$f(x)=x^n+a_1x^{n-1}+\cdots+a_n$$

having roots $\alpha_1, \alpha_2, \cdots, \alpha_n$, the discriminant of $f(x)$ (denoted by $Disc(f)$) is defined as

$$Disc(f) = \prod_{i>j} (\alpha_i - \alpha_j)^2$$

Let

$$A = \begin{bmatrix} 1 & \alpha_1 & \alpha_1^2 & \cdots & \cdots & \alpha_1^{n-1} \\ 1 & \alpha_2 & \alpha_2^2 & \cdots & \cdots & \alpha_2^{n-1} \\ \vdots & \vdots & \vdots & \cdots & \cdots & \vdots \\ 1 & \alpha_n & \alpha_n^2 & \cdots & \cdots & \alpha_n^{n-1} \end{bmatrix}$$

be the Vandermonde determinant. Let $det(A)$ denote the determinant of the square matrix A. We know that $det(A) = \prod_{i>j}(\alpha_i - \alpha_j)$. So $(det(A))^2 = \prod_{i>j}(\alpha_i - \alpha_j)^2 = Disc(f)$. We know that $det(A) = det(A^t)$, where A^t denotes the transpose of the matrix A. We have

$$Disc(f) = (det(A))^2 = det(A) \cdot det(A) = det(A^t) \cdot det(A) = det(A^t \cdot A)$$

Now

$$A^t \cdot A = \begin{bmatrix} 1 & 1 & 1 & \cdots & \cdots & 1 \\ \alpha_1 & \alpha_2 & \alpha_3 & \cdots & \cdots & \alpha_n \\ \alpha_1^2 & \alpha_2^2 & \alpha_3^2 & \cdots & \cdots & \alpha_n^2 \\ \vdots & \vdots & \vdots & \cdots & \cdots & \vdots \\ \alpha_1^{n-1} & \alpha_2^{n-1} & \alpha_3^{n-1} & \cdots & \cdots & \alpha_n^{n-1} \end{bmatrix} \cdot \begin{bmatrix} 1 & \alpha_1 & \alpha_1^2 & \cdots & \cdots & \alpha_1^{n-1} \\ 1 & \alpha_2 & \alpha_2^2 & \cdots & \cdots & \alpha_2^{n-1} \\ 1 & \alpha_3 & \alpha_3^2 & \cdots & \cdots & \alpha_3^{n-1} \\ \vdots & \vdots & \vdots & \cdots & \cdots & \vdots \\ 1 & \alpha_n & \alpha_n^2 & \cdots & \cdots & \alpha_n^{n-1} \end{bmatrix}$$

$$= \begin{bmatrix} n & s_1 & s_2 \cdots & \cdots & s_{n-1} \\ s_1 & s_2 & s_3 \cdots & \cdots & s_n \\ s_2 & s_3 & s_4 \cdots & \cdots & s_{n+1} \\ \vdots & \vdots & \vdots & \vdots & \vdots \\ s_{n-1} & s_n & s_{n+1} \cdots & \cdots & s_{2n-2} \end{bmatrix}.$$

In particular if $n = 3$ then $Disc(f)$ is given by the determinant

$$\begin{bmatrix} 3 & s_1 & s_2 \\ s_1 & s_2 & s_3 \\ s_2 & s_3 & s_4 \end{bmatrix}.$$

Example 1. *Discriminant of the cubic* $x^3 + px + q$.

As an example we calculate the discriminant or the cubic polynomial $x^3 + px + q$. Assuming its roots to be α_1, α_2, and α_3, we have

$$\begin{aligned} \alpha_1 + \alpha_2 + \alpha_3 &= 0 \\ \alpha_1\alpha_2 + \alpha_2\alpha_3 + \alpha_3\alpha_1 &= p \\ \alpha_1\alpha_2\alpha_3 &= -q. \end{aligned}$$

Using the Newton's Identities, we have (here $a_1 = 0, a_2 = p$, and $a_3 = q$)

$$\begin{aligned} s_1 + a_1 &= 0 \\ s_2 + a_1 s_1 + 2a_2 &= 0. \end{aligned}$$

So $s_1 = 0$, and $s_2 = -2p$. Also

$$\begin{aligned} s_3 + a_1 s_2 + a_2 s_1 + 3a_3 &= 0 \\ s_4 + a_1 s_3 + a_2 s_2 + a_3 s_1 &= 0. \end{aligned}$$

So $s_3 = -3q$, and $s_4 = 2p^2$. Substituting for s_1, s_2, s_3, and s_4 in the expression for $Disc(f)$, we get the following value of the discriminant of the cubic $x^3 + px + q$.

$$\begin{vmatrix} 3 & s_1 & s_2 \\ s_1 & s_2 & s_3 \\ s_2 & s_3 & s_4 \end{vmatrix} = \begin{vmatrix} 3 & 0 & -2p \\ 0 & -2p & -3q \\ -2p & -3q & 2p^2 \end{vmatrix} = 3(-4p^3 - 9q^2) - 2p(-4p^2) = -4p^3 - 27q^2.$$

CHAPTER 3

Solution in Radicals

1. Solving a Cubic Equation (Cardano's Formula)

Consider the following cubic equation whose coefficients a, b, and c are complex numbers:

$$y^3 + ay^2 + by + c = 0.$$

Suppose that the roots of the above equations are β_1, β_2, and β_3. The substitution $y = x - \frac{a}{3}$ transforms the above equation into

$$x^3 + px + q = 0,$$

where

$$p = b - \frac{a^2}{3} \qquad \text{and} \qquad q = \frac{2a^3}{27} - \frac{ab}{3} + c,$$

The roots of the new equation are α_1, α_2, and α_3 where

$$\alpha_i = \beta_i + \frac{a}{3}, \qquad 1 \leq i \leq 3,$$

So in order to solve a general cubic equation it is sufficient to give a method of solving a cubic equation of the type $x^3 + px + q = 0$, that is one in which the coefficient of x^2 is zero. The roots α_1, α_2, and α_3 of the cubic equation $x^3 + px + q = 0$ satisfy the relations

$$\begin{aligned}
\alpha_1 + \alpha_2 + \alpha_3 &= 0 \\
\alpha_1\alpha_2 + \alpha_2\alpha_3 + \alpha_3\alpha_1 &= p \\
\alpha_1\alpha_2\alpha_3 &= -q \\
\alpha_1^2 + \alpha_2^2 + \alpha_3^2 &= -2p \\
\textstyle\sum_{i\neq j} \alpha_i^2\alpha_j = \alpha_1^2\alpha_2 + \alpha_2^2\alpha_1 + \alpha_1^2\alpha_3 + \alpha_3^2\alpha_1 + \alpha_2^2\alpha_3 + \alpha_3^2\alpha_2 &= 3q \\
\alpha_1^3 + \alpha_2^3 + \alpha_3^3 &= -3q,
\end{aligned}$$

The last two of the above relations can be seen to be true as follows

$$0 = (\alpha_1 + \alpha_2 + \alpha_3)(\alpha_1\alpha_2 + \alpha_2\alpha_3 + \alpha_3\alpha_1) = \sum_{i\neq j} \alpha_i^2\alpha_j + 3\alpha_1\alpha_2\alpha_3$$

and so $\sum_{i\neq j}\alpha_i^2\alpha_j = -3\alpha_1\alpha_2\alpha_3 = 3q$. Also

$$\alpha_1^3+\alpha_2^3+\alpha_3^3 = (\alpha_1+\alpha_2+\alpha_3)(\alpha_1^2+\alpha_2^2+\alpha_3^2) - \sum_{i\neq j}\alpha_i^2\alpha_j = -3q.$$

Let $\omega = e^{\frac{2\pi i}{3}} = \cos\frac{2\pi}{3} + i\sin\frac{2\pi}{3} = \cos 120^\circ + i\sin 120^\circ = \dfrac{-1+\sqrt{-3}}{2}$, a primitive third root of unity and

$$\begin{aligned} L &= \alpha_1 + \omega\alpha_2 + \omega^2\alpha_3, \\ M &= \alpha_1 + \omega^2\alpha_2 + \omega\alpha_3, \end{aligned}$$

using the fact that $\alpha_1+\alpha_2+\alpha_3 = 0$, we have

$$L+M = 3\alpha_1, \qquad \omega^2L+\omega M = 3\alpha_2, \qquad \omega L + \omega^2 M = 3\alpha_3$$

or

$$\alpha_1 = \frac{1}{3}(L+M), \qquad \alpha_2 = \frac{1}{3}(\omega^2L+\omega M), \qquad \alpha_3 = \frac{1}{3}(\omega L+\omega^2 M),$$

So if we could express L and M in terms of the coefficients p and q of the given equation, we will have radical expressions for the roots of the given cubic equation in terms of its coefficients. By direct multiplication we see that

$$L\cdot M = \alpha_1^2+\alpha_2^2+\alpha_3^2 + (\omega+\omega^2)(\alpha_1\alpha_2+\alpha_2\alpha_3+\alpha_3\alpha_1) = -2p - p = -3p,$$

and so $L^3\cdot M^3 = -27p^3$. Observing that $L^3+M^3 = (L+M)(\omega^2L+\omega M)(\omega L+\omega^2M)$ and multiplying the following three equations

$$L+M = 3\alpha_1, \qquad \omega^2L+\omega M = 3\alpha_2, \qquad \omega L + \omega^2 M = 3\alpha_3,$$

we get

$$L^3+M^3 = (L+M)(\omega^2L+\omega M)(\omega L+\omega^2M) = 27\alpha_1\alpha_2\alpha_3 = -27q.$$

So L^3 and M^3 are the roots of the quadratic equation

$$z^2+27qz-27p^3 = 0,$$

Therefore we have

$$L^3, M^3 = \frac{-27q \pm \sqrt{(27q)^2+4\cdot 27p^3}}{2} = 27\left(-\frac{q}{2} \pm \sqrt{\frac{q^2}{4}+\frac{p^3}{27}}\right).$$

Since we have the additional requirement $L \cdot M = -3p$, we get only three pairs of values of L and M. These give the following values for the three roots α_1, α_2, and α_3 of the given cubic:

$$\alpha_1 = \frac{1}{3}(L+M) = \left(-\frac{q}{2} + \sqrt{\frac{q^2}{4} + \frac{p^3}{27}}\right)^{\frac{1}{3}} + \left(-\frac{q}{2} - \sqrt{\frac{q^2}{4} + \frac{p^3}{27}}\right)^{\frac{1}{3}}$$

$$\alpha_2 = \frac{1}{3}(\omega^2 L + \omega M) = \omega^2\left(-\frac{q}{2} + \sqrt{\frac{q^2}{4} + \frac{p^3}{27}}\right)^{\frac{1}{3}} + \omega\left(-\frac{q}{2} - \sqrt{\frac{q^2}{4} + \frac{p^3}{27}}\right)^{\frac{1}{3}}$$

$$\alpha_3 = \frac{1}{3}(\omega L + \omega^2 M) = \omega\left(-\frac{q}{2} + \sqrt{\frac{q^2}{4} + \frac{p^3}{27}}\right)^{\frac{1}{3}} + \omega^2\left(-\frac{q}{2} - \sqrt{\frac{q^2}{4} + \frac{p^3}{27}}\right)^{\frac{1}{3}}.$$

Writing $A = \dfrac{L}{3}$ and $B = \dfrac{M}{3}$, we have

$$A^3, B^3 = \left(-\frac{q}{2} \pm \sqrt{\frac{q^2}{4} + \frac{p^3}{27}}\right), \qquad A \cdot B = -\frac{p}{3}$$

and

$$\alpha_1 = A + B = \left(-\frac{q}{2} + \sqrt{\frac{q^2}{4} + \frac{p^3}{27}}\right)^{\frac{1}{3}} + \left(-\frac{q}{2} - \sqrt{\frac{q^2}{4} + \frac{p^3}{27}}\right)^{\frac{1}{3}}$$

$$\alpha_2 = \omega^2 A + \omega B = \omega^2\left(-\frac{q}{2} + \sqrt{\frac{q^2}{4} + \frac{p^3}{27}}\right)^{\frac{1}{3}} + \omega\left(-\frac{q}{2} - \sqrt{\frac{q^2}{4} + \frac{p^3}{27}}\right)^{\frac{1}{3}}$$

$$\alpha_3 = \omega A + \omega^2 B = \omega\left(-\frac{q}{2} + \sqrt{\frac{q^2}{4} + \frac{p^3}{27}}\right)^{\frac{1}{3}} + \omega^2\left(-\frac{q}{2} - \sqrt{\frac{q^2}{4} + \frac{p^3}{27}}\right)^{\frac{1}{3}}.$$

The above expressions for the roots of a cubic equation are known as Cardano's formula. Cardano is known to have derived these in sixteenth century in a somewhat different manner.

2. Cubic with Real Coefficients

Suppose that the coefficients p and q of the cubic $x^3 + px + q = 0$ are real numbers. In such a case we shall refer to the given cubic as a *real cubic*. The discriminant D of this cubic as seen earlier is given by

$$D = (\alpha_1 - \alpha_2)^2(\alpha_2 - \alpha_3)^2(\alpha_3 - \alpha_1)^2 = -(4p^3 + 27q^2)$$

We note that in the Cardano's formula for the roots of the cubic the expression under the square root radical is

$$\frac{q^2}{4}+\frac{p^3}{27}=\frac{4p^3+27q^2}{108}.$$

Its numerator (except for the sign) is the discriminant of the cubic. Since the given cubic polynomial is assumed to have real coefficients either all of its roots are real or one of its root is real and remaining two complex conjugates or each other. Therefore if $D<0$ then given cubic cannot have three real roots and if $D>0$ it cannot have two nonreal roots. In other words we have the following conclusion about the roots of the cubic equation $x^3+px+q=0$ whose coefficients p and q are real:

- If $4p^3+27q^2\leq 0$ then it has three real roots.
- If $4p^3+27q^2>0$ then one of its root is real and other two nonreal (complex conjugates of each other).

We shall now discuss several examples of cubic equation whose roots can be easily found out, but the Cardano's formula gives a seemingly complicated expression for the roots. As is well known the attempts to explain this phenomenon led to the introduction and development of complex numbers.

Example 1.

Consider the cubic equation

$$x^3+x-2=0$$

Here $p=1$, $q=-2$, so the discriminant of this cubic is given be

$$D=-(4p^3+27q^2)=-(4+108)=-112$$

So the given cubic has just one real root—which is evidently 1. Since

$$\sqrt{\frac{q^2}{4}+\frac{p^3}{27}}=\sqrt{1+\frac{1}{27}}=\frac{2}{3}\sqrt{\frac{7}{3}}$$

Cardano's formula gives the following expressions for the (only) real root (namely 1) of the above equation

$$\left(1+\frac{2}{3}\sqrt{\frac{7}{3}}\right)^{\frac{1}{3}}+\left(1-\frac{2}{3}\sqrt{\frac{7}{3}}\right)^{\frac{1}{3}}$$

Therefore

$$\left(1+\frac{2}{3}\sqrt{\frac{7}{3}}\right)^{\frac{1}{3}}+\left(1-\frac{2}{3}\sqrt{\frac{7}{3}}\right)^{\frac{1}{3}}=1.$$

Example 2.

Consider the cubic equation

$$x^3-15x-4=0$$

In this case $p=-15$, $q=-4$ and we have

$$D=-(4p^3+27q^2)=13068$$

and all the three roots of this equation are real. These roots are easily seen to be 4 and $-2\pm\sqrt{3}$. Using Cardano's formula we have the following expression for the three roots

$$\begin{aligned}
(2+\sqrt{-121})^{\frac{1}{3}} &+ (2-\sqrt{-121})^{\frac{1}{3}} &= 4\\
\omega\,(2+\sqrt{-121})^{\frac{1}{3}} &+ \omega^2\,(2-\sqrt{-121})^{\frac{1}{3}} &= -2-\sqrt{3}\\
\omega^2\,(2+\sqrt{-121})^{\frac{1}{3}} &+ \omega\,(2-\sqrt{-121})^{\frac{1}{3}} &= -2+\sqrt{3}.
\end{aligned}$$

These can be verified as follows.

$$(2+\sqrt{-121})^{\frac{1}{3}}+(2-\sqrt{-121})^{\frac{1}{3}}=2\times\text{ Real Part of }(2+\sqrt{-121})^{\frac{1}{3}}$$

and

$$(2+\sqrt{-121})^{\frac{1}{3}}=(2+11i)^{\frac{1}{3}}=\left[\sqrt{125}\left(\frac{2}{\sqrt{125}}+\frac{11}{\sqrt{125}}\right)\right]^{\frac{1}{3}}=\sqrt{5}\left(\frac{2}{5\sqrt{5}}+\frac{11}{5\sqrt{5}}i\right)^{\frac{1}{3}}.$$

Now

$$\sqrt{5}\left(\frac{2}{5\sqrt{5}}+\frac{11}{5\sqrt{5}}i\right)^{\frac{1}{3}}=\sqrt{5}\,(\cos\theta+i\sin\theta)^{\frac{1}{3}}=\sqrt{5}\left(\cos\frac{\theta}{3}+i\sin\frac{\theta}{3}\right).$$

where

$$\cos\theta=\frac{2}{5\sqrt{5}},\ \text{and so }\theta\approx 79.695^o,\quad \frac{\theta}{3}\approx 26.565^o,\quad \cos\frac{\theta}{3}=0.8944,\quad \sin\frac{\theta}{3}\approx 0.4472.$$

Therefore

$$2\times\text{ Real Part of }(2+11i)^{\frac{1}{3}}=2\times\sqrt{5}\times\cos\frac{\theta}{3}=4.$$

Likewise as $\omega\,(2+\sqrt{-121})^{\frac{1}{3}}$ and $\omega^2\,(2-\sqrt{-121})^{\frac{1}{3}}$ are complex conjugate of each other we have

$$\omega\,(2+\sqrt{-121})^{\frac{1}{3}}+\omega^2\,(2-\sqrt{-121})^{\frac{1}{3}}=2\times\text{ Real Part of }\quad\omega\,(2+\sqrt{-121})^{\frac{1}{3}}.$$

Writing $\omega\left(2+\sqrt{-121}\right)^{\frac{1}{3}}$ as $\omega\left(2+\sqrt{-121}\right)^{\frac{1}{3}}$ we have

$$\begin{aligned}
\omega\left(2+\sqrt{-121}\right)^{\frac{1}{3}}+\omega^2\left(2-\sqrt{-121}\right)^{\frac{1}{3}} &= 2\times \text{ Real Part of } \quad \omega\left(2+\sqrt{-121}\right)^{\frac{1}{3}} \\
&= 2\times \text{ Real Part of } \quad \left(-\tfrac{1}{2}+\tfrac{3}{2}i\right)(2+11i)^{\frac{1}{3}} \\
&= 2\times \text{ Real Part of } \quad \left(-\tfrac{1}{2}+\tfrac{3}{2}i\right)\sqrt{5}\left(\cos\tfrac{\theta}{3}+i\sin\tfrac{\theta}{3}\right) \\
&= 2\times\sqrt{5}\left(-\tfrac{1}{2}\cos\tfrac{\theta}{3}-\tfrac{\sqrt{3}}{2}\sin\tfrac{\theta}{3}\right) \\
&= -\sqrt{5}\left(0.8944+\sqrt{3}\times 0.4472\right) \\
&= -3.7319\approx -2-\sqrt{3}.
\end{aligned}$$

Example 3.

Consider the cubic equation

$$x^3-19x+30=0$$

Here $p=-19$, $q=30$. The above cubic has three real roots namely 2, 3, and -5. However Cardano's formula gives the following expression for the three roots:

$$\begin{aligned}
\alpha_1 &= \left(-15+\frac{28\sqrt{-3}}{9}\right)^{\frac{1}{3}}+\left(-15-\frac{28\sqrt{-3}}{9}\right)^{\frac{1}{3}} &= 3 \\
\alpha_2 &= \omega\left(-15+\frac{28\sqrt{-3}}{9}\right)^{\frac{1}{3}}+\omega^2\left(-15-\frac{28\sqrt{-3}}{9}\right)^{\frac{1}{3}} &= 2 \\
\alpha_3 &= \omega^2\left(-15+\frac{28\sqrt{-3}}{9}\right)^{\frac{1}{3}}+\omega\left(-15-\frac{28\sqrt{-3}}{9}\right)^{\frac{1}{3}} &= -5
\end{aligned}$$

These can be verified as was done in the previous example.

3. Radical Expressions for $\cos\frac{2\pi}{n}$

Example 4.

In this example we shall find a radical expression for $\cos\frac{2\pi}{18}=\cos\frac{\pi}{9}=\cos 20^o$. It is easily seen that the cyclotomic polynomial $\Phi_{18}=x^6-x^3+1$. Its roots are $\alpha,\alpha^5,\alpha^7,\alpha^{11},\alpha^{13}$, and α^{17}, where

$$\alpha=\cos\frac{2\pi}{18}+i\sin\frac{2\pi}{18}=\cos 20^o+i\sin 20^o$$

To solve $x^6 - x^3 + 1 - 0$, we divide it by x^3 getting $x^3 - 1 + \dfrac{1}{x^3} = 0$. The substitution $x + \dfrac{1}{x} = y$ transforms this into the equation $y^3 - 3y - 1 = 0$, whose roots are

$$\alpha + \alpha^{17} = 2\cos 20^o, \qquad \alpha^5 + \alpha^{13} = 2\cos 100^o, \qquad \alpha^7 + \alpha^{11} = 2\cos 140^o.$$

To solve $y^3 - 3y - 1 = 0$ by Cardano's formula we note that $p = -3$ and $q = -1$. This gives

$$A = \left(-\frac{q}{2} + \sqrt{\frac{q^2}{4} + \frac{p^3}{27}}\right)^{\frac{1}{3}} = \left(\frac{1}{2} + \frac{\sqrt{-3}}{2}\right)^{\frac{1}{3}}, \quad B = \left(-\frac{q}{2} - \sqrt{\frac{q^2}{4} + \frac{p^3}{27}}\right)^{\frac{1}{3}} = \left(\frac{1}{2} - \frac{\sqrt{-3}}{2}\right)^{\frac{1}{3}}.$$

and we have

$$\begin{aligned}
2\cos 20^o &= A + B &&= \left(\frac{1}{2} + \frac{\sqrt{-3}}{2}\right)^{\frac{1}{3}} + \left(\frac{1}{2} - \frac{\sqrt{-3}}{2}\right)^{\frac{1}{3}} &&\approx 1.879385242 \\
2\cos 140^o &= \omega A + \omega^2 B &&= \omega\left(\frac{1}{2} + \frac{\sqrt{-3}}{2}\right)^{\frac{1}{3}} + \omega^2\left(\frac{1}{2} - \frac{\sqrt{-3}}{2}\right)^{\frac{1}{3}} &&\approx -1.53208886 \\
2\cos 100^o &= \omega^2 A + \omega B &&= \omega^2\left(\frac{1}{2} + \frac{\sqrt{-3}}{2}\right)^{\frac{1}{3}} + \omega\left(\frac{1}{2} - \frac{\sqrt{-3}}{2}\right)^{\frac{1}{3}} &&\approx -0.347296355
\end{aligned}$$

Example 5. Radical expression for $2\cos\frac{2\pi}{7}$.

Since $z^7 - 1 = (z-1)(z^6 + z^5 + z^4 + z^3 + z^2 + z + 1)$, the roots of the equation $z^6 + z^5 + z^4 + z^3 + z^2 + z + 1 = 0$ are $\alpha, \alpha^2, \alpha^3, \alpha^4, \alpha^5$, and α^6, where $\alpha = e^{\frac{2\pi i}{7}} = \cos\frac{2\pi}{7} + i\sin\frac{2\pi}{7}$. After performing the transformations as indicated in the following we arrive at a cubic equation whose roots are $2\cos\frac{2\pi}{7}, 2\cos\frac{4\pi}{7}$, and $2\cos\frac{6\pi}{7}$.

$$z^6 + z^5 + z^4 + z^3 + z^2 + z + 1 = 0.$$

Dividing the above equation by z^3 we get

$$z^3 + z^2 + z + 1 + \frac{1}{z} + \frac{1}{z^2} + \frac{1}{z^3} = 0$$

Let $z + \frac{1}{z} = y$. Then we have $z^2 + \frac{1}{z^2} = y^2 - 2$ and $z^3 + \frac{1}{z^3} = y^3 - 3y$.
Therefore substitution $y = z + \dfrac{1}{z}$ transforms the above equation into

$$y^3 + y^2 - 2y - 1 = 0.$$

The transformation $u = y + \frac{1}{3}$ transforms the above equation into

$$u^3 - \frac{7}{3}u - \frac{7}{27} = 0.$$

The transformation $x = 3u$ transforms the above equation into

$$x^3 - 21x - 7 = 0.$$

The following table gives the roots of the successive equations appearing above.

Equation	Roots
$z^6 + z^5 + z^4 + z^3 + z^2 + z + 1 = 0$	$\alpha^i, 1 \leq i \leq 6, \quad \alpha = \cos\frac{2\pi}{7} + i\sin\frac{2\pi}{7}$
$y^3 + y^2 - 2y - 1 = 0$	$2\cos\frac{2\pi}{7},\ \ 2\cos\frac{4\pi}{7},\ \ 2\cos\frac{6\pi}{7}$
$u^3 - \frac{7}{3}u - \frac{7}{27} = 0$	$2\cos\frac{2\pi}{7} + \frac{1}{3},\ \ 2\cos\frac{4\pi}{7} + \frac{1}{3},\ \ 2\cos\frac{6\pi}{7} + \frac{1}{3}$
$x^3 - 21x - 7 = 0$	$6\cos\frac{2\pi}{7} + 1, 6\cos\frac{4\pi}{7} + 1,$ and $6\cos\frac{6\pi}{7} + 1$

Solving the equation $x^3 - 21x - 7 = 0$ by using Cardano's formula we find the following expression for the values of $2\cos\frac{2\pi}{7}, 2\cos\frac{4\pi}{7}$, and $2\cos\frac{6\pi}{7}$ (i.e., roots of the equation $y^3 + y^2 - 2y - 1 = 0$).

$$\begin{aligned}
2\cos\frac{2\pi}{7} &= \frac{\sqrt{7}}{3}\left(\frac{1+3\sqrt{-3}}{2\sqrt{7}}\right)^{\frac{1}{3}} + \frac{\sqrt{7}}{3}\left(\frac{1-3\sqrt{-3}}{2\sqrt{7}}\right)^{\frac{1}{3}} - \frac{1}{3} &\approx 1.2469796\\
2\cos\frac{4\pi}{7} &= \frac{\sqrt{7}}{3}\left(\frac{1+3\sqrt{-3}}{2\sqrt{7}}\right)^{\frac{1}{3}}\omega^2 + \frac{\sqrt{7}}{3}\left(\frac{1-3\sqrt{-3}}{2\sqrt{7}}\right)^{\frac{1}{3}}\omega - \frac{1}{3} &\approx -0.4450418\\
2\cos\frac{6\pi}{7} &= \frac{\sqrt{7}}{3}\left(\frac{1+3\sqrt{-3}}{2\sqrt{7}}\right)^{\frac{1}{3}}\omega + \frac{\sqrt{7}}{3}\left(\frac{1-3\sqrt{-3}}{2\sqrt{7}}\right)^{\frac{1}{3}}\omega^2 - \frac{1}{3} &\approx -1.8019377
\end{aligned}$$

Example 6. Radical expression for $2\cos\frac{2\pi}{13}$.

Let $\alpha = e^{\frac{2\pi i}{13}} = \cos\frac{2\pi}{13} + i\sin\frac{2\pi}{13}$. Then $\alpha^i,\ \ 1 \leq i \leq 12$ are the roots of the equation

$$z^{12} + z^{11} + z^{10} + z^9 + z^8 + z^7 + z^6 + z^5 + z^4 + z^3 + z^2 + z + 1 = 0$$

Let

$$\begin{aligned}
\beta_1 &= \alpha + \alpha^5 + \alpha^{12} + \alpha^8 &= 2\cos\tfrac{2\pi}{13} + 2\cos\tfrac{10\pi}{13}\\
\beta_2 &= \alpha^2 + \alpha^{10} + \alpha^{11} + \alpha^3 &= 2\cos\tfrac{4\pi}{13} + 2\cos\tfrac{6\pi}{13}\\
\beta_3 &= \alpha^4 + \alpha^7 + \alpha^9 + \alpha^6 &= 2\cos\tfrac{8\pi}{13} + 2\cos\tfrac{14\pi}{13}
\end{aligned}$$

Then using the fact that $\alpha^{13} = 1$ we see by direct calculation that

$$\beta_1 + \beta_2 + \beta_3 = -1, \qquad \beta_1\beta_2 + \beta_2\beta_3 + \beta_3\beta_1 = -4, \qquad \beta_1\beta_2\beta_3 = -1$$

So the equation with roots β_1, β_2, and β_3 is

$$y^3 + y^2 - 4y + 1 = 0$$

The substitution $u = y + \frac{1}{3}$ transforms the above equation into

$$27u^3 - 117u + 65 = 0$$

whose roots are $\beta_1 + \frac{1}{3}$, $\beta_2 + \frac{1}{3}$, and $\beta_3 + \frac{1}{3}$. The transformation $x = 3u$ transforms the previous equation into the equation

$$x^3 - 39x + 65 = 0$$

whose roots are $3\beta_1 + 1$, $3\beta_2 + 1$ and $3\beta_3 + 1$.

Let

$$\gamma_1 = \alpha + \alpha^{12} = 2\cos\frac{2\pi}{13} \quad \text{and} \quad \gamma_2 = \alpha^5 + \alpha^8 = 2\cos\frac{10\pi}{13}$$

It is easily seen that $\gamma_1 + \gamma_2 = \beta_1$ and $\gamma_1\gamma_2 = \beta_3$. In other words γ_1 and γ_2 are the roots of the equation

$$t^2 - \beta_1 t + \beta_3 = 0$$

and so

$$\gamma_1, \gamma_2 = \frac{\beta_1 \pm \sqrt{\beta_1^2 - 4\beta_3}}{2}$$

So all we need is radical expressions for β_1 and β_3. We can get these by solving the cubic equation.

$$x^3 - 39x + 65 = 0,$$

whose roots are, as mentioned earlier, $3\beta_1 + 1$, $3\beta_2 + 1$, and $3\beta_3 + 1$. We solve this cubic equation by using Cardano's formula, we have $p = -39$, $q = 65$ and so

$$\frac{q^2}{4} + \frac{p^3}{27} = -\frac{4563}{3} = -\left(\frac{39}{2}\right)^2 \cdot 3.$$

Therefore

$$A = \left(-\frac{q}{2} + \sqrt{\frac{q^2}{4} + \frac{p^3}{27}}\right)^{\frac{1}{3}} = \left(-\frac{65}{2} + \frac{39\sqrt{-3}}{2}\right)^{\frac{1}{3}},$$

$$B = \left(-\frac{q}{2} - \sqrt{\frac{q^2}{4} + \frac{p^3}{27}}\right)^{\frac{1}{3}} = \left(-\frac{65}{2} - \frac{39\sqrt{-3}}{2}\right)^{\frac{1}{3}}.$$

We have

$$A=\left(-\frac{65}{2}+\frac{39\sqrt{-3}}{2}\right)^{\frac{1}{3}}=\sqrt{13}\,(\cos\theta+i\sin\theta)^{\frac{1}{3}}=\sqrt{13}\left(\cos\frac{\theta}{3}+i\sin\frac{\theta}{3}\right)$$

and

$$B=\left(-\frac{65}{2}-\frac{39\sqrt{-3}}{2}\right)^{\frac{1}{3}}=\sqrt{13}\,(\cos\theta-i\sin\theta)^{\frac{1}{3}}=\sqrt{13}\left(\cos\frac{\theta}{3}-i\sin\frac{\theta}{3}\right)$$

where θ is in the second quadrant and $\tan\theta=-\frac{3\sqrt{3}}{5}$. This gives $\theta\approx 133.8978862$ degrees and so $\frac{\theta}{3}\approx 44.63262875$ degrees. Therefore $\cos\frac{\theta}{3}\approx 0.711626069$ and $\frac{\theta}{3}\approx 0.702558423$.

We have

$$\begin{array}{lclcl} A+B & = & 3\beta_2+1 & = & 6\cos\frac{4\pi}{13}+6\cos\frac{6\pi}{13}+1 \approx 5.131553736\\ \omega A+\omega^2 B & = & 3\beta_3+1 & = & 6\cos\frac{8\pi}{13}+6\cos\frac{14\pi}{13}+1 \approx -6.953280226\\ \omega^2 A+\omega B & = & 3\beta_1+1 & = & 6\cos\frac{2\pi}{13}+6\cos\frac{6\pi}{13}+1 \approx 1.821671665 \end{array}$$

we finally get the following radical expression for $2\cos\frac{2\pi}{13}$ and $2\cos\frac{10\pi}{13}$

$$\gamma_1=2\cos\frac{2\pi}{13}=\frac{1}{2}\left(\beta_1+\sqrt{\beta_1^2-4\beta_3}\right)$$

$$=\frac{1}{2}\left[\frac{1}{3}\left(\omega^2A+\omega B-1\right)+\sqrt{\frac{1}{9}\left(\omega^2A+\omega B-1\right)^2-\frac{4}{3}\left(\omega A+\omega^2B-1\right)}\right]$$

$$\gamma_2=2\cos\frac{10\pi}{13}=\frac{1}{2}\left(\beta_1-\sqrt{\beta_1^2-4\beta_3}\right)$$

$$=\frac{1}{2}\left[\frac{1}{3}\left(\omega^2A+\omega B-1\right)-\sqrt{\frac{1}{9}\left(\omega^2A+\omega B-1\right)^2-\frac{4}{3}\left(\omega A+\omega^2B-1\right)}\right]$$

where

$$\omega=\frac{-1+\sqrt{-3}}{2},\qquad \omega^2=\frac{-1-\sqrt{-3}}{2}$$

and A and B are as defined above. The following table gives the various equations and their roots involved in the previous example.

Equation	Roots
$z^{12}+z^{11}+z^{10}+z^9+z^8+z^7+z^6+z^5+z^4+z^3+z^2+z+1=0$	$\alpha^i,\quad 1\le i\le 12,\quad \alpha=e^{\frac{2\pi i}{13}}$
$y^3+y^2-4y+1=0$	$\beta_1=\alpha+\alpha^5+\alpha^{12}+\alpha^8$ $\beta_2=\alpha^2+\alpha^{10}+\alpha^{11}+\alpha^3$ $\beta_3=\alpha^4+\alpha^7+\alpha^9+\alpha^6$
$27u^3-117u+65=0$	$\beta_1+\frac{1}{3},\beta_2+\frac{1}{3},\beta_3+\frac{1}{3}$
$x^3-39x+65=0$	$3\beta_1+1,3\beta_2+1,3\beta_3+1$
$t^2-\beta_1 t+\beta_3=0$	$\gamma_1=\alpha+\alpha^{12}=2\cos\frac{2\pi}{13}$ and $\gamma_2=\alpha^5+\alpha^8=2\cos\frac{10\pi}{13}$

The complete radical expression for $\cos\frac{2\pi}{13}$ is as follows:

A Radical Expression for $\cos\dfrac{2\pi}{13}$

$$\cos\frac{2\pi}{13}=\frac{1}{4}\left[\frac{1}{3}\left\{\frac{-1-\sqrt{-3}}{2}\left(-\frac{65}{2}+\frac{39\sqrt{-3}}{2}\right)^{\frac{1}{3}}+\frac{-1+\sqrt{-3}}{2}\left(-\frac{65}{2}-\frac{39\sqrt{-3}}{2}\right)^{\frac{1}{3}}-1\right\}\right]+$$

$$\frac{1}{4}\left[\sqrt{\begin{array}{l}\frac{1}{9}\left\{\frac{-1-\sqrt{-3}}{2}\left(-\frac{65}{2}+\frac{39\sqrt{-3}}{2}\right)^{\frac{1}{3}}+\frac{-1+\sqrt{-3}}{2}\left(-\frac{65}{2}-\frac{39\sqrt{-3}}{2}\right)^{\frac{1}{3}}-1\right\}^2-\\ \frac{4}{3}\left\{\frac{-1+\sqrt{-3}}{2}\left(-\frac{65}{2}+\frac{39\sqrt{-3}}{2}\right)^{\frac{1}{3}}+\frac{-1-\sqrt{-3}}{2}\left(-\frac{65}{2}-\frac{39\sqrt{-3}}{2}\right)^{\frac{1}{3}}-1\right\}\end{array}}\right]$$

Example 7. Radical expression for $\cos\dfrac{2\pi}{19}$.
In this case we shall indicate only the outlines of the derivation of the value of $\cos\frac{2\pi}{19}$ and leave the details to the readers as these are very similar to the previous example. We begin by observing that the roots of $z^{19}-1=0$ are $\alpha^i,\quad 0\le i\le 18$, where $\alpha=e^{\frac{2\pi i}{19}}=\cos\frac{2\pi}{19}+i\cos\frac{2\pi}{19}$. So the equation

$$z^{18}+z^{17}+z^{16}+z^{15}+z^{14}+z^{13}+z^{12}+z^{11}+z^{10}+z^9+z^8+z^7+z^6+z^5+z^4+z^3+z^2+z+1=0$$

has roots $\alpha^i,\quad 1\le i\le 18$. Since the order of 8 modulo 19 is 6 we let

$$\begin{aligned}\beta_1&=\alpha+\alpha^8+\alpha^7+\alpha^{18}+\alpha^{11}+\alpha^{12}\\ \beta_2&=\alpha^2+\alpha^{16}+\alpha^{14}+\alpha^{17}+\alpha^3+\alpha^5\\ \beta_3&=\alpha^4+\alpha^{13}+\alpha^9+\alpha^{15}+\alpha^6+\alpha^{10}\end{aligned}$$

so the equation with roots β_1, β_2, and β_3 is

$$y^3 + y^2 - 6y - 7 = 0$$

The substitution $y = u - \frac{1}{3}$ transforms the above equation into the equation

$$u^3 - \frac{19}{3}u - \frac{133}{27} = 0$$

whose roots are $\beta_1 + \frac{1}{3}$, $\beta_2 + \frac{1}{3}$, and $\beta_3 + \frac{1}{3}$. Finally the substitution $u = \frac{x}{3}$ transforms the previous equation into the equation

$$x^3 - 57x - 133 = 0,$$

whose roots are $3\beta_1 + 1$, $3\beta_2 + 1$, and $3\beta_3 + 1$. If we let

$$\begin{aligned}
\gamma_1 &= \alpha + \alpha^{18} &&= 2\cos\tfrac{2\pi}{19} \\
\gamma_2 &= \alpha^8 + \alpha^{11} &&= 2\cos\tfrac{16\pi}{19} \\
\gamma_3 &= \alpha^7 + \alpha^{12} &&= 2\cos\tfrac{14\pi}{19}.
\end{aligned}$$

Then we can easily see by direct calculation that

$$\begin{aligned}
\gamma_1 + \gamma_2 + \gamma_3 &= \beta_1 \\
\gamma_1\gamma_2 + \gamma_2\gamma_3 + \gamma_3\gamma_1 &= \beta_1 + \beta_3 \\
\gamma_1\gamma_2\gamma_3 &= \beta_2 + 2.
\end{aligned}$$

So the equation with roots γ_1, γ_2, and γ_3 is

$$x^3 - \beta_1 x^2 + (\beta_1 + \beta_3)x - (\beta_2 + 2) = 0.$$

The substitution $x = y + \frac{\beta_1}{3}$ transforms the above equation into

$$27y^3 + 9\left[3(\beta_1 + \beta_3) - \beta_1^2\right] y + 9\beta_1(\beta_1 + \beta_3) - 2\beta_1^3 - 27\beta_2 - 54 = 0.$$

whose roots are $\gamma_1 - \frac{\beta_1}{3}$, $\gamma_2 - \frac{\beta_1}{3}$, and $\gamma_3 - \frac{\beta_1}{3}$. Finally the substitution $y = \frac{z}{3}$ transforms the previous equation into the equation

$$z^3 + 3\left[3(\beta_1 + \beta_3) - \beta_1^2\right] z + 9\beta_1(\beta_1 + \beta_3) - 2\beta_1^3 - 27\beta_2 - 54 = 0$$

whose roots are $3\gamma_1 - \beta_1$, $3\gamma_2 - \beta_1$, and $3\gamma_3 - \beta_1$. The following table illustrates the transformation used and the roots of the polynomials at each step of the above procedure.

Equation	Roots
$z^{18}+z^{17}+\cdots+z^3+z^2+z+1=0$	$\alpha^i,\quad 1\le i\le 18,\quad \alpha=e^{\frac{2\pi i}{19}}$
$y^3+y^2-6y-7=0$	$\beta_1=\alpha+\alpha^8+\alpha^7+\alpha^{18}+\alpha^{11}+\alpha^{12}$ $\beta_2=\alpha^2+\alpha^{16}+\alpha^{14}+\alpha^{17}+\alpha^3+\alpha^5$ $\beta_3=\alpha^4+\alpha^{13}+\alpha^9+\alpha^{15}+\alpha^6+\alpha^{10}$
$27u^3-171u-133=0$	$\beta_1+\frac{1}{3},\ \beta_2+\frac{1}{3},\ \beta_3+\frac{1}{3}$
$x^3-57x-133=0$	$3\beta_1+1,\ 3\beta_2+1,\ 3\beta_3+1$
$x^3-\beta_1x^2+(\beta_1+\beta_3)x-(\beta_2+2)=0$	$\gamma_1=\alpha+\alpha^{18}=2\cos\frac{2\pi}{19}$, $\gamma_2=\alpha^8+\alpha^{11}=2\cos\frac{16\pi}{19}$, $\gamma_3=\alpha^7+\alpha^{12}=2\cos\frac{14\pi}{19}$
$27y^3+9\left[3(\beta_1+\beta_3)-\beta_1^2\right]y+9\beta_1(\beta_1+\beta_3)-2\beta_1^3-27\beta_2-54=0$	$\gamma_1-\frac{\beta_1}{3},\ \gamma_2-\frac{\beta_1}{3},\ \gamma_3-\frac{\beta_1}{3}$
$z^3+3\left[3(\beta_1+\beta_3)-\beta_1^2\right]z+9\beta_1(\beta_1+\beta_3)-2\beta_1^3-27\beta_2-54=0$	$3\gamma_1-\beta_1,\ 3\gamma_2-\beta_1,\ 3\gamma_3-\beta_1$

The following provides a summary of the calculation for $\cos\dfrac{2\pi}{19}$ carried out above.

$$
\begin{aligned}
\omega &= \tfrac{-1+i\sqrt{3}}{2}, \qquad i=\sqrt{-1}\\
A &= \left(\tfrac{133+57\sqrt{-3}}{2}\right)^{\frac{1}{3}}\\
B &= \left(\tfrac{133-57\sqrt{-3}}{2}\right)^{\frac{1}{3}}, \qquad A\cdot B=19\\
\beta_1 &= \tfrac{1}{3}\left(\omega^2A+\omega B-1\right)\\
\beta_2 &= \tfrac{1}{3}\left(A+B-1\right)\\
\beta_3 &= \tfrac{1}{3}\left(\omega A+\omega^2B-1\right)\\
p &= 3\left[3(\beta_1+\beta_2)-\beta_1^2\right]\\
q &= 9\beta_1(\beta_1+\beta_3)-2\beta_1^3-27\beta_2-54\\
P &= \left(-\tfrac{q}{2}+\sqrt{\tfrac{q^2}{4}+\tfrac{p^3}{27}}\right)^{\frac{1}{3}}\\
Q &= \left(-\tfrac{q}{2}-\sqrt{\tfrac{q^2}{4}+\tfrac{p^3}{27}}\right)^{\frac{1}{3}}, \qquad P\cdot Q=-\tfrac{p}{3}\\
\cos\frac{2\pi}{19} &= \tfrac{1}{6}\left(P+Q+\beta_1\right).
\end{aligned}
$$

4. Gauss' Construction of Regular 17-gon

A Radical Expression of $\cos\frac{2\pi}{17}$

As is well known Gauss found a radical expression for $\cos\dfrac{2\pi}{17}$ involving only (real) square root radicals and so proved that a regular 17-gon is constructible. In fact, Gauss proved that this can be done for any integer N such that $N = 2^k \cdot p_1 \cdot p_2 \cdots p_r$, where $p_1, p_2, \cdots p_r$ are distinct Fermat prime (i.e., primes of the form $2^n + 1$). The basic idea behind Gauss's proof can be easily understood by observing that if p is a Fermat prime then the set of nonzero residue modulo p form a cyclic group G of order $p - 1 = 2^n$. In such a group there is a (unique) tower of subgroup in which each subgroup (other than G) is of index two in the preceding group. For instance in the case of $p = 17$, such a tower of subgroups is

$$G = \langle\bar{3}\rangle \supset \langle\bar{9}\rangle \supset \langle\bar{13}\rangle \supset \langle\bar{16}\rangle \supset \langle\bar{1}\rangle$$

Here the bars indicate the residue class modulo 17 and $\langle\bar{x}\rangle$ denotes the subgroup of G generated by $\bar{x}$. Gauss' precise calculation of $\cos\frac{2\pi}{17}$ can be more easily explained by means of the following diagram. In this diagram $\alpha = e^{\frac{2\pi i}{17}} = \cos\frac{2\pi}{17} + i\sin\frac{2\pi}{17}$.

Diagram illustrating the calculation of $2\cos\dfrac{2\pi}{17}$

$$\alpha = e^{\frac{2\pi i}{17}} = \cos\frac{2\pi}{17} + i\sin\frac{2\pi}{17}, \alpha^k, 0 \le k \le 16 \text{ are roots of } z^{17} - 1 = 0$$

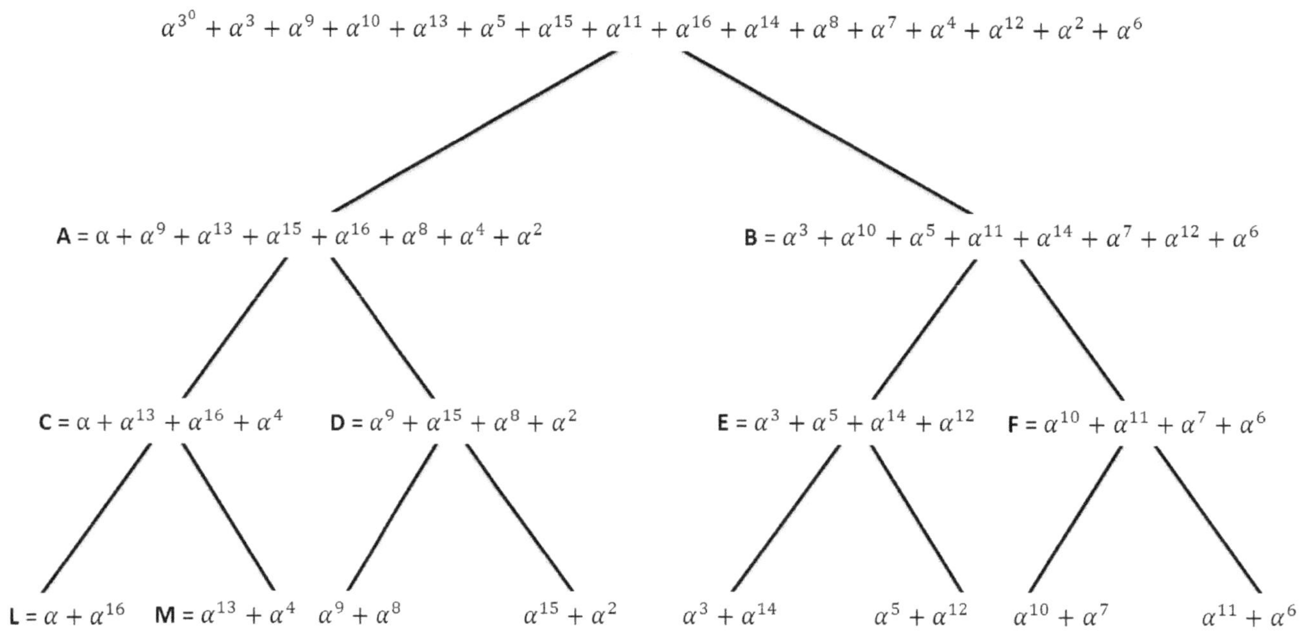

$$\begin{aligned}
\mathbf{A} &= \alpha+\alpha^2+\alpha^4+\alpha^8+\alpha^{16}+\alpha^{15}+\alpha^{13}+\alpha^9 \\
\mathbf{B} &= \alpha^3+\alpha^6+\alpha^{12}+\alpha^7+\alpha^{14}+\alpha^{11}+\alpha^5+\alpha^{10} \\
\mathbf{C} &= \alpha+\alpha^4+\alpha^{16}+\alpha^{13} \\
\mathbf{D} &= \alpha^2+\alpha^8+\alpha^{15}+\alpha^9 \\
\mathbf{E} &= \alpha^3+\alpha^{12}+\alpha^{14}+\alpha^5 \\
\mathbf{G} &= \alpha^6+\alpha^7+\alpha^{11}+\alpha^{10} \\
\mathbf{L} &= \alpha+\alpha^{16} \\
\mathbf{M} &= \alpha^4+\alpha^{13}
\end{aligned}$$

We note that

$$\begin{aligned}
&A+B=-1, \quad A\cdot B=-4 \\
&C+D=A, \quad C\cdot D=-1 \\
&E+G=B, \quad E\cdot G=-1 \\
&L+M=C, \quad L\cdot M=E \\
&L=2\cos\frac{2\pi}{17}
\end{aligned}$$

Caution:

Here A, B, L, and M do not have the same meaning as they did in the first section of this chapter.

The following steps lead to a radical expression (containing only square root radicals) for $\cos\frac{2\pi}{17}$. The notation is that of the previous diagram.

Step 1.

$A+B=-1$, $A\cdot B=-4$, so A and B are the roots of the equation $x^2+x-4=0$, therefore

$$A=\frac{-1+\sqrt{17}}{2}, \qquad B=\frac{-1-\sqrt{17}}{2}$$

Step 2.

$C+D=A$, $C\cdot D=-1$, so C and D are the roots of the equation $x^2-Ax-1=0$, therefore

$$C=\frac{A+\sqrt{A^2+4}}{2}, \qquad D=\frac{A-\sqrt{A^2+4}}{2}$$

Step 3.

$E + G = B$, $E \cdot G = -1$, so E and G are the roots of the equation $x^2 - Bx - 1 = 0$, therefore

$$E = \frac{B + \sqrt{B^2+4}}{2}, \qquad G = \frac{B - \sqrt{B^2+4}}{2}$$

Step 4.

$L + M = C$, $L \cdot M = E$, so L and M are the roots of the equation $x^2 - Cx + E = 0$, therefore

$$L = \frac{C + \sqrt{C^2-4E}}{2}, \qquad M = \frac{C - \sqrt{C^2-4E}}{2}$$

Squaring A and B gives us

$$A^2 = \frac{9-\sqrt{17}}{2}, \qquad A^2 + 4 = \frac{17-\sqrt{17}}{2}$$

$$B^2 = \frac{9+\sqrt{17}}{2}, \qquad B^2 + 4 = \frac{17+\sqrt{17}}{2}$$

$$L = \frac{C+\sqrt{C^2-4E}}{2} = \frac{1}{4}\left[A + \sqrt{A^2+4} + \sqrt{(A+\sqrt{A^2+4})^2 - 8(B+\sqrt{B^2+4})}\right].$$

Substituting the values $A = \frac{-1+\sqrt{17}}{2}$, $B = \frac{-1-\sqrt{17}}{2}$ in above and simplifying we get the following expression for $2\cos\frac{2\pi}{17}$

$$L = 2\cos\frac{2\pi}{17} = \frac{1}{8}\left\{-1+\sqrt{17}+\sqrt{(34-2\sqrt{17})}\right\}$$
$$+\frac{1}{8}\sqrt{\left\{68+12\sqrt{17}-16\sqrt{(34+2\sqrt{17})}+2(\sqrt{17}-1)\sqrt{(34-2\sqrt{17})}\right\}}.$$

Therefore

$$\cos\frac{2\pi}{17} = \frac{1}{16}\left\{-1+\sqrt{17}+\sqrt{(34-2\sqrt{17})}\right\}$$
$$+\frac{1}{16}\sqrt{\left\{68+12\sqrt{17}-16\sqrt{(34+2\sqrt{17})}+2(\sqrt{17}-1)\sqrt{(34-2\sqrt{17})}\right\}}.$$

Remark 1 It should be noted that expressing roots of an equation in terms of radicals (whenever it is possible) is only of academic interest and has little practical implication. This is made much clearer from the observation that in order to calculate square roots, cube roots and so on of complex numbers one is forced to take recourse to De Moiver's theorem and thus take recourse to values of trigonometric functions. Nevertheless the problem of finding a radical expression for the roots of equation is of much historical importance as its study has contributed to the development of modern mathematics especially in the field of algebra.

Remark 2 In a later chapter we shall discuss the computation of a radical expression for $\cos\frac{2\pi}{11}$, which was first done by Vandermonde (1771) over 200 years ago.

5. Solution of the Quartic Equation

As discussed in the previous section the roots α, β, and γ of the cubic equation

$$x^3 + px + q = 0,$$

are given by

$$\alpha = A + B; \quad \beta = \omega^2 A + \omega B; \quad \gamma = \omega A + \omega^2 B,$$

where $\omega = e^{\frac{2\pi i}{3}}$ is a primitive cube root of unity and

$$A, B = \left[-\frac{q}{2} \pm \sqrt{\frac{q^2}{4} + \frac{p^3}{27}} \right]^{\frac{1}{3}}.$$

Here the cube roots appearing in the above expression are so chosen that $A \cdot B = -\frac{p}{3}$.

A general monic quartic equation is of the form $x^4 + ax^3 + bx^2 + cx + d = 0$. In this equation if we substitute $x = a - \frac{a}{4}$ then in the resulting quartic equation in y the coefficient of y^3 is zero. So while trying to find a radical expression of a quartic equation we may assume that the coefficient of the cubic term is zero.

Consider the quartic equation $x^4 + px^2 + qx + r = 0$. Suppose its roots are $\alpha_1, \alpha_2, \alpha_3$, and α_4. Then we have

$$\begin{aligned}
\alpha_1 + \alpha_2 + \alpha_3 + \alpha_4 &= 0 \\
\alpha_1\alpha_2 + \alpha_1\alpha_3 + \alpha_1\alpha_4 + \alpha_2\alpha_3 + \alpha_2\alpha_4 + \alpha_3\alpha_4 &= p \\
\alpha_1\alpha_2\alpha_3 + \alpha_1\alpha_2\alpha_4 + \alpha_1\alpha_3\alpha_4 + \alpha_2\alpha_3\alpha_4 &= -q \\
\alpha_1\alpha_2\alpha_3\alpha_4 &= r.
\end{aligned}$$

Let

$$\begin{aligned}\Theta_1 &= (\alpha_1+\alpha_2)(\alpha_3+\alpha_4) &= -(\alpha_1+\alpha_2)^2 &= -(\alpha_3+\alpha_4)^2\\ \Theta_2 &= (\alpha_1+\alpha_3)(\alpha_2+\alpha_4) &= -(\alpha_1+\alpha_3)^2 &= -(\alpha_2+\alpha_4)^2\\ \Theta_3 &= (\alpha_1+\alpha_4)(\alpha_2+\alpha_3) &= -(\alpha_1+\alpha_4)^2 &= -(\alpha_2+\alpha_3)^2.\end{aligned}$$

We shall show that

$$\begin{aligned}\Theta_1+\Theta_2+\Theta_3 &= 2p\\ \Theta_1\Theta_2+\Theta_2\Theta_3+\Theta_3\Theta_1 &= p^2-4r\\ \Theta_1\Theta_2\Theta_3 &= -q^2.\end{aligned}$$

We have

$$\Theta_1+\Theta_2+\Theta_3 = (\alpha_1+\alpha_2)(\alpha_3+\alpha_4)+(\alpha_1+\alpha_3)(\alpha_2+\alpha_4)+(\alpha_1+\alpha_4)(\alpha_2+\alpha_3) = 2\sum_{i\neq j}\alpha_i\alpha_j = 2p.$$

To prove that $\Theta_1\Theta_2+\Theta_2\Theta_3+\Theta_3\Theta_1 = p^2-4r$ we first note that

$$\left(\sum_{i\neq j}\alpha_i\alpha_j\right)^2 = \sum_{i\neq j}\alpha_i^2\alpha_j^2 + 2\sum\alpha_i^2\alpha_j\alpha_k + 6\alpha_1\alpha_2\alpha_3\alpha_4.$$

In the sums above the subscripts i, j, k vary from 1 to 4 and are all different. We further note that

$$(\alpha_1+\alpha_2+\alpha_3+\alpha_4)(\alpha_1\alpha_2\alpha_3+\alpha_1\alpha_2\alpha_4+\alpha_1\alpha_3\alpha_4+\alpha_2\alpha_3\alpha_4) = \sum\alpha_i^2\alpha_j\alpha_k + 4\alpha_1\alpha_2\alpha_3\alpha_4$$

Since $\alpha_1+\alpha_2+\alpha_3+\alpha_4 = 0$, we have

$$\sum\alpha_i^2\alpha_j\alpha_k = -4\alpha_1\alpha_2\alpha_3\alpha_4 = -4r.$$

Now

$$\begin{aligned}&\Theta_1\Theta_2+\Theta_2\Theta_3+\Theta_3\Theta_1\\ &= (\alpha_1+\alpha_2)(\alpha_3+\alpha_4)(\alpha_1+\alpha_3)(\alpha_2+\alpha_4)+(\alpha_1+\alpha_3)(\alpha_2+\alpha_4)(\alpha_1+\alpha_4)(\alpha_2+\alpha_3)\\ &\quad +(\alpha_1+\alpha_4)(\alpha_2+\alpha_3)(\alpha_1+\alpha_2)(\alpha_3+\alpha_4)\\ &= 3\sum\alpha_i^2\alpha_j\alpha_k+\sum\alpha_i^2\alpha_j^2+6\alpha_1\alpha_2\alpha_3\alpha_4 = \sum\alpha_i^2\alpha_j\alpha_k+\left(\sum\alpha_i\alpha_j\right)^2\\ &= -4r+p^2.\end{aligned}$$

Finally we have

$$\begin{aligned}&(\alpha_1+\alpha_2)(\alpha_1+\alpha_3)(\alpha_1+\alpha_4) =\\ &\alpha_1^3+\alpha_1^2\alpha_2+\alpha_1^2\alpha_3+\alpha_1^2\alpha_4+\alpha_1\alpha_2\alpha_3+\alpha_1\alpha_2\alpha_4+\alpha_1\alpha_3\alpha_4+\alpha_2\alpha_3\alpha_4 =\\ &\alpha_1^2(\alpha_1+\alpha_2+\alpha_3+\alpha_4)-q = -q.\end{aligned}$$

Therefore

$$\Theta_1\Theta_2\Theta_3 = -(\alpha_1+\alpha_2)^2(\alpha_1+\alpha_3)^2(\alpha_1+\alpha_4)^2 = -q^2$$

and

$$\Theta_1 = -(\alpha_1+\alpha_2)^2 \Longrightarrow \alpha_1+\alpha_2 = \pm\sqrt{-\Theta_1}$$
$$\Theta_2 = -(\alpha_1+\alpha_3)^2 \Longrightarrow \alpha_1+\alpha_3 = \pm\sqrt{-\Theta_2}$$
$$\Theta_3 = -(\alpha_1+\alpha_4)^2 \Longrightarrow \alpha_1+\alpha_4 = \pm\sqrt{-\Theta_3}\,.$$

As noted above

$$(\alpha_1+\alpha_2)(\alpha_1+\alpha_3)(\alpha_1+\alpha_4) = -q.$$

Therefore

$$\sqrt{-\Theta_1}\sqrt{-\Theta_2}\sqrt{-\Theta_3} = (\alpha_1+\alpha_2)(\alpha_1+\alpha_3)(\alpha_1+\alpha_4) = -q.$$

Thus in

$$\alpha_1+\alpha_2 = \pm\sqrt{-\Theta_1}, \qquad \alpha_1+\alpha_3 = \pm\sqrt{-\Theta_2}, \qquad \alpha_1+\alpha_4 = \pm\sqrt{-\Theta_3}$$

the signs have to be chosen so they satisfy

$$(\alpha_1+\alpha_2)(\alpha_1+\alpha_3)(\alpha_1+\alpha_4) = -q.$$

If we let

$$\alpha_1+\alpha_2 = \sqrt{-\Theta_1},\ \ \alpha_1+\alpha_3 = \sqrt{-\Theta_2},\ \ \alpha_1+\alpha_4 = \sqrt{-\Theta_3}$$

then

$$\alpha_1 = \frac{1}{2}\left(\sqrt{-\boldsymbol{\Theta}_1} + \sqrt{-\boldsymbol{\Theta}_2} + \sqrt{-\boldsymbol{\Theta}_3}\right)$$

and as

$$\alpha_2+\alpha_1 = \sqrt{-\Theta_1},\ \ \alpha_2+\alpha_3 = -\sqrt{-\Theta_3},\ \ \alpha_2+\alpha_4 = -\sqrt{-\Theta_2}$$

so

$$\alpha_2 = \frac{1}{2}\left(\sqrt{-\boldsymbol{\Theta}_1} - \sqrt{-\boldsymbol{\Theta}_2} - \sqrt{-\boldsymbol{\Theta}_3}\right)$$

and as

$$\alpha_3+\alpha_1 = \sqrt{-\Theta_2},\ \ \alpha_3+\alpha_2 = -\sqrt{-\Theta_3},\ \ \alpha_3+\alpha_4 = -\sqrt{-\Theta_1}$$

so

$$\alpha_3 = \frac{1}{2}\left(-\sqrt{-\boldsymbol{\Theta}_1} + \sqrt{-\boldsymbol{\Theta}_2} - \sqrt{-\boldsymbol{\Theta}_3}\right)$$

and finally as

$$\alpha_4+\alpha_1=\sqrt{-\Theta_3},\quad \alpha_4+\alpha_2=-\sqrt{-\Theta_2},\quad \alpha_4+\alpha_3=-\sqrt{-\Theta_1}$$

so

$$\alpha_4=\frac{1}{2}\left(-\sqrt{-\Theta_1}-\sqrt{-\Theta_2}+\sqrt{-\Theta_3}\right).$$

This completes the solution of the quartic equation.

An Example

Let us solve the quartic equation

$$x^4-x^3-4x^2+4x+1=0$$

by the method discussed above. Let the roots of the above equation be denoted by $\alpha,\beta,\gamma,$ and δ. Then the equation whose roots are

$$\alpha_1=4\alpha-1,\quad \alpha_2=4\beta-1,\quad \alpha_3=4\gamma-1,\quad \text{and}\quad \alpha_4=4\delta-1$$

is

$$x^4-70x^2+120x+445=0.$$

So in the notation that we used for solving the quartic above $p=-70,\ \ q=120$, and $r=445$.

Let

$$\Theta_1=(\alpha_1+\alpha_2)(\alpha_3+\alpha_4),\quad \Theta_2=(\alpha_1+\alpha_3)(\alpha_2+\alpha_4),\quad \Theta_3=(\alpha_1+\alpha_4)(\alpha_2+\alpha_3)$$

Then as we have seen before

$$\begin{aligned}\Theta_1+\Theta_2+\Theta_3 &= 2p &&= -140\\ \Theta_1\Theta_2+\Theta_2\Theta_3+\Theta_3\Theta_1 &= p^2-4r &&= 3120\\ \Theta_1\Theta_2\Theta_3 &= -q^2 &&= -14400.\end{aligned}$$

Therefore the cubic equation whose roots are Θ_1, Θ_2, and Θ_3 is

$$x^3+140x^2+3120x+14400=0$$

This equation can be solved easily as one of its roots is -20 and $x^3+140x^2+3120x+14400=(x+20)\times(x^2+120x+720)$. So we let

$$\Theta_1=-20\ \text{ and }\ \Theta_2,\Theta_3=-60\pm 24\sqrt{5}.$$

Since $\sqrt{-\Theta_1} \cdot \sqrt{-\Theta_2} \cdot \sqrt{-\Theta_3} = -q = -120$, we let

$$\sqrt{-\Theta_1} = -\sqrt{20}, \quad \sqrt{-\Theta_2} = \sqrt{60 - 24\sqrt{5}}, \quad \sqrt{-\Theta_3} = \sqrt{60 + 24\sqrt{5}}.$$

We recall from the preceding discussion of the solution of the quartic equation that

$$\begin{aligned}
\alpha_1 &= \tfrac{1}{2}(\sqrt{-\Theta_1} + \sqrt{-\Theta_2} + \sqrt{-\Theta_3}) \\
\alpha_2 &= \tfrac{1}{2}(\sqrt{-\Theta_1} - \sqrt{-\Theta_2} - \sqrt{-\Theta_3}) \\
\alpha_3 &= -\tfrac{1}{2}(\sqrt{-\Theta_1} + \sqrt{-\Theta_2} - \sqrt{-\Theta_3}) \\
\alpha_4 &= -\tfrac{1}{2}(\sqrt{-\Theta_1} - \sqrt{-\Theta_2} + \sqrt{-\Theta_3}).
\end{aligned}$$

So we have

$$\begin{aligned}
\alpha_1 = 4\alpha - 1 &= \tfrac{1}{2}\left(-\sqrt{20} + \sqrt{60 - 24\sqrt{5}} + \sqrt{60 + 24\sqrt{5}}\right) \\
&= -\sqrt{5} + \sqrt{15 - 6\sqrt{5}} + \sqrt{15 - 6\sqrt{5}}
\end{aligned}$$

$$\begin{aligned}
\alpha_2 = 4\beta - 1 &= \tfrac{1}{2}\left(-\sqrt{20} - \sqrt{60 - 24\sqrt{5}} - \sqrt{60 + 24\sqrt{5}}\right) \\
&= -\sqrt{5} - \sqrt{15 - 6\sqrt{5}} - \sqrt{15 - 6\sqrt{5}}
\end{aligned}$$

$$\begin{aligned}
\alpha_3 = 4\gamma - 1 &= \tfrac{1}{2}\left(+\sqrt{20} + \sqrt{60 - 24\sqrt{5}} - \sqrt{60 + 24\sqrt{5}}\right) \\
&= \sqrt{5} + \sqrt{15 - 6\sqrt{5}} - \sqrt{15 - 6\sqrt{5}}
\end{aligned}$$

$$\begin{aligned}
\alpha_4 = 4\delta - 1 &= \tfrac{1}{2}\left(+\sqrt{20} - \sqrt{60 - 24\sqrt{5}} + \sqrt{60 + 24\sqrt{5}}\right) \\
&= \sqrt{5} - \sqrt{15 - 6\sqrt{5}} + \sqrt{15 - 6\sqrt{5}}.
\end{aligned}$$

Therefore

$$\alpha = \tfrac{1}{4}\left(1 - \sqrt{5} + \sqrt{15 - 6\sqrt{5}} + \sqrt{15 + 6\sqrt{5}}\right)$$

$$\beta = \tfrac{1}{4}\left(1 - \sqrt{5} - \sqrt{15 - 6\sqrt{5}} - \sqrt{15 + 6\sqrt{5}}\right)$$

$$\gamma = \tfrac{1}{4}\left(1 + \sqrt{5} + \sqrt{15 - 6\sqrt{5}} - \sqrt{15 + 6\sqrt{5}}\right)$$

$$\delta = \tfrac{1}{4}\left(1 + \sqrt{5} - \sqrt{15 - 6\sqrt{5}} + \sqrt{15 + 6\sqrt{5}}\right).$$

The values of α, β, γ, and δ given above should not come as a surprise. In fact we know that 15-th cyclotomic polynomial $\Phi_{15}(x)$ is $x^8 - x^7 + x^5 - x^4 + x^3 - x + 1$. So if after dividing the equation $x^8 - x^7 + x^5 - x^4 + x^3 - x + 1 = 0$, by x^4 we substitute $x + \dfrac{1}{x} = y$, we get the equation $y^4 - y^3 - 4y^2 + 4y + 1 = 0$—which is the same equation we solved above. As the eight roots of $\Phi_{15}(x) = 0$ are $e^{\frac{2k\pi i}{15}}$ for $k = 1, 2, 4, 7, 8, 11, 13, 14$, the four roots of $y^4 - y^3 - 4y^2 + 4y + 1 = 0$ are $2\cos\frac{2\pi}{15}$, $2\cos\frac{4\pi}{15}$, $2\cos\frac{8\pi}{15}$, and $2\cos\frac{14\pi}{15}$. It can be checked by a calculator that

$$\alpha = 2\cos\frac{4\pi}{15}, \quad \beta = 2\cos\frac{14\pi}{15}, \quad \gamma = 2\cos\frac{8\pi}{15}, \quad \delta = 2\cos\frac{2\pi}{15}.$$

CHAPTER 4

Field Automorphisms and Galois Groups

We shall first give some more facts about field extensions. Although much of what we write is true about fields in general but unless otherwise specified we shall continue to assume that all fields considered have characteristic zero.

Theorem 0.1. *(Kronecker) Let $\mathbb{K}$ be a field and $f(x) \in \mathbb{K}[x]$ is a nonconstant irreducible polynomial. Then there is an extension field $\mathbb{L}$ of $\mathbb{K}$ which has a root of $f(x)$.*

Before starting the proof of the above theorem we shall make a comment. The field $\mathbb{L}$ that we are about to construct (containing a root of $f(x)$) *will not* contain $\mathbb{K}$ as a subfield as such but will have an isomorphic copy of $\mathbb{K}$ inside it. Without loss of generality we can identify this isomorphic copy of $\mathbb{K}$ (lying in $\mathbb{L}$) with $\mathbb{K}$ itself. It is in this sense that $\mathbb{L}$ will contain $\mathbb{K}$ as a subfield. Likewise the polynomial $f(x)$ will be replaced by the polynomial obtained by replacing the coefficients of $f(x)$ by their images under the embedding of $\mathbb{K}$ into $\mathbb{L}$ we just talked about.

Proof. Without loss of generality we may assume that the irreducible polynomial $f(x) \in \mathbb{K}[x]$ is a monic polynomial. Let $f(x) = x^n + a_1x^{n-1} + a_2x^{n-2} + \cdots a_{n-1}x + a_n \in \mathbb{K}[x]$. As $f(x) \in \mathbb{K}[x]$ is irreducible the principal ideal generated by $f(y) \in \mathbb{K}[y]$ is a maximal ideal, and the quotient $\mathbb{K}[y]/\langle f(y)\rangle$ of $\mathbb{K}[y]$ by the maximal ideal $\langle f(y)\rangle$ is a field. We shall denote this field by $\mathbb{L}$. The natural ring homomorphism

$$\eta : \mathbb{K}[y] \longrightarrow \mathbb{K}[y]/\langle f(y)\rangle = \mathbb{L}$$

is such that its restriction to $\mathbb{K}$ is one-to-one. Therfore the image $\eta(\mathbb{K})$ of $\mathbb{K}$ in the above map is a subfield of the field $\mathbb{L} = \mathbb{K}[y]/\langle f(y)\rangle$. With abuse of notation we denote this embedding of $\mathbb{K}$ into $\mathbb{L}$ by η as well. So we have the embedding

$$\eta : \mathbb{K} \longrightarrow \mathbb{L} = \mathbb{K}[y]/\langle f(y)\rangle$$

in this mapping

$$a \longrightarrow \bar{a} = a + \langle f(y) \rangle$$

The above mapping can be extended to the following ring homomorphism.

$$\eta : \mathbb{K}[x] \longrightarrow \mathbb{L}[x], \qquad \text{where } \mathbb{L} = \mathbb{K}[y]/\langle f(y) \rangle.$$

In this mapping the image of the polynomial

$$f(x) = x^n + a_1 x^{n-1} + a_2 x^{n-2} + \cdots a_{n-1} x + a_n \in \mathbb{K}[x]$$

is the polynomial

$$\eta\left(f(x)\right) = \bar{f}(x) = x^n + \bar{a}_1 x^{n-1} + \bar{a}_2 x^{n-2} + \cdots \bar{a}_{n-1} x + \bar{a}_n \in \mathbb{L}[x].$$

We claim that the element $\bar{y} = y + \langle f(y) \rangle \in \mathbb{L} = \mathbb{K}[y]/\langle f(y) \rangle$ is a zero of the polynomial $\bar{f}(x)$. In fact $\bar{f}(\bar{y})$ is equal to

$$\bar{y}^n + \bar{a}_1 \bar{y}^{(n-1)} + \cdots + \bar{a}_n = \overline{y^n + a_1 y^{n-1} + \cdots + a_n} = \overline{f(y)} = f(y) + \langle f(y) \rangle = \langle f(y) \rangle$$

which is the zero element of the field $\mathbb{L} = \mathbb{K}[y]/\langle f(y) \rangle$. □

Note 1. The above proof giving the construction of an extension field having a zero of the irreducible polynomial with coefficients in the base field may appear to be rather confusing to a beginner. If we follow the construction of the extension field of the preceding theorem then field of complex numbers is nothing but the field $\mathbb{R}[x]/\langle x^2+1 \rangle$. However if we identify (or rename) the element $x + \langle x^2+1 \rangle$ of this field with the symbol i and identify $a + \langle x^2+1 \rangle$, $(a \in \mathbb{R})$ with a then

$$\begin{aligned} \mathbb{R}[x]/\langle x^2+1 \rangle &= \{a + bx + \langle x^2+1 \rangle : a, b \in \mathbb{R}\} \\ &= \{a + \langle x^2+1 \rangle + b(x + \langle x^2+1 \rangle) : a, b \in \mathbb{R}\} \\ &= \{a + bi : a, b \in \mathbb{R}\}. \end{aligned}$$

giving us the usual representation of the complex numbers. The point here is that the symbol $b \cdot i$ gets the meaning only in above context. It is only our excessive familiarity in dealing with complex numbers that in a way deprives us of the appreciation that we otherwise would have for the above construction of complex numbers—which can be thought of extension of the field of real numbers in which the polynomial x^2+1 has a zero. After having constructed the field $\mathbb{L}$—an extension of $\mathbb{K}$ having a zero of the irreducible polynomial $f(x) \in \mathbb{K}[x]$, we do not have to continue to write $\mathbb{L}$ as a residue classes modulo the maximal ideal $\langle f(x) \rangle$. We regard $\mathbb{K}$ as a subfield of the field $\mathbb{L}$ in as much as $\mathbb{R}$ is the subfield of the field of complex numbers. The readers may recall the similar situation in the course of construction of real numbers by means of Dedekind cuts or as equivalence classes of Cauchy sequences of rational numbers. The second note below may help to see this similarity better.

Note 2. After having shown that given a nonconstant irreducible polynomial $f(x) \in \mathbb{K}[x]$ how to construct an extension of the field $\mathbb{K}$ containing a root of $f(x)$, it may help the readers to think of the extension field in the following manner. Let $\mathbb{L}$ consist of the formal sums

$$\mathbb{L} = \left\{b_0 + b_1\alpha + b_2\alpha^2 + \cdots + b_{n-1}\alpha^{n-1} : b_i \in \mathbb{K}\right\}$$

where α is some "unknown" which will constitute a root of the irreducible polynomial $f(x) \in \mathbb{K}[x]$ in an extension of the field $\mathbb{K}$. The operations of addition and multiplication are defined in $\mathbb{L}$ subject to the condition that

$$\alpha^n = -\left(a_1\alpha^{n-1} + a_2\alpha^{n-2} + \cdots + a_{n-1}\alpha + a_n\right)$$

It can be shown that $\mathbb{L}$ is a field containing $\mathbb{K}$ as a subfield and contains a root α of the polynomial $f(x) \in \mathbb{K}[x]$.

We also note that the extension field $\mathbb{L}$ constructed above is generated over the field $\mathbb{K}$ by a zero of the irreducible polynomial $f(x)$ and has dimension equal to degree $(f(x))$ when regarded as a vector space over the field $\mathbb{K}$.

By continuing the above process of constructing an extension field of given field having a zero of the given irreducible polynomial, repeatedly if necessary we can find an extension $\mathbb{L}$ of a field $\mathbb{K}$ in which the polynomial $f(x) \in \mathbb{K}[x]$ splits completely and $\mathbb{L}$ is generated over $\mathbb{K}$ by the zeros of the given polynomial. Such an extension field $\mathbb{L}$ is called the *splitting field* of the polynomial $f(x)$. The splitting field of a polynomial plays an important role in Galois Theory.

A field $\mathbb{L}$ which is such that every polynomial $f(x) \in \mathbb{L}[x]$ has zero in $\mathbb{L}$ (and therefore splits completely in $\mathbb{L}$) is called an algebraically closed field. The field $\mathbb{C}$ of complex numbers is an example of an algebraically closed field. In fact the set of all algebraic numbers in $\mathbb{C}$ is an algebraically closed field. With the help of Zorn's Lemma (or its equivalent Axiom of Choice) it can be shown that given a field $\mathbb{K}$ we can find a field $\mathbb{L}$ such that $\mathbb{L}/\mathbb{K}$ is an algebraic extension and the field $\mathbb{L}$ is an algebraically closed field. Such a field $\mathbb{L}$ is called *algebraic closure* of the field $\mathbb{K}$. It can also be shown that if the fields $\mathbb{L}_1$ and $\mathbb{L}_2$ are algebraic closures

of the same field $\mathbb{K}$ then $\mathbb{L}_1$ and $\mathbb{L}_2$ are isomorphic fields. The field $\mathbb{C}$ of complex numbers is algebraic closure of the field $\mathbb{R}$ of real numbers and the field consisting of all algebraic numbers is the algebraic closure of the field $\mathbb{Q}$ of rational numbers and is denoted as $\overline{\mathbb{Q}}$. In the case of characteristic zero the splitting field of a polynomial (or a set of polynomials) will also be referred to as a normal extension or a Galois extension of the base field.
As mentioned above a splitting field of $f(x) \in \mathbb{K}[x]$ is also called a normal extension or a Galois extension. The reason that we are a bit cautious about it is that a Galois extension is generally defined as an extension which is *normal and separable.* Although we have not defined what is meant by a separable extension (we shall do so soon) but in the case of characteristic zero all (algebraic) extensions are separable. In fact the case of so called inseparable extensions can only arise while dealing with infinite fields of finite characteristic.

1. Separable Elements and Separable Extensions

Let $\mathbb{L}/\mathbb{K}$ be an algebraic extension. An element $\alpha \in \mathbb{L}$ is called separable if the minimal polynomial of α over $\mathbb{K}$ has distinct roots in any extension of $\mathbb{K}$. The extension $\mathbb{L}/\mathbb{K}$ is said to be separable if every element of $\mathbb{L}$ is separable. An algebraic extension which is not separable is said to be inseparable.
As mentioned before it can be shown that inseparable extensions can only appear in case of infinite fields of finite characteristic. We shall discuss finite fields later, here we shall simply provide an example of an inseparable extension.
Let $\mathbb{F}_p$ denote the finite field of p elements. Let $\mathbb{F}_p[x]$ denote the polynomial ring in one variable x over $\mathbb{F}_p$ and $\mathbb{F}_p(x)$ the corresponding quotient field. Consider the following diagram of fields:

$$\begin{array}{c} \mathbb{F}_p(x) \\ | \\ \mathbb{F}_p(x^p) \\ | \\ \mathbb{F}_p \end{array}$$

The field extension $\mathbb{F}_p(x)$ is an extension of the field $\mathbb{F}_p(x^p)$ of degree p, as the element $x \in \mathbb{F}_p(x)$ is algebraic of degree p over $\mathbb{F}_p(x^p)$, since x is a root of the polynomial $y^p - x^p \in \mathbb{F}(x^p)[y]$. The set $\{1, x, x^2, \cdots, x^{p-1}\}$ constitutes a basis of $\mathbb{F}_p(x)$ over $\mathbb{F}_p(x^p)$. The minimal polynomial of $x \in \mathbb{F}_p(x)$ over $\mathbb{F}_p(x^p)$ is $y^p - x^p \in \mathbb{F}_p(x)[y]$. It is easily seen that $y^p - x^p$, as a polynomial in y, is irreducible over the field $\mathbb{F}_p(x^p)$. Since $y^p - x^p = (y-x)^p$ the polynomial $y^p - x^p$ has only one zero, namely x, of multiplicity p. So $x \in \mathbb{F}_p(x)$ is not a separable element over $\mathbb{F}_p(x^p)$. Therefore the extension $\mathbb{F}_p(x)/\mathbb{F}_p(x^p)$ is inseparable.

As mentioned earlier practically all the extensions that we shall consider in this book will be separable. So for us a finite normal extension is same as splitting field of a polynomial which is in turn is same as a finite Galois extension.

2. Some Properties of Normal Extensions

Let $\mathbb{L}/\mathbb{K}$ be a finite normal extension of degree n—that is $\mathbb{L}$ is the splitting field of some polynomial in $\mathbb{K}[x]$. We mention without proof the following properties of such extensions

(1) If $f(x) \in \mathbb{K}[x]$ is an irreducible polynomial having one roots in $\mathbb{L}$ then it has all of its root in $\mathbb{L}$—in other words $f(x)$ splits completely in $\mathbb{L}$.

(2) Let $\overline{\mathbb{L}}$ be an algebraic closure of $\mathbb{L}$, and

$$\phi : \mathbb{L} \longrightarrow \overline{\mathbb{L}}$$

is an embedding of $\mathbb{L}$ into $\overline{\mathbb{L}}$ over $\mathbb{K}$ (that is it fixes every element of $\mathbb{K}$). Then $\phi(\mathbb{L}) = \mathbb{L}$ and ϕ is an automorphism of $\mathbb{L}$ over $\mathbb{K}$.

(3) An embedding

$$\phi : \mathbb{K} \longrightarrow \mathbb{L}$$

of $\mathbb{K}$ into $\mathbb{L}$ can be extended to exactly n (the degree of $\mathbb{L}$ over $\mathbb{K}$) embedding of $\mathbb{L}$ into itself.

3. Automorphism of Fields—Galois Group

Let $\mathbb{K}$ be any field (of characteristic zero) and $f(x) \in \mathbb{K}[x]$ be a nonconstant polynomial. Let $\mathbb{L}$ be the splitting field of $f(x)$ over $\mathbb{K}$. So $\mathbb{L}/\mathbb{K}$ is a finite normal extension. Let $Aut(\mathbb{L}/\mathbb{K})$ denote the set of all automorphism of $\mathbb{L}$ over $\mathbb{K}$ (that is all automorphism of $\mathbb{L}$ which fix every element of $\mathbb{K}$). Then the set $Aut(\mathbb{L}/\mathbb{K})$ forms a group under the binary operation of the composition of mappings and the order of $Aut(\mathbb{L}/\mathbb{K})$ is equal to the dimension of $\mathbb{L}$ as a vector space over $\mathbb{K}$, namely $|\mathbb{L} : \mathbb{K}|$. The group $Aut(\mathbb{L}/\mathbb{K})$ is called the Galois group of the polynomial $f(x)$ over $\mathbb{K}$.

4. Fundamental Theorem of Galois Theory

Let $\mathbb{K}$ be field and $f(x) \in \mathbb{K}[x]$ be a nonconstant polynomial and $\mathbb{L}$ be its splitting field over $\mathbb{K}$. Let $G = Aut(\mathbb{L}/\mathbb{K})$ be the Galois group of $f(x)$ over $\mathbb{K}$. Further let

$$\begin{aligned} \Omega &= \text{Set of all subgroups of the group } G\text{, and} \\ \Lambda &= \text{Set of all subfields of } \mathbb{L} \text{ which contain } \mathbb{K}. \end{aligned}$$

Then we have the following one-to-one correspondence between Ω and Λ, called the Galois correspondence. (Here H denotes a subgroup of the group G and $\mathbb{F}$ denotes a subfield of $\mathbb{L}$ containing $\mathbb{K}$.)

$$\begin{aligned} H \longrightarrow \mathbb{L}_H &= \{x \in \mathbb{L} : \sigma(x) = x, \quad \text{for all} \quad \sigma \in H\} \\ \mathbb{F} \longrightarrow Aut(\mathbb{L}/\mathbb{F}) &= \{\sigma \in Aut(\mathbb{L}/\mathbb{K}) : \sigma(x) = x, \quad \text{for all} \quad x \in \mathbb{F}\}. \end{aligned}$$

So $\mathbb{L}_H$ is a subfield of the field $\mathbb{L}$ consisting of those elements which are fixed by every element of the subgroup H of the group $G = Aut(\mathbb{L}/\mathbb{K})$.

The above correspondence satisfies the following properties:

(1) $Aut(\mathbb{L}/\mathbb{L}_H) = H$

(2) $\mathbb{L}_{Aut(\mathbb{L}/\mathbb{F})} = \mathbb{F}$

(3) A subgroup $H \subseteq G$ is a normal subgroup if and only if $\mathbb{L}_H/\mathbb{K}$ is a normal extension—in other words the field $\mathbb{L}_H$ is a splitting field of certain polynomial in $\mathbb{K}[x]$. In such a case we have the following isomorphism of groups:

$$Aut(\mathbb{L}_H) = Aut(\mathbb{L}_H/\mathbb{K}) \approx \frac{Aut(\mathbb{L}/\mathbb{K})}{Aut(\mathbb{L}/\mathbb{L}_H)}$$

Note 1. The mappings comprising Galois correspondence between the sets Ω and Λ are inverse of each other. One can also interpret Galois correspondence as setting up an "anti-isomorphism" (that is order reversing) between the lattice of subgroups of G and the lattice of subfields of $\mathbb{L}$ containing the field $\mathbb{K}$. The last remark is often helpful in setting up the Galois correspondence between the subgroups of Galois group G and the subfields of the splitting field $\mathbb{L}$ of the polynomial $f(x) \in \mathbb{K}[x]$.

Note 2. An important application of Fundamental Theorem of Galois Theory is that the group $G = Aut(\mathbb{L}/\mathbb{K})$ is solvable if and only if the polynomial equation $f(x) = 0$ is solvable by means of radicals—in other words its roots can be expressed in terms of coefficients of $f(x)$ using basic arithmetical operations and radicals only. In particular, this implies that roots of a general equation of degree five or more cannot be expressed in terms of radicals.

5. Computing Galois Group of Polynomials over $\mathbb{Q}$

Now we discuss the computation of the Galois group of polynomials in $\mathbb{Q}[x]$. Generally we shall begin with an irreducible polynomial in $f(x) \in \mathbb{Q}[x]$, find its splitting field and its degree over $\mathbb{Q}$ as a vector space. Some knowledge of the groups of small orders will be helpful in identifying the group of automorphism of the splitting field of the polynomial $f(x)$ over $\mathbb{Q}$. First we shall discuss some criterion for the irreducibility of polynomials over $\mathbb{Q}$ which would be helpful for identifying the Galois group of a given polynomial. After this we shall be in a position to discuss the computation of Galois group of a polynomial in $\mathbb{Q}[x]$ and illustrate the Galois correspondence (i.e., exhibit the $1-1$ mapping between the subgroups of the Galois group and subfields of the splitting field for a given polynomial $f(x) \in \mathbb{Q}[x]$). We shall provide several examples of the Galois correspondence in the next chapter.

6. Polynomial over Finite Fields

Just as starting with the polynomial $x^2+1 \in \mathbb{R}[x]$ (or any other irreducible polynomial of degree two in $\mathbb{R}[x]$) we can construct the field $\mathbb{C}$ of complex numbers, if we start with the finite field $\mathbb{F}_2$ of order two and irreducible polynomial $x^2+x+1 \in \mathbb{F}_2[x]$, we can construct a finite field of order four whose elements can be described as $\{a+b\alpha, \;\; a,b \in \mathbb{F}_2, \alpha^2+\alpha+1=0\}$. Starting with any irreducible polynomial of degree n with coefficients in the finite field $\mathbb{F}_q$ we can construct a finite field of order q^n. It can be shown that the number of monic irreducible polynomials of degree n with coefficients in the finite field $\mathbb{F}_q$ is given by $\frac{1}{n}\sum_{d|n}\mu(d)q^{\frac{n}{d}}$ where μ is the Mobius μ-function. All finite fields of same order are isomorphic. For example, there are two distinct monic irreducible polynomials over the finite field $\mathbb{F}_3$ of order three. These are x^2+x+2 and x^2+2x+2. We can use any one of these two polynomials to construct a field of order nine and the two resulting fields will be isomorphic. It will be helpful to have some information about irreducible polynomials of low degree over finite fields of small orders. This will be useful in proving the irreducibility of certain polynomials in $\mathbb{Q}[x]$ whose Galois groups we may be interested in computing. In the table on the following page we list the monic irreducible polynomials of degrees two, three, and four over the fields $\mathbb{F}_2$ and $\mathbb{F}_3$. For a comprehensive list of irreducible polynomials over finite fields see Lidl and Niederreiter (2008).

n	Base Field	Monic Irreducible Polynomials of Degree n
2	$\mathbb{F}_2$	x^2+x+1
	$\mathbb{F}_3$	$x^2+x+2,\ \ x^2+2x+2$
3	$\mathbb{F}_2$	$x^3+x+1,\ \ x^3+x^2+1$
	$\mathbb{F}_3$	$x^3+2x^2+2x+2,\ \ x^3+2x^2+x+1,\ \ x^3+2x^2+1,\ \ x^3+x^2+2x+1,$
		$x^3+x^2+x+2,\ \ x^3+x^2+2,\ \ x^3+2x+2,\ \ x^3+2x+1$
4	$\mathbb{F}_2$	$x^4+x^3+1,\ \ x^4+x+1,\ \ x^4+x^3+x^2+x+1,$
	$\mathbb{F}_3$	$x^4+2x^3+2,\ \ x^4+x^3+2,\ \ x^4+2x^2+2,$
		$x^4+x^2+2,\ \ x^4+2x+2,\ \ x^4+x+2,$
		$x^4+2x^3+x^2+1,\ \ x^4+2x^3+x+1,$
		$x^4+x^3+x^2+1,\ \ x^4+x^3+2x+1,$
		$x^4+x^2+2x+1,\ \ x^4+x^2+x+1,$
		$x^4+2x^3+2x^2+x+2,\ \ x^4+2x^3+x^2+2x+1,$
		$x^4+2x^3+x^2+x+2,\ \ x^4+x^3+2x^2+2x+2,$
		$x^4+x^3+x^2+2x+2,\ \ x^4+x^3+x^2+x+1$

7. Automorphisms of Finite Fields

Any field of characteristic zero contains the field of rational numbers $\mathbb{Q}$ as a subfield and any field of characteristic p contains the field $\mathbb{F}_p = \mathbb{Z}_p = GF(p)$ as a subfield. For this reason the fields $\mathbb{Q}$ and $\mathbb{F}_p$ are called *prime fields.* To construct a finite field $\mathbb{F}_{p^n}$ of order p^n we start with an irreducible polynomial $f(x) \in \mathbb{F}_p[x]$ of degree n. If $\beta \in \mathbb{F}_{p^n} — \mathbb{F}_p$ is any root of the polynomial $f(x)$ then so are β^{p^i}, for $1 \le i \le n$. If the polynomial $f(x)$ is monic then

$$f(x) = \prod_{1 \le i \le n} (x - \beta^{p^i})$$

We know that the multiplicative group $\mathbb{F}_{p^n}^*$ of a finite field $\mathbb{F}_{p^n}$ is cyclic. A root of $f(x)$ may or may not be a generator of the cyclic group $\mathbb{F}_{p^n}^*$. If $\alpha \in \mathbb{F}_{p^n}^*$ is a generator of $\mathbb{F}_{p^n}^*$ then $\mathbb{F}_{p^n}^* = \langle \alpha \rangle$, and so $\mathbb{F}_{p^n} = \mathbb{F}_p[\alpha]$. The group $Aut(\mathbb{F}_{p^n}/\mathbb{F}_p)$ is cyclic of order n. A generating automorphism of this group being $\sigma : \alpha \longrightarrow \alpha^p$. This automorphism is called *Frobenius automorphism.* In fact if q is any power of a prime number then the group $Aut(\mathbb{F}_{q^n}/\mathbb{F}_q)$ is cyclic of order n and a generating automorphism maps every element of $\mathbb{F}_{q^n}$ to its q-th power.

8. Irreducibility of Polynomials over $\mathbb{Q}$

We shall state some important facts which are helpful in proving irreducibility of polynomials in $\mathbb{Q}[x]$.

(1) Let $f(x) \in \mathbb{Z}[x]$. If $f(x)$ does factors in $\mathbb{Q}[x]$ then it also factors in $\mathbb{Z}[x]$. In other words in order to check the irreducibility of $f(x) \in \mathbb{Z}[x]$ over $\mathbb{Q}$ we only need to prove that it is irreducible over $\mathbb{Z}$.

(2) *Rational root theorem* The rational zeros (if any) of the polynomial

$$f(x) = a_0x^n + a_1x^{n-1} + \cdots + a_{n-1}x + a_n$$

are of the type $\pm\frac{r}{s}$ where r is some divisor of a_n and s is some divisor of a_0.

(3) *Eisenstein's Irreducibility Criterion* Let

$$f(x) = a_0x^n + a_1x^{n-1} + \cdots + a_{n-1}x + a_n \in \mathbb{Z}[x]$$

Let p be a prime number such that p does not divide a_0 but it divides a_i, $1 \leq i \leq n$, and p^2 does not divide a_n. Then $f(x)$ is irreducible in $\mathbb{Z}[x]$ (and therefore in $\mathbb{Q}[x]$).

This criterion is very useful in proving irreducibility of certain polynomials in $\mathbb{Z}[x]$. Using this criterion we can show that if p is any prime then the polynomial

$$f(x) = x^{p-1} + x^{p-2} + \cdots + x + 1$$

is irreducible in $\mathbb{Z}[x]$ (and hence in $\mathbb{Q}[x]$). It is clear that if $f(x)$ is irreducible then so is the polynomial $f(x+1)$ and vice versa. Now

$$f(x) = x^{p-1} + x^{p-2} + \cdots + x + 1 = \frac{x^p - 1}{x - 1}$$

and

$$f(x+1) = \frac{(x+1)^p - 1}{x+1-1} = x^{p-1} + px^{p-2} + \frac{p(p-1)}{1 \cdot 2}x^{p-3} + \cdots + \frac{p(p-1)}{1 \cdot 2}x + p.$$

The last polynomial satisfies the hypothesis of the Eisenstein's criterion for the prime p and so the polynomial $f(x+1)$ (and therefore the polynomial $f(x)$) is irreducible in $\mathbb{Z}[x]$ and so in $\mathbb{Q}[x]$.

(4) Let $f(x) \in \mathbb{Z}[x]$. Let p be a prime and $f_p(x)$ denote the polynomial in $\mathbb{Z}_p[x]$ obtained from $f(x)$ by reducing its coefficients modulo prime p. In other words, if we extend the mapping $\mathbb{Z} \longrightarrow \mathbb{Z}_p$ which maps $a \longrightarrow \bar{a} = a$ (modulo p) to the ring homomorphism $\mathbb{Z}[x] \longrightarrow \mathbb{Z}_p[x]$ then $f_p(x) \in \mathbb{Z}_p[x]$ is the image of $f(x) \in \mathbb{Z}[x]$. Then if $f(x)$ is reducible in $\mathbb{Z}[x]$, $f_p(x)$ is reducible in $\mathbb{Z}_p[x]$. In other words, if $f_p(x) \in \mathbb{Z}_p[x]$ is irreducible then so is $f(x) \in \mathbb{Z}[x]$.

The above result is very useful in proving irreducibility of certain polynomials $f(x) \in \mathbb{Z}[x]$. Often we are able to find a prime p such that $f_p(x)$ is irreducible in $\mathbb{Z}_p[x]$. Since in $\mathbb{Z}_p[x]$ there are only finite number of irreducible polynomials often (especially for small degree) it is easy to check whether $f_p(x) \in \mathbb{Z}[x]$ is irreducible or not. Sometimes a polynomial $f(x) \in \mathbb{Z}[x]$ is such that the factorization of $f_p(x)$ for two different primes is inconsistent with any possible factorization of $f(x)$ in $\mathbb{Z}[x]$. For example, if a polynomial $f(x) \in \mathbb{Z}[x]$ of degree six is such that $f_p(x)$ (for some prime p) factors in $\mathbb{Z}_p[x]$ as a product two irreducible cubic polynomials and $f_q(x)$ (for some prime $q \neq p$) factors as a product of three second degree irreducible polynomials in $\mathbb{Z}_q[x]$ then $f(x)$ must be irreducible in $\mathbb{Z}[x]$. For no factorization of $f(x)$ in $\mathbb{Z}[x]$ can give rise to such factorization of $f_p(x)$ and $f_q(x)$. The Example 3 in the following illustrates this.

Example 1. $f(x) = x^5 - x^2 + 1$.

Consider this polynomial modulo 2. We can write $f(x)$ (modulo 2) as $x^5 + x^2 + 1$. This polynomial does not have any linear factor over $\mathbb{Z}_2$. There is only one irreducible polynomial of degree two in $\mathbb{Z}_2[x]$, namely $x^2 + x + 1$. By direct calculation we can see that $x^2 + x + 1$ does not divide $x^5 + x^2 + 1$ in $\mathbb{Z}_2[x]$. Hence $f(x)$ is irreducible in $\mathbb{Z}[x]$.

Example 2. $f(x) = x^4 + 3x^3 + 3x^2 - 5$.

We note that $f(x) \equiv x^4 + x^3 + x^2 + 1 \equiv (x+1)(x^3 + x + 1)$ (modulo 2). This shows that $f(x)$ cannot have any factor of degree two. Since $f(x)$ has no linear factors in $\mathbb{Z}[x]$ (as none of the numbers ± 1, ± 5 is a zero of $f(x)$ and so by rational root theorem $f(x)$ has no rational zeros). Therfore $f(x)$ is irreducible in $\mathbb{Z}[x]$.

Example 3. $f(x) = x^6 + 3x^5 + 6x^4 + 3x^3 + 9x + 9$.
Here $f(x) \equiv x^6 + x^5 + x^3 + x + 1 \equiv (x^2 + x + 1)^3$ (modulo 2) and $f(x) \equiv (x^3 + 2x^2 + 6x + 4)(x^3 + x^2 + 5x + 4)$ (modulo 7). Therefore $f(x)$ is irreducible over $\mathbb{Z}$ and therefore over $\mathbb{Q}$.

Note. Since modulo 3, $f(x)$ factors as a product of linear factors we could not use that. However instead of factoring $f(x)$ modulo 2 we could have factored it modulo 5, namely

$$f(x) \equiv (x^2 + 4x + 1)(x^2 + 4x + 2)(x^2 + 2) \text{ (modulo 5)}$$

and reached the same conclusion.

Example 4. $f(x) = x^5 - x - 1$
In this case, as we can easily see, reducing the polynomial $f(x)$ modulo prime $p = 2$, that it does not have any linear factor but it does have the factor $x^2 + x + 1$. So in $\mathbb{Z}_2[x]$, $f(x)$ factors as a product of degree two and a degree three irreducible polynomials. This leads us nowhere. We try to factor $f(x)$ modulo 3. We see that $f(x)$ has no linear factors as can be seen by calculating $f(0)$, $f(1)$, and $f(2)$ modulo 3. There are only three monic irreducible polynomials of degree two in $\mathbb{Z}_3[x]$. These are $x^2 + 1$, $x^2 + x + 2$, and $x^2 + 2x + 2$. By direct calculation we see that none of these is a factor of $f(x)$ modulo 3. Hence $f(x)$ is irreducible in $\mathbb{Z}_3[x]$ and therefore in $\mathbb{Z}[x]$.

Note. The irreducible polynomial $x^5 - x - 1$ has three real roots and two nonreal roots. We shall see later that this implies that the Galois group of this polynomial is isomorphic to S_5, the symmetric group on five letters.
Alternately we could have used the following important theorem to prove that the polynomial $x^5 - x - 1$ is irreducible in $\mathbb{Z}_5[x]$. This theorem is due to Artin and Schreier.

Theorem 8.1: *(Artin-Schreier)*

Let p be a prime and $f(x) = x^p - x - a \in \mathbb{Z}_p[x]$, where $a \neq 0$ (i.e., p does not divide a). Then $f(x)$ is irreducible in $\mathbb{Z}_p[x]$.

Proof. We first recall the Fermat's Little Theorem which states that for a prime p, $x^p \equiv x$ (modulo p) for any $x \in \mathbb{Z}$. Since $a \neq 0$ the polynomial $f(x)$ cannot have any root in $\mathbb{Z}_p$. Let $\mathbb{L}$ denote the splitting field of $f(x) \in \mathbb{Z}_p[x]$. So $\mathbb{Z}_p \subseteq \mathbb{L}$ and $f(x)$ has all of its roots in $\mathbb{L}$. Let $\alpha \in \mathbb{L}$ be one of these roots. So $\alpha^p - \alpha - a = 0$. Again in view of Fermat's Little Theorem we have, for $0 \leq i \leq p-1$,

$$f(\alpha + i) = (\alpha + i)^p - (\alpha + i) - a \equiv \alpha^p + i^p - \alpha - i - a \equiv \alpha^p - \alpha - a \equiv 0 \text{ (modulo } p)$$

and therefore

$$x^p - x - a = \prod_{i=0}^{p-1} (x - \alpha - i) \quad \in \mathbb{L}[x].$$

Now suppose that $f(x)$ is reducible in $\mathbb{Z}_p[x]$ and has a factor $g(x)$ or degree n, where $0 < n < p$. Let

$$g(x) = x^n + b_1 x^{n-1} + \cdots + b_n \in \mathbb{Z}_p[x].$$

Then we can find $i_1, i_2, \cdots, i_n \in \mathbb{Z}_p$ such that

$$g(x) = (x - \alpha - i_1)(x - \alpha - i_2) \cdots (x - \alpha - i_n).$$

Equating the coefficients of x^{n-1} in the two expressions for $g(x)$ we have

$$b_1 = -(\alpha + i_1 + \alpha + i_2 + \cdots + \alpha + i_n) = -n\alpha - (i_1 + i_2 + \cdots + i_n).$$

So

$$\alpha = -\frac{1}{n}(b_1 + i_1 + i_2 + \cdots + i_n) \in \mathbb{Z}_p.$$

But then $\alpha^p \equiv \alpha$ and $\alpha^p - \alpha - a \equiv -a \neq 0$ (modulo p). This contradiction proves that $x^p - x - a$ is irreducible in $\mathbb{Z}_p[x]$ □

Note. In the above proof we have assumed that $f(x) \in \mathbb{Z}_p[x]$ has a splitting field. Its proof is similar to what we discussed in case of polynomials over $\mathbb{Q}$.

The following properties of the Galois groups will be helpful in the determination of the Galois group of the irreducible polynomials over $\mathbb{Q}$. The properties listed below are true for any field of characteristic zero. In what follows we shall assume that $f(x)$ is an irreducible polynomial over $\mathbb{Q}$, with splitting field $\mathbb{K}$ and $G = Gal(f) = Aut(\mathbb{K}/\mathbb{Q})$ is the Galois group of $f(x)$ over $\mathbb{Q}$. If degree $f(x) = n$ and $\alpha_1, \alpha_2, \cdots, \alpha_n$ are the roots of $f(x)$ in an algebraic closure of $\mathbb{Q}$ then we shall write $\mathbb{K} = \mathbb{Q}(\alpha_1, \alpha_2, \cdots, \alpha_n)$.

(1) The group G acts transitively on the set $\{\alpha_1, \alpha_2, \cdots, \alpha_n\}$ that is given any two elements α_i and α_j in $\{\alpha_1, \alpha_2, \cdots, \alpha_n\}$ there exists an element $\sigma \in G$ which maps α_i to α_j. In particular, the order of G is divisible by the degree n of the irreducible polynomial $f(x)$, and G is isomorphic to a subgroup of S_n, the symmetric group on n letters.

(2) Let

$$f = f(x) = a_0x^n + a_1x^{n-1} + \cdots + a_n \in \mathbb{Z}[x]$$

be an irreducible polynomial. Let $\alpha_1, \alpha_2, \cdots, \alpha_n$ be its zeros in some extension of $\mathbb{Q}$. Then the discriminant $Disc(f)$ of f is defined as

$$Disc(f) = a_0^{2(n-1)} \prod_{i \neq j} (\alpha_i - \alpha_j)^2.$$

Here the factor $a_0^{2(n-1)}$ ensures that $Disc(f)$ is a rational integer. Then $Disc(f)$ is a square in $\mathbb{Z}$ if and only if the Galois group of f is a subgroup of the alternating group A_n. In particular for the case $n = 3$, the Galois group of an irreducible polynomial of degree three is cyclic group of order three if and only if $Disc(f)$ is a square otherwise it is isomorphic to S_3—the symmetric group on three letters.

(3) Let p be a prime which does not divide the discriminant of the polynomial $f(x)$. In other words the n roots of $\bar{f}(x) \in \mathbb{Z}_p[x]$ (the polynomial obtained by reducing the coefficients of $f(x)$ modulo p) are distinct in any algebraic closure of the field $\mathbb{Z}_p$. Suppose that over $\mathbb{Z}_p[x]$ the polynomial $\bar{f}(x)$ factors as

$$\bar{f}(x) = \bar{f}_1(x)\bar{f}_2(x)\cdots \bar{f}_k(x)$$

where the polynomials $\bar{f}_i(x) \in \mathbb{Z}_p[x]$, $1 \leq i \leq k$ are irreducible. The polynomial $\bar{f}_i(x)$, is of degree n_i for $1 \leq i \leq k$, and $n_1 + n_2 + \cdots + n_k = n$. Then the Galois group G of $f(x)$ over $\mathbb{Q}$ has elements whose effect on the set of roots of $f(x)$, namely, $\{\alpha_1, \alpha_2, \cdots, \alpha_n\}$ as a permutation is of type $n_1{\cdot}n_2 \cdots n_k$ (that is a permutation which is a product of disjoint cycles of lengths n_1, n_2, $\cdots$ and n_k). In other words there are elements $\sigma \in G = Aut(\mathbb{K}/\mathbb{Q})$ such that as a permutation on $\{\alpha_1, \alpha_2, \cdots, \alpha_n\}$, σ breaks up into disjoint cycles of lengths n_i, $1 \leq i \leq k$. This theorem is due to Dedekind and we shall give a proof of this in one of the Appendix.

The above properties are helpful in identifying the Galois group of polynomials over the field of rational numbers and we will use these in later chapters.

Note. In describing the Galois group $\mathbb{G}$ we shall multiply the permutation from left to right. Likewise if $\sigma, \eta \in Aut(\mathbb{K}/\mathbb{Q})$ then in determining the effect of the composition $\sigma \circ \eta$, on the elements of the fields $\mathbb{K}$, we shall assume that σ is applied first and is then followed by η—we realize that it is somewhat unconventional but while looking at a Galois group of a polynomial as a permutation group on its roots we find it more convenient.

CHAPTER 5

Examples of Galois Groups and Galois Correspondence

In this chapter we shall compute the Galois group of over $\mathbb{Q}$ of several polynomials and exhibit the Galois correspondence between the subgroups of the Galois group and the subfields of the splitting field of these polynomials. We shall begin with an irreducible polynomial $f(x) \in \mathbb{Q}[x]$ and describe its splitting field. In most cases from the nature of the roots of $f(x)$ we will be able to determine the degree of its splitting field over $\mathbb{Q}$ which will be same as the order of the Galois group of $f(x)$ over $\mathbb{Q}$. By looking at the roots of $f(x)$ along with some information about the groups of small orders we shall be able to characterize its Galois group. In all of our examples $\mathbb{K}$ will denote the splitting field of the irreducible polynomial $f(x) \in \mathbb{Q}[x]$ and G its Galois group over $\mathbb{Q}$. In each case we provide the following information:

(1) The irreducible polynomial $f(x) \in \mathbb{Q}[x]$ and its Galois group G over $\mathbb{Q}$.

(2) The roots of the polynomial $f(x)$.

(3) The splitting field $\mathbb{K}$ of $f(x)$.

(4) The description of the group G in terms of generators and relations.

(5) A table giving the actions of the elements of G on some elements of the field $\mathbb{K}$.

(6) A table giving the Galois correspondence between the subgroups of G and subfields of the field $\mathbb{K}$.

(7) In some cases we will have separate tables giving the subgroups of the Galois group G and their order along with the (fixed) subfields of $\mathbb{K}$ corresponding to these subgroups of G.

(8) In first few cases we will draw the diagrams exhibiting the lattice of subgroups of the Galois group and the lattice of subfields of the splitting field of the polynomial $f(x) \in \mathbb{Q}[x]$.

1. The Galois Group of $x^3 - 2$

The Galois groups of x^3-2 is dihedral group of order 6 and is isomorphic to S_3, the symmetric group on three letters.

Polynomial $x^3 - 2$.

Zeros $2^{\frac{1}{3}}, \omega 2^{\frac{1}{3}}, \omega^2 2^{\frac{1}{3}}$; $\omega = e^{\frac{2\pi i}{3}} = \dfrac{-1+\sqrt{-3}}{2} = \dfrac{-1+i\sqrt{3}}{2}$.

Splitting Field $\mathbb{K} = \mathbb{Q}(2^{\frac{1}{3}}, \omega) = \mathbb{Q}(2^{\frac{1}{3}} + \omega)$.

Galois Group $G = Aut(\mathbb{K}/\mathbb{Q}) = \langle \sigma, \eta : \sigma^3 = \eta^2 = e, \eta\sigma = \sigma^2\eta \rangle = \{e, \sigma, \sigma^2, \eta, \sigma\eta, \sigma^2\eta\}$.

Action of G on $\mathbb{K}$

	e	σ	σ^2	η	$\sigma\eta$	$\sigma^2\eta$
$2^{\frac{1}{3}} \longrightarrow$	$2^{\frac{1}{3}}$	$\omega 2^{\frac{1}{3}}$	$\omega^2 2^{\frac{1}{3}}$	$2^{\frac{1}{3}}$	$\omega^2 2^{\frac{1}{3}}$	$\omega 2^{\frac{1}{3}}$
$\omega \longrightarrow$	ω	ω	ω^2	ω^2	ω^2	ω^2

Galois Correspondence for $x^3 - 2$

Subgroup	Order	Fixed Field
$\{e\}$	1	$\mathbb{K} = \mathbb{Q}(2^{\frac{1}{3}}, \omega)$
$\langle\sigma\rangle$	3	$\mathbb{Q}(\omega)$
$\langle\eta\rangle$	2	$\mathbb{Q}(2^{\frac{1}{3}})$
$\langle\sigma\eta\rangle$	2	$\mathbb{Q}(\omega 2^{\frac{1}{3}})$
$\langle\sigma^2\eta\rangle$	2	$\mathbb{Q}(\omega^2 2^{\frac{1}{3}})$
G	6	$\mathbb{Q}$

Here $\sigma\eta$ means σ followed by η

Galois Group of $x^3 - 2$ - Lattice of subfields of $\mathbb{K} = \mathbb{Q}(2^{\frac{1}{3}}, \omega)$

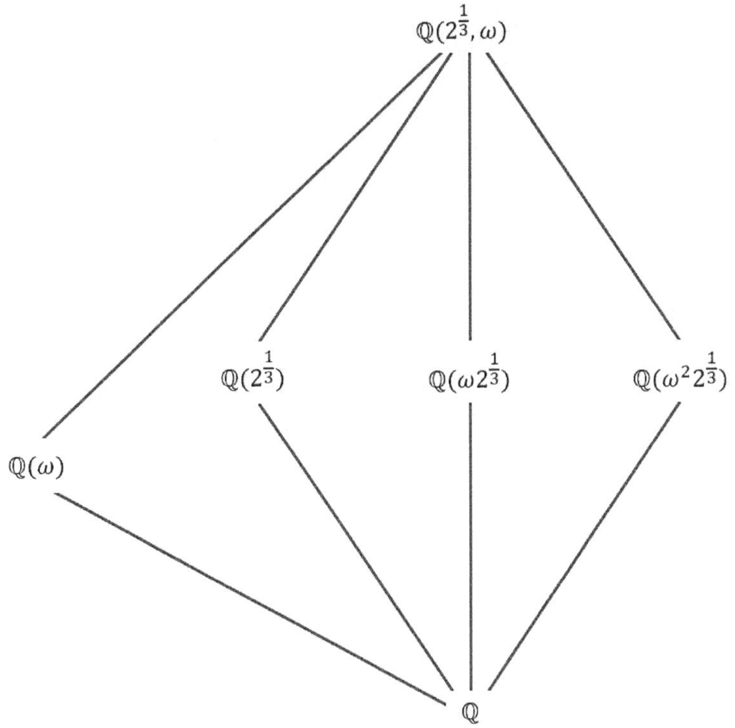

Galois Group of $x^3 - 2$ - Lattice of subgroups of $G = Aut(\mathbb{K}/\mathbb{Q}) = \langle \sigma, \eta : \sigma^3 = \eta^2 = e,\ \eta^{-1}\sigma\eta = \sigma^2 \rangle$

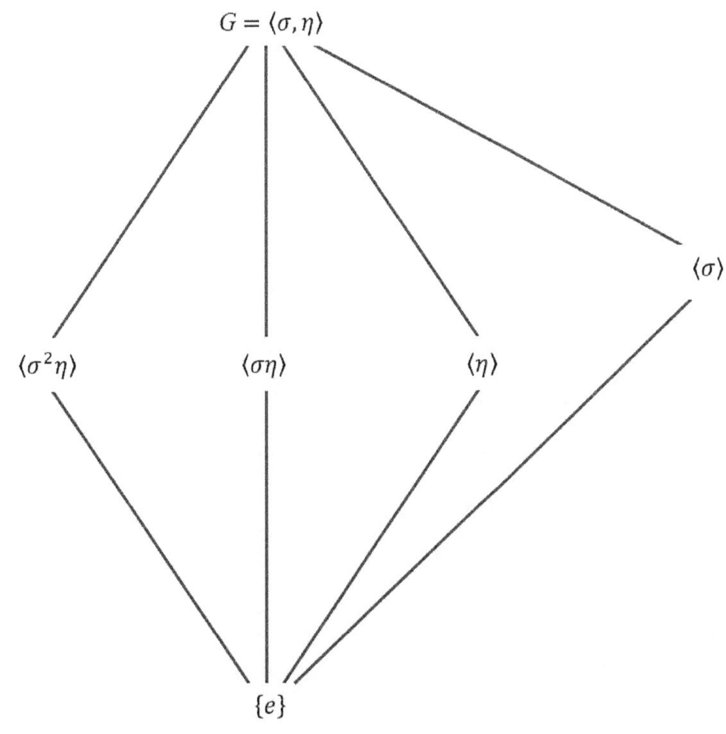

2. The Galois Groups of $x^4 - 2$

The Galois groups of $x^4 - 2$ is dihedral group of order 8.

Polynomial $x^4 - 2$.

Zeros $\pm 2^{\frac{1}{4}}, \pm i2^{\frac{1}{4}}; \quad i = \sqrt{-1}$.

Splitting Field $\mathbb{K} = \mathbb{Q}(2^{\frac{1}{4}}, i) = \mathbb{Q}(2^{\frac{1}{4}} + i)$.

Galois Group $G = Aut(\mathbb{K}/\mathbb{Q}) = \langle \sigma, \eta : \sigma^4 = \eta^2 = e, \eta\sigma = \sigma^3\eta \rangle = \{e, \sigma, \sigma^2, \sigma^3, \eta, \sigma\eta, \sigma^2\eta, \sigma^3\eta\}$.

Action of $\mathbb{G}$ on $\mathbb{K}$

	e	σ	σ^2	σ^3	η	$\sigma\eta$	$\sigma^2\eta$	$\sigma^3\eta$
$2^{\frac{1}{4}} \longrightarrow$	$2^{\frac{1}{4}}$	$i2^{\frac{1}{4}}$	$-2^{\frac{1}{4}}$	$-i2^{\frac{1}{4}}$	$2^{\frac{1}{4}}$	$-i2^{\frac{1}{4}}$	$-2^{\frac{1}{4}}$	$i2^{\frac{1}{4}}$
$i \longrightarrow$	i	i	i	i	$-i$	$-i$	$-i$	$-i$

Galois Correspondence for $x^4 - 2$

Subgroup	Order	Fixed Field
$\{e\}$	1	$\mathbb{K} = \mathbb{Q}(2^{\frac{1}{4}}, i)$
$\langle \sigma^2 \rangle$	2	$\mathbb{Q}(i, \sqrt{2})$
$\langle \eta \rangle$	2	$\mathbb{Q}(2^{\frac{1}{4}})$
$\langle \sigma^2\eta \rangle$	2	$\mathbb{Q}(i2^{\frac{1}{4}})$
$\langle \sigma\eta \rangle$	2	$\mathbb{Q}\left((1-i)2^{\frac{1}{4}}\right) = \mathbb{Q}\left(\frac{1+i}{\sqrt{2}} \cdot 2^{\frac{1}{4}}\right)$
$\langle \sigma^3\eta \rangle$	2	$\mathbb{Q}\left((1+i)2^{\frac{1}{4}}\right) = \mathbb{Q}\left(\frac{1-i}{\sqrt{2}} \cdot 2^{\frac{1}{4}}\right)$
$\langle \sigma \rangle$	4	$\mathbb{Q}(i)$
$\langle \sigma^2, \eta \rangle$	4	$\mathbb{Q}(2^{\frac{1}{2}})$
$\langle \sigma^2, \sigma\eta \rangle$	4	$\mathbb{Q}(i2^{\frac{1}{2}})$
$G = \langle \sigma, \eta \rangle$	8	$\mathbb{Q}$

Here $\sigma\eta$ means σ followed by η

Galois Group of $x^4 - 2$ - Lattice of subgroups of $G = Aut(\mathbb{K}/\mathbb{Q}) = \langle \sigma, \eta : \sigma^4 = \eta^2 = e,\ \eta^{-1}\sigma\eta = \sigma^3 \rangle$

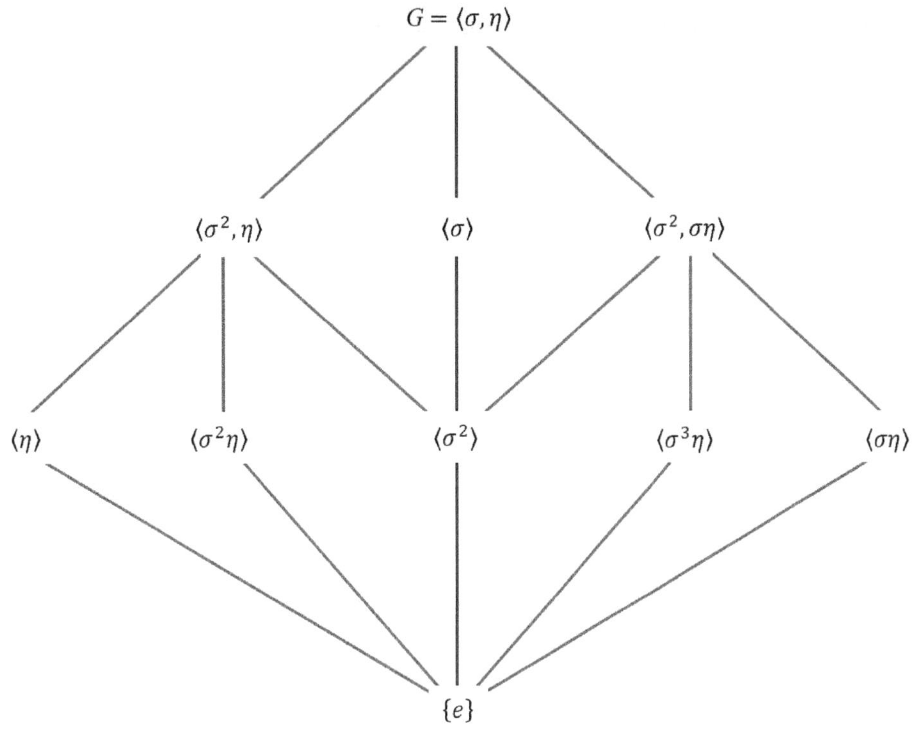

Galois Group of $x^4 - 2$ - Lattice of subfields of $\mathbb{K} = \mathbb{Q}(2^{\frac{1}{4}}, i)$

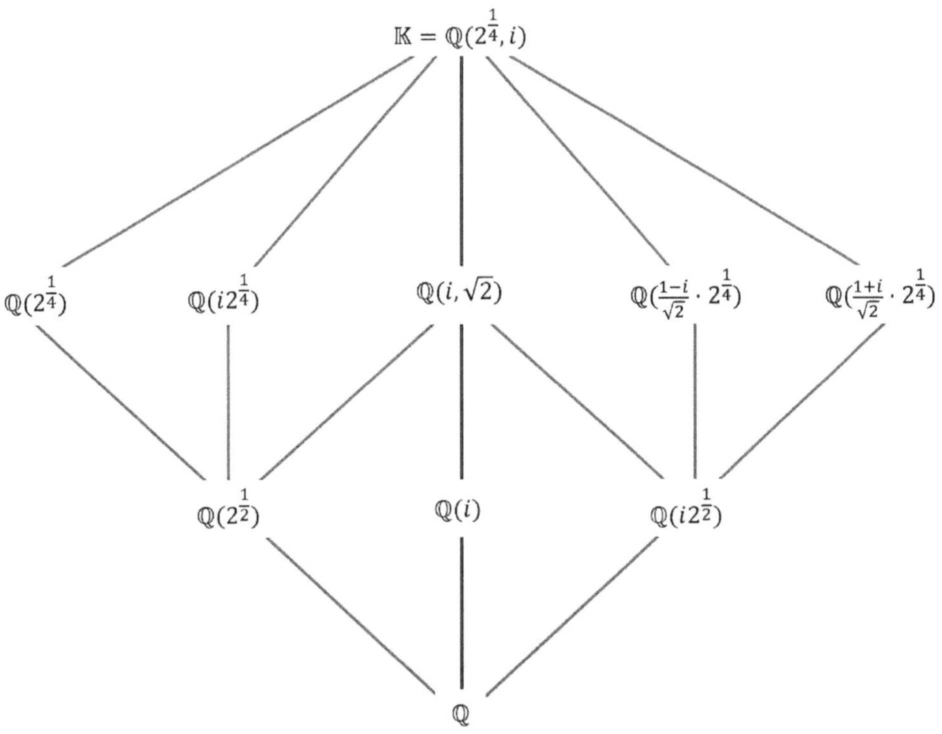

3. Galois Group of $x^8 - 24x^6 + 144x^4 - 288x^2 + 144$

In this section we shall show that the Galois group of the above polynomial is isomorphic to the quaternion group of order eight. (This example has been taken from the book by Dummit and Foote [2004].)

Let $\alpha = \sqrt{(2+\sqrt{2})(3+\sqrt{3})}$. It can be easily seen that $\mathbb{Q}\left((2+\sqrt{2})(3+\sqrt{3})\right) = \mathbb{Q}(\sqrt{2},\sqrt{3})$. Therefore α has degree eight over $\mathbb{Q}$. By squaring the expression for α and by further simplifying we find that α is a zero of an irreducible polynomial of degree eight. This can be seen as follows:

$$\begin{aligned}
\alpha &= \sqrt{(2+\sqrt{2})(3+\sqrt{3})} \quad \text{squaring this we get}\\
\alpha^2 &= 6+2\sqrt{3}+3\sqrt{2}+\sqrt{6} \quad \text{or}\\
\alpha^2-6-\sqrt{6} &= 2\sqrt{3}+3\sqrt{2} \quad \text{squaring this we get}\\
\alpha^4-12\alpha^2+12 &= 2\alpha^2\sqrt{6} \quad \text{finally squaring this we get}
\end{aligned}$$

$$\alpha^8 - 24\alpha^6 + 144\alpha^4 - 288\alpha^2 + 144 = 0.$$

Therefore α is a zero of the polynomial $f(x) = x^8 - 24x^6 + 144x^4 - 288x^2 + 144$ and α^2 is a zero of the polynomial $g(x) = x^4 - 24x^3 + 144x^2 - 288x + 144$. We will show that $f(x)$ is irreducible over $\mathbb{Q}$. It will be enough to show that α is an algebraic number of degree 8 over $\mathbb{Q}$. As the field

$$\mathbb{Q}(\alpha^2) = \mathbb{Q}\left((2+\sqrt{2})(3+\sqrt{3})\right) = \mathbb{Q}(\sqrt{2},\sqrt{3})$$

is of degree four over $\mathbb{Q}$, it follows that $g(x)$ is irreducible over $\mathbb{Q}$ and so α^2 is of degree four over $\mathbb{Q}$ and α is of degree eight over $\mathbb{Q}$. The four roots of $g(x)$ (which are squares of the roots of $f(x)$) are as follows

$$\begin{aligned}
\alpha^2 &= (2+\sqrt{2})(3+\sqrt{3})\\
\beta^2 &= (2+\sqrt{2})(3-\sqrt{3})\\
\gamma^2 &= (2-\sqrt{2})(3+\sqrt{3})\\
\delta^2 &= (2-\sqrt{2})(3-\sqrt{3}).
\end{aligned}$$

So we have

$$\mathbb{Q}(\sqrt{2},\sqrt{3}) = \mathbb{Q}(\alpha^2) = \mathbb{Q}(\beta^2) = \mathbb{Q}(\gamma^2) = \mathbb{Q}(\delta^2).$$

The eight zeros of $f(x)$ are: $\pm\alpha, \pm\beta, \pm\gamma$, and $\pm\delta$. Consider the field diagram:

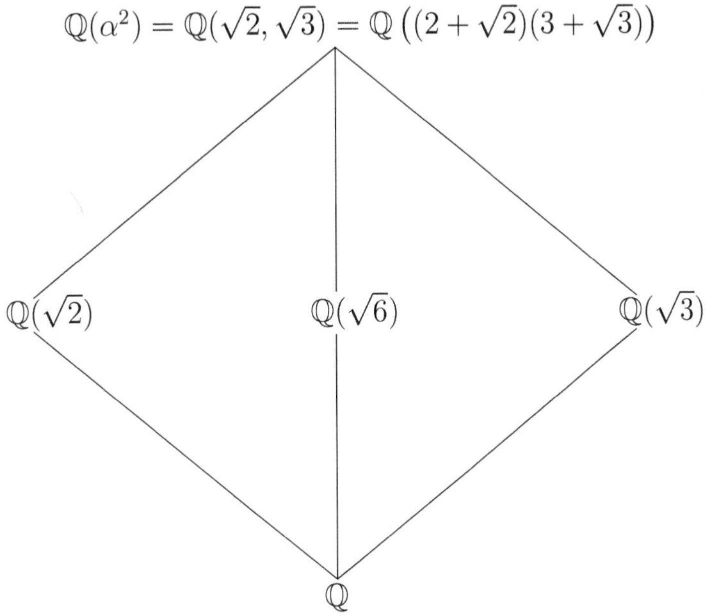

We claim that $\alpha \notin \mathbb{Q}(\alpha^2)$. This would imply that α is of degree two over $\mathbb{Q}(\alpha^2)$, so α is an algebraic number of degree 8 and $f(x)$ is its minimal polynomial over $\mathbb{Q}$. From this we would conclude that $f(x)$ is irreducible over $\mathbb{Q}$.

Suppose that $\alpha \in \mathbb{Q}(\alpha^2) = \mathbb{Q}(\sqrt{2}, \sqrt{3})$. The group $Aut\left(\mathbb{Q}(\sqrt{2}, \sqrt{3})/\mathbb{Q}(\sqrt{3})\right) = \langle\sigma\rangle$ has order two and its nontrivial automorphism σ maps $\sqrt{2}$ to $-\sqrt{2}$ and $\mathbb{Q}(\sqrt{3})$ is its fixed field. Therefore σ maps $\alpha\sigma(\alpha)$ to itself and so $\alpha\sigma(\alpha) \in \mathbb{Q}(\sqrt{3})$. Now

$$(\alpha \cdot \sigma(\alpha))^2 = \alpha^2 \cdot \sigma(\alpha^2) = (2+\sqrt{2})(3+\sqrt{3})(2-\sqrt{2})(3+\sqrt{3}) = 2(3+\sqrt{3})^2.$$

Therefore $\alpha\sigma(\alpha) = \pm\sqrt{2}(3+\sqrt{3}) \in \mathbb{Q}(\sqrt{3})$ - which is not true.

Now the eight roots of the equation $f(x) = x^8 - 24x^6 + 144x^4 - 288x^2 + 144 = 0$ are

$$\begin{aligned}
\pm\alpha &= \pm\sqrt{(2+\sqrt{2})(3+\sqrt{3})} \\
\pm\beta &= \pm\sqrt{(2+\sqrt{2})(3-\sqrt{3})} \\
\pm\gamma &= \pm\sqrt{(2-\sqrt{2})(3+\sqrt{3})} \\
\pm\delta &= \pm\sqrt{(2-\sqrt{2})(3-\sqrt{3})}
\end{aligned}$$

So if we denote its splitting field by $\mathbb{K}$ then

$$\mathbb{K} = \mathbb{Q}(\alpha, \beta, \gamma, \delta) = \mathbb{Q}(\alpha) = \mathbb{Q}(\beta) = \mathbb{Q}(\gamma) = \mathbb{Q}(\delta).$$

We observe that the polynomial $f(x)$ is such that each of its zero is a rational function of any one of them. We have the following diagram of subfields of $\mathbb{K}$.

Lattice of subfields of the field $\mathbb{K} = \mathbb{Q}\left(\sqrt{(2+\sqrt{2})(3+\sqrt{3})}\right)$.

We note that $\mathbb{K} = \mathbb{Q}(\alpha) = \mathbb{Q}(\beta) = \mathbb{Q}(\gamma) = \mathbb{Q}(\delta) = \mathbb{Q}(\alpha, \beta, \gamma, \delta) = \mathbb{Q}\left(\sqrt{(2+\sqrt{2})(3+\sqrt{3})}\right)$.

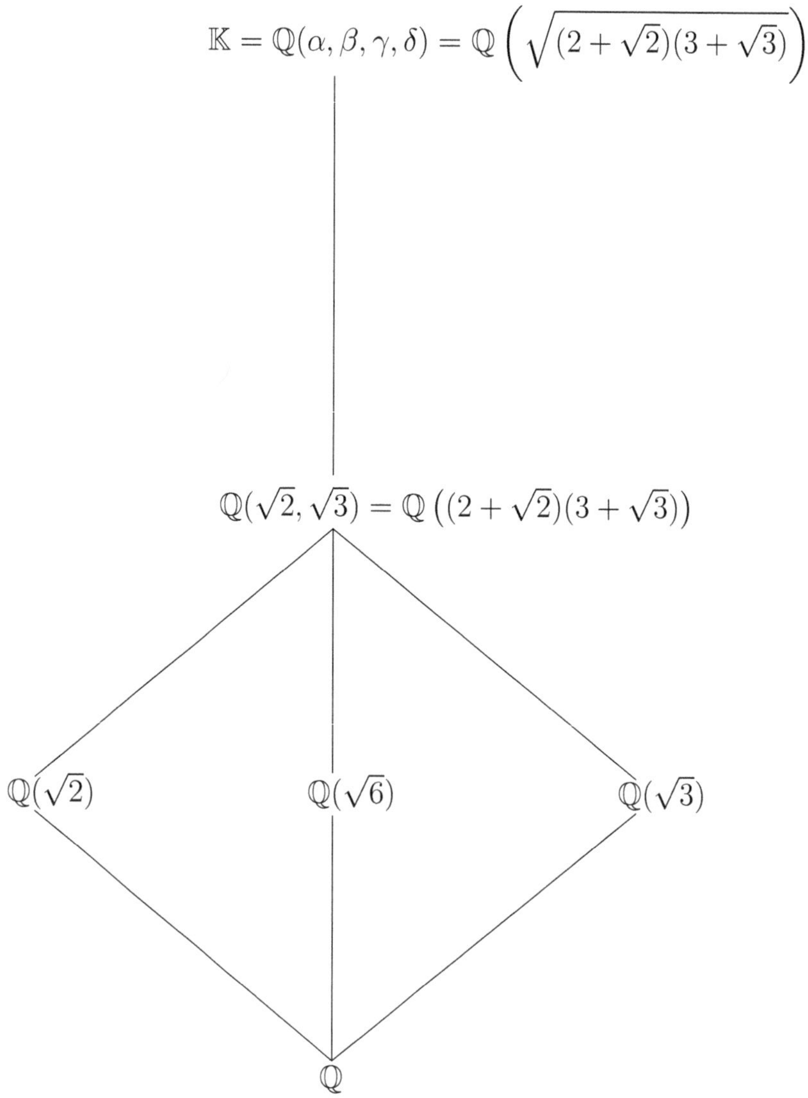

Let $G = Aut(\mathbb{K}/\mathbb{Q})$, then G has order 8. Let σ and η be the following two automorphisms of $\mathbb{K}$, $\sigma : \alpha \longrightarrow \beta$ and $\quad \eta : \alpha \longrightarrow \gamma$.

Then we have

$$G = \langle \sigma, \eta : \sigma^4 = \eta^4 = e, \sigma^2 = \eta^2, \eta\sigma = \sigma^3\eta \rangle = \{e, \sigma, \sigma^2, \sigma^3, \eta, \sigma\eta, \sigma^2\eta, \sigma^3\eta\}.$$

The following table gives the effect of each automorphism of $\mathbb{K}$. The following observations would be useful in completing the entering in the table that follows:

- $\alpha^2\beta^2 = 6(2+\sqrt{2})^2, \quad \alpha\beta = \sqrt{6}(2+\sqrt{2}), \quad \beta = \dfrac{\sqrt{6}(2+\sqrt{2})}{\alpha}.$
- $\alpha^2\gamma^2 = 2(3+\sqrt{3})^2, \quad \alpha\gamma = \sqrt{2}(3+\sqrt{3}), \quad \gamma = \dfrac{\sqrt{2}(3+\sqrt{3})}{\alpha}.$
- $\alpha^2\delta^2 = 12, \quad \delta^2 = \dfrac{12}{\alpha^2}, \quad \delta = \dfrac{2\sqrt{3}}{\alpha}.$
- $\beta^2\gamma^2 = 12, \quad \beta\gamma = 2\sqrt{3}, \quad \gamma = \dfrac{2\sqrt{3}}{\beta}.$
- $\beta^2\delta^2 = 2(3-\sqrt{3})^2, \quad \beta\delta = \sqrt{2}(3-\sqrt{3}), \quad \delta = \dfrac{\sqrt{2}(3-\sqrt{3})}{\beta}.$
- $\gamma^2\delta^2 = 6(2-\sqrt{2})^2, \quad \gamma\delta = \sqrt{6}(2-\sqrt{2}), \quad \delta = \dfrac{\sqrt{6}(2-\sqrt{2})}{\gamma}.$

We note that $\sigma : \alpha \longrightarrow \beta$ implies that $\alpha^2 \longrightarrow \beta^2$ and therefore effect of the automorphism σ on $\sqrt{2}$, $\sqrt{3}$, $\sqrt{6}$, and β, γ, δ can be described as follows:

$$\begin{aligned}
&\sqrt{2} \longrightarrow \sqrt{2}, \quad \sqrt{3} \longrightarrow -\sqrt{3}, \quad \sqrt{6} \longrightarrow -\sqrt{6} \\
&\beta = \frac{\sqrt{6}(2+\sqrt{2})}{\alpha} \longrightarrow -\frac{\sqrt{6}(2+\sqrt{2})}{\beta} = -\alpha \\
&\gamma = \frac{\sqrt{2}(3+\sqrt{3})}{\alpha} \longrightarrow \frac{\sqrt{2}(3-\sqrt{3})}{\beta} = \delta \\
&\delta = \frac{2\sqrt{3}}{\alpha} \longrightarrow -\frac{2\sqrt{3}}{\beta} = -\gamma.
\end{aligned}$$

Also as $\eta : \alpha \longrightarrow \gamma$ implies that $\alpha^2 \longrightarrow \gamma^2$ the effect of the automorphism η on $\sqrt{2}$, $\sqrt{3}$, $\sqrt{6}$, and β, γ, δ can be described as follows:

$$\begin{aligned}
&\sqrt{2} \longrightarrow -\sqrt{2}, \quad \sqrt{3} \longrightarrow \sqrt{3}, \quad \sqrt{6} \longrightarrow -\sqrt{6} \\
&\beta = \frac{\sqrt{6}(2+\sqrt{2})}{\alpha} \longrightarrow -\frac{\sqrt{6}(2-\sqrt{2})}{\gamma} = -\delta \\
&\gamma = \frac{\sqrt{2}(3+\sqrt{3})}{\alpha} \longrightarrow \frac{-\sqrt{2}(3+\sqrt{3})}{\beta} = -\alpha \\
&\delta = \frac{2\sqrt{3}}{\alpha} \longrightarrow \frac{2\sqrt{3}}{\gamma} = \beta.
\end{aligned}$$

Action of G on $\mathbb{K}$

	e	σ	σ^2	σ^3	η	$\sigma\eta$	$\sigma^2\eta$	$\sigma^3\eta$
$\alpha \longrightarrow$	α	β	$-\alpha$	$-\beta$	γ	$-\delta$	$-\gamma$	δ
$\beta \longrightarrow$	β	$-\alpha$	$-\beta$	α	$-\delta$	$-\gamma$	δ	γ
$\gamma \longrightarrow$	γ	δ	$-\gamma$	$-\delta$	$-\alpha$	β	α	$-\beta$
$\delta \longrightarrow$	δ	$-\gamma$	$-\delta$	γ	β	α	$-\beta$	$-\alpha$
$\sqrt{2} \longrightarrow$	$\sqrt{2}$	$\sqrt{2}$	$\sqrt{2}$	$\sqrt{2}$	$-\sqrt{2}$	$-\sqrt{2}$	$-\sqrt{2}$	$-\sqrt{2}$
$\sqrt{3} \longrightarrow$	$\sqrt{3}$	$-\sqrt{3}$	$\sqrt{3}$	$-\sqrt{3}$	$\sqrt{3}$	$-\sqrt{3}$	$\sqrt{3}$	$-\sqrt{3}$

Here $\sigma\eta$ means σ followed by η

Galois Correspondence

Subgroup	Order	Fixed Field
$\{e\}$	1	$\mathbb{K} = \mathbb{Q}\left(\sqrt{(2+\sqrt{2})(3+\sqrt{3})}\right)$
$\langle\sigma^2\rangle$	2	$\mathbb{Q}(\sqrt{2}, \sqrt{3})$
$\langle\sigma\rangle$	4	$\mathbb{Q}(\sqrt{2})$
$\langle\eta\rangle$	4	$\mathbb{Q}(\sqrt{3})$
$\langle\sigma\eta\rangle$	4	$\mathbb{Q}(\sqrt{6})$
$\langle\sigma, \eta\rangle$	8	$\mathbb{Q}$

Lattice of subgroups of the Galois group $G = Aut\left(\mathbb{Q}\left(\sqrt{(2+\sqrt{2})(3+\sqrt{3})}\right) / \mathbb{Q}\right)$

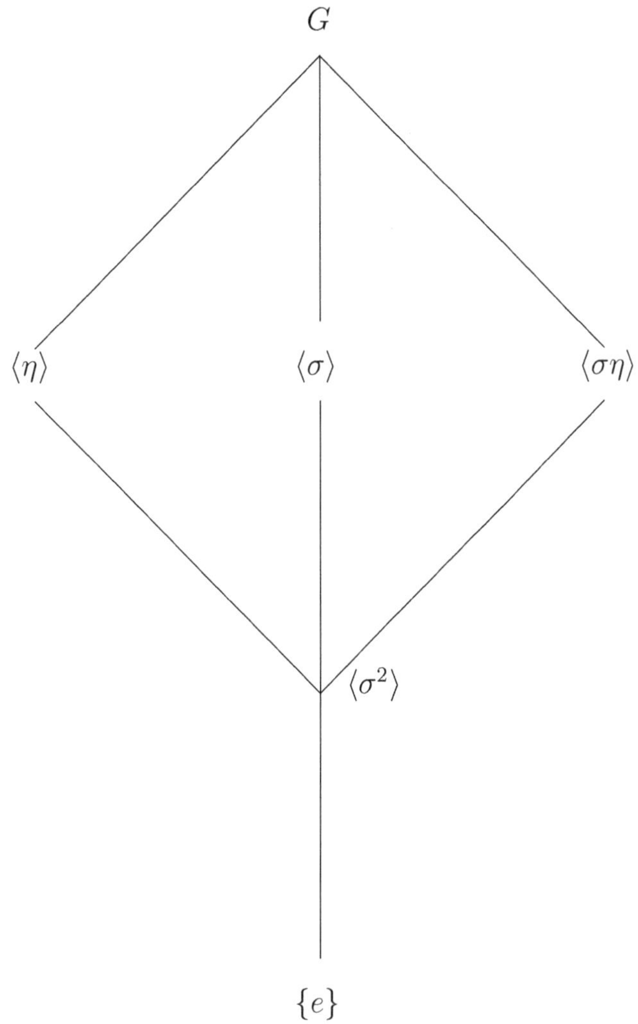

Note. The polynomial $f(x) = x^8 - 24x^6 + 144x^4 - 288x^2 + 144$ has discriminant $2^{52} \times 3^{14}$ and it factors modulo every prime. As the primes 2 and 3 divide its discriminant, the polynomial $f(x)$ has repeated roots modulo $p = 2$ and $p = 3$ and we would expect $f(x)$ to not be irreducible modulo 2 and modulo 3. However, even for primes $p > 3$ the polynomial $f(x)$ cannot be irreducible as this would imply that the Galois group of the polynomial $f(x)$

over $\mathbb{Q}$ has an element of order 8 which is not true. It is also interesting to note that modulo prime $p > 3$, the polynomial $f(x)$ factors in one of the three ways:

(1) Two factors of degree four each.
(2) Four factors of degree two each.
(3) Eight linear factors.

This is so because $\sigma \in Aut(\mathbb{K})$, $\sigma \neq e$ cannot fix any root of $f(x)$ as the field $\mathbb{K}$ can be generated by any of its eight roots.

It can be easily seen that if for a prime p the polynomial $f(x)$ factors into eight linear factors then 2 and 3 must be quadratic residues modulo p—which in turn implies that $p \equiv \pm 1$ (modulo 24). The converse of this is not true as

$$x^8 - 24x^6 + 144x^4 - 288x^2 + 144 \equiv (x^2+2)(x^2+6)(x^2+28)(x^2+34) \text{ (modulo 47)}.$$

It appears difficult to decide for which primes $p \equiv \pm 1$ (modulo 24) the polynomial $f(x)$ factors into eight linear factors modulo p. We will study polynomials in $\mathbb{Z}[x]$ which are irreducible over $\mathbb{Q}$ but factor modulo every prime in more detail in a later chapter.

Note. The polynomial $f(x) = x^8 - 24x^6 + 144x^4 - 288x^2 + 144$ could also have been obtained by multiplying $(x^2-\alpha^2)(x^2-\beta^2)(x^2-\gamma^2)(x^2-\delta^2)$ as follows:

$$\begin{aligned}
&(x^2-\alpha^2)(x^2-\beta^2)(x^2-\gamma^2)(x^2-\delta^2)\\
&= \left[x^2-(2+\sqrt{2})(3+\sqrt{3})\right]\left[x^2-(2+\sqrt{2})(3-\sqrt{3})\right]\left[x^2-(2-\sqrt{2})(3+\sqrt{3})\right]\left[x^2-(2-\sqrt{2})(3-\sqrt{3})\right]\\
&= \left[x^4-6x^2(2+\sqrt{2})+6(2+\sqrt{2})^2\right]\left[x^4-6x^2(2-\sqrt{2})+6(2-\sqrt{2})^2\right]\\
&= \left[x^4-12x^2-6\sqrt{2}x^2+6(6+4\sqrt{2})\right]\left[x^4-12x^2+6\sqrt{2}x^2+6(6-4\sqrt{2})\right]\\
&= \left[x^4-12x^2+36-6\sqrt{2}(x^2-4)\right]\left[x^4-12x^2+36+6\sqrt{2}(x^2-4)\right]\\
&= (x^4-12x^2+36)^2-72(x^2-4)^2\\
&= x^8+144x^2+1296-24x^6+72x^4-864x^2-72(x^4-8x^2+16) = x^8-24x^6+144x^4-288x^2+144.
\end{aligned}$$

4. The Galois Group of $x^5 - 2$

The polynomial x^5-2 is easily seen to be irreducible over $\mathbb{Q}$. Its five zeros are $2^{\frac{1}{5}}$, $\varepsilon 2^{\frac{1}{5}}$, $\varepsilon^2 2^{\frac{1}{5}}$, $\varepsilon^3 2^{\frac{1}{5}}$, $\varepsilon^4 2^{\frac{1}{5}}$, where $\varepsilon = e^{\frac{2\pi i}{5}} = \cos\frac{2\pi}{5} + i\sin\frac{2\pi}{5}$. We note that $\varepsilon+\varepsilon^4 = \dfrac{-1+\sqrt{5}}{2}$ and $\varepsilon^2+\varepsilon^3 = \dfrac{-1-\sqrt{5}}{2}$. The splitting field of the polynomial x^5-2 is $\mathbb{K} = \mathbb{Q}(2^{\frac{1}{5}}, \varepsilon)$.
Let σ and η be the following two automorphisms of $\mathbb{K}$.

$$\sigma : \begin{cases} 2^{\frac{1}{5}} \longrightarrow \varepsilon 2^{\frac{1}{5}} \\ \varepsilon \longrightarrow \varepsilon \end{cases} \qquad \eta : \begin{cases} 2^{\frac{1}{5}} \longrightarrow 2^{\frac{1}{5}} \\ \varepsilon \longrightarrow \varepsilon^2 \end{cases}.$$

It can be easily checked that $\sigma^5 = \eta^4 = e$ (identity) and $\eta^{-1}\sigma\eta = \sigma^2$. The group $G = Aut\,(\mathbb{K}/\mathbb{Q}) = \langle \sigma, \eta \rangle$ is of order 20 and is (so called) a Frobenius group. The following table shows the effect of its generating elements (namely σ and η) on $2^{\frac{1}{5}}$, ε, and $\sqrt{5}$.

Action of $G = Aut\,(\mathbb{K}/\mathbb{Q}) = \langle \sigma, \eta \rangle$ on $\mathbb{K} = \mathbb{Q}(2^{\frac{1}{5}}, \varepsilon)$

	e	σ	σ^2	σ^3	σ^4	η	$\sigma\eta$	$\sigma^2\eta$	$\sigma^3\eta$	$\sigma^4\eta$
$2^{\frac{1}{5}} \longrightarrow$	$2^{\frac{1}{5}}$	$\varepsilon 2^{\frac{1}{5}}$	$\varepsilon^2 2^{\frac{1}{5}}$	$\varepsilon^3 2^{\frac{1}{5}}$	$\varepsilon^4 2^{\frac{1}{5}}$	$2^{\frac{1}{5}}$	$\varepsilon^2 2^{\frac{1}{5}}$	$\varepsilon^4 2^{\frac{1}{5}}$	$\varepsilon 2^{\frac{1}{5}}$	$\varepsilon^3 2^{\frac{1}{5}}$
$\varepsilon \longrightarrow$	ε	ε	ε	ε	ε	ε^2	ε^2	ε^2	ε^2	ε^2
$\sqrt{5} \longrightarrow$	$\sqrt{5}$	$\sqrt{5}$	$\sqrt{5}$	$\sqrt{5}$	$\sqrt{5}$	$-\sqrt{5}$	$-\sqrt{5}$	$-\sqrt{5}$	$-\sqrt{5}$	$-\sqrt{5}$

	η^2	$\sigma\eta^2$	$\sigma^2\eta^2$	$\sigma^3\eta^2$	$\sigma^4\eta^2$	η^3	$\sigma\eta^3$	$\sigma^2\eta^3$	$\sigma^3\eta^3$	$\sigma^4\eta^3$
$2^{\frac{1}{5}} \longrightarrow$	$2^{\frac{1}{5}}$	$\varepsilon^4 2^{\frac{1}{5}}$	$\varepsilon^3 2^{\frac{1}{5}}$	$\varepsilon^2 2^{\frac{1}{5}}$	$\varepsilon 2^{\frac{1}{5}}$	$2^{\frac{1}{5}}$	$\varepsilon^3 2^{\frac{1}{5}}$	$\varepsilon 2^{\frac{1}{5}}$	$\varepsilon^4 2^{\frac{1}{5}}$	$\varepsilon^2 2^{\frac{1}{5}}$
$\varepsilon \longrightarrow$	ε^4	ε^4	ε^4	ε^4	ε^4	ε^3	ε^3	ε^3	ε^3	ε^3
$\sqrt{5} \longrightarrow$	$\sqrt{5}$	$\sqrt{5}$	$\sqrt{5}$	$\sqrt{5}$	$\sqrt{5}$	$-\sqrt{5}$	$-\sqrt{5}$	$-\sqrt{5}$	$-\sqrt{5}$	$-\sqrt{5}$

Here $\sigma\eta$ means σ followed by η

Subfields of $\mathbb{K} = \mathbb{Q}\left(2^{\frac{1}{5}}, \varepsilon\right)$

Subfield	Degree over $\mathbb{Q}$
$\mathbb{Q}$	1
$\mathbb{Q}(\varepsilon + \varepsilon^4) = \mathbb{Q}(\cos\frac{2\pi}{5}) = \mathbb{Q}(\sqrt{5})$	2
$\mathbb{Q}(\varepsilon)$	4
$\mathbb{Q}(2^{\frac{1}{5}})$	5
$\mathbb{Q}(\varepsilon 2^{\frac{1}{5}})$	5
$\mathbb{Q}(\varepsilon^2 2^{\frac{1}{5}})$	5
$\mathbb{Q}(\varepsilon^3 2^{\frac{1}{5}})$	5
$\mathbb{Q}(\varepsilon^4 2^{\frac{1}{5}})$	5
$\mathbb{Q}(2^{\frac{1}{5}}, \sqrt{5})$	10
$\mathbb{Q}(\varepsilon 2^{\frac{1}{5}}, \sqrt{5})$	10
$\mathbb{Q}(\varepsilon^2 2^{\frac{1}{5}}, \sqrt{5})$	10
$\mathbb{Q}(\varepsilon^3 2^{\frac{1}{5}}, \sqrt{5})$	10
$\mathbb{Q}(\varepsilon^4 2^{\frac{1}{5}}, \sqrt{5})$	10
$\mathbb{K} = \mathbb{Q}(2^{\frac{1}{5}}, \varepsilon)$	20

Subgroups of $G = Aut(\mathbb{K}/\mathbb{Q})$

Subgroup	Order
$G = \langle \sigma, \eta \rangle = \{\sigma^i \eta^j : 0 \leq i \leq 4, 0 \leq j \leq 3\}$	20
$\langle \sigma, \eta^2 \rangle = \{e, \sigma, \sigma^2, \sigma^3, \sigma^4, \eta^2, \sigma\eta^2, \sigma^2\eta^2, \sigma^3\eta^2, \sigma^4\eta^2\}$	10
$\langle \sigma \rangle = \{e, \sigma, \sigma^2, \sigma^3, \sigma^4\}$	5
$\langle \eta \rangle = \{e, \eta, \eta^2, \eta^3\}$	4
$\langle \sigma\eta \rangle = \{e, \sigma\eta, \sigma^4\eta^2, \sigma^3\eta^3\}$	4
$\langle \sigma^2\eta \rangle = \{e, \sigma^2\eta, \sigma^3\eta^2, \sigma\eta^3\}$	4
$\langle \sigma^3\eta \rangle = \{e, \sigma^3\eta, \sigma^3\eta^2, \sigma\eta^4\}$	4
$\langle \sigma^4\eta \rangle = \{e, \sigma^4\eta, \sigma\eta^2, \sigma\eta^2\}$	4
$\langle \eta^2 \rangle$	2
$\langle \sigma^4\eta^2 \rangle$	2
$\langle \sigma^3\eta^2 \rangle$	2
$\langle \sigma^2\eta^2 \rangle$	2
$\langle \sigma\eta^2 \rangle$	2
$\langle e \rangle$	1

Galois Correspondence for $x^5 - 2$

Subgroup	Order	Fixed Field
$\{e\}$	1	$\mathbb{K} = \mathbb{Q}(2^{\frac{1}{5}}, \varepsilon)$
$\langle \eta^2 \rangle$	2	$\mathbb{Q}(2^{\frac{1}{5}}, \varepsilon + \varepsilon^4) = \mathbb{Q}(2^{\frac{1}{5}}, \sqrt{5})$
$\langle \sigma\eta^2 \rangle$	2	$\mathbb{Q}(\varepsilon^2 2^{\frac{1}{5}}, \sqrt{5})$
$\langle \sigma^2\eta^2 \rangle$	2	$\mathbb{Q}(\varepsilon^4 2^{\frac{1}{5}}, \sqrt{5})$
$\langle \sigma^3\eta^2 \rangle$	2	$\mathbb{Q}(\varepsilon 2^{\frac{1}{5}}, \sqrt{5})$
$\langle \sigma^4\eta^2 \rangle$	2	$\mathbb{Q}(\varepsilon^3 2^{\frac{1}{5}}, \sqrt{5})$
$\langle \eta \rangle$	4	$\mathbb{Q}(2^{\frac{1}{5}})$
$\langle \sigma\eta \rangle$	4	$\mathbb{Q}(\varepsilon^3 2^{\frac{1}{5}})$
$\langle \sigma^2\eta \rangle$	4	$\mathbb{Q}(\varepsilon 2^{\frac{1}{5}})$
$\langle \sigma^3\eta \rangle$	4	$\mathbb{Q}(\varepsilon^4 2^{\frac{1}{5}})$
$\langle \sigma^4\eta \rangle$	4	$\mathbb{Q}(\varepsilon^2 2^{\frac{1}{5}})$
$\langle \sigma \rangle$	5	$\mathbb{Q}(\varepsilon)$
$\langle \sigma, \eta^2 \rangle$	10	$\mathbb{Q}(\varepsilon + \varepsilon^4) = \mathbb{Q}(\sqrt{5})$
$G = \langle \sigma, \eta \rangle$	20	$\mathbb{Q}$

Galois Group of $x^5 - 2$ - Lattice of subfields of $\mathbb{K} = \mathbb{Q}(2^{\frac{1}{5}}, \varepsilon)$

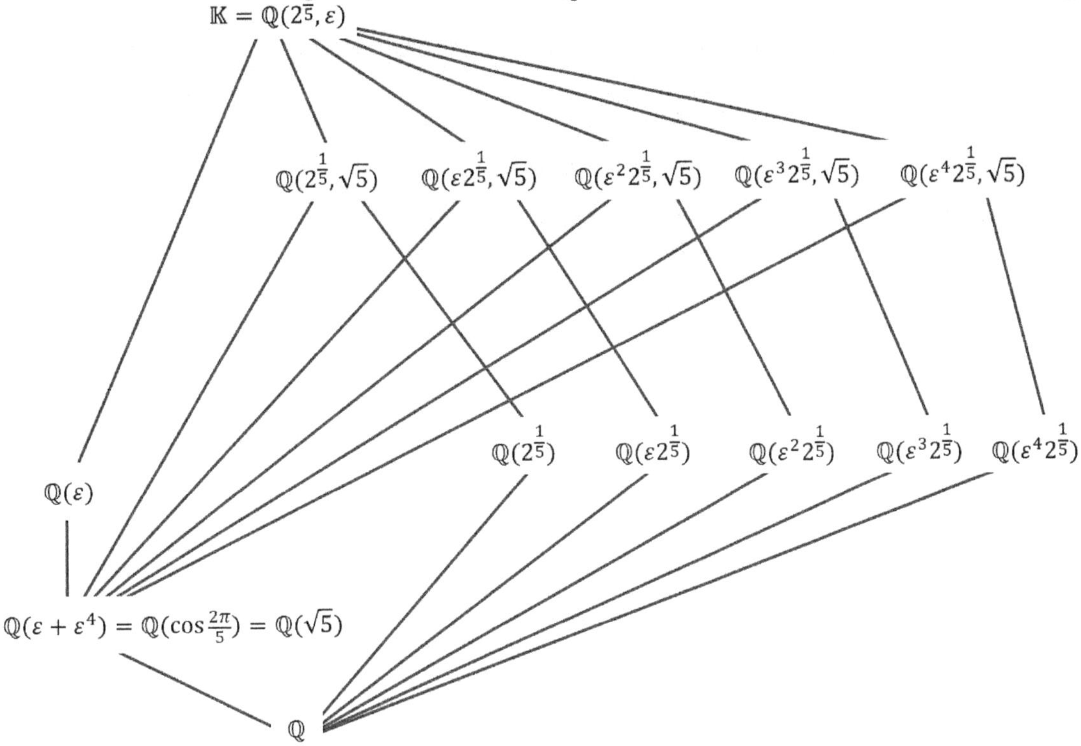

Galois Group of $x^5 - 2$ - Lattice of subgroups of $G = Aut(\mathbb{K}/\mathbb{Q}) = \langle \sigma, \eta : \sigma^5 = \eta^4 = e,\ \eta^{-1}\sigma\eta = \sigma^2 \rangle$

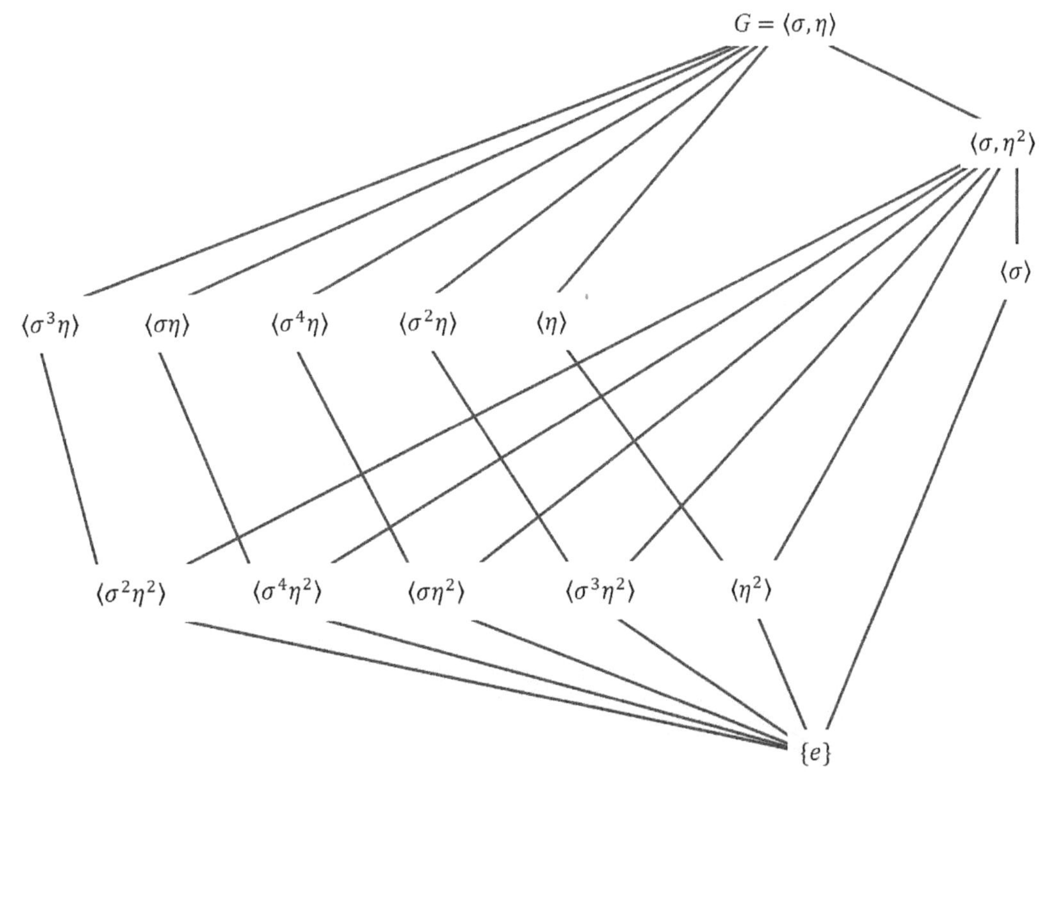

5. The Galois Group of $x^6 - 2$

The polynomial $x^6 - 2$ is easily seen to be irreducible over $\mathbb{Q}$. Its six zeros are $\pm 2^{\frac{1}{6}}$, $\pm\omega 2^{\frac{1}{6}}$, $\pm\omega^2 2^{\frac{1}{6}}$, where $\omega = e^{\frac{2\pi i}{3}} = \cos\frac{2\pi}{3} + i\sin\frac{2\pi}{3} = \dfrac{-1+\sqrt{-3}}{2}$. The splitting field of the polynomial $x^6 - 2$ is $\mathbb{K} = \mathbb{Q}(2^{\frac{1}{6}}, \omega)$. Evidently it has degree 12 over $\mathbb{Q}$ and so its Galois group $G = Aut(\mathbb{K}/\mathbb{Q})$ has order 12. Let σ and η be the following two automorphisms of $\mathbb{K}$.

$$\sigma : \begin{cases} 2^{\frac{1}{6}} & \longrightarrow & -\omega 2^{\frac{1}{6}} \\ \omega & \longrightarrow & \omega \end{cases} \qquad\qquad \eta : \begin{cases} 2^{\frac{1}{6}} & \longrightarrow & 2^{\frac{1}{6}} \\ \omega & \longrightarrow & \omega^2 \end{cases}.$$

It can be easily checked that $\sigma^6 = \eta^2 = e$ (identity), and $\eta^{-1}\sigma\eta = \sigma^5$ and so the group $G = Aut\,(\mathbb{K}/\mathbb{Q}) = \langle\sigma, \eta\rangle$ is a dihedral group of order 12. We have

$$G = \langle\sigma, \eta : \sigma^6 = \eta^2 = e, \eta^{-1}\sigma\eta = \sigma^5\rangle = \{e, \sigma, \sigma^2, \sigma^3, \sigma^4, \sigma^5, \eta, \sigma\eta, \sigma^2\eta, \sigma^3\eta, \sigma^4\eta, \sigma^5\eta\}.$$

In this and subsequent examples we shall not draw the diagrams exhibiting the lattice of subfields of the splitting field $\mathbb{K}$ and the lattice of subgroups of its Galois group $G = Aut(\mathbb{K}/\mathbb{Q})$. In this and subsequent examples we shall only give a list of subfields of the splitting field and a list of subgroups of the Galois group together with the action of the Galois group on certain elements of $\mathbb{K}$ (namely $2^{\frac{1}{6}}, \omega, \omega 2^{\frac{1}{6}}, \omega^2 2^{\frac{1}{6}}, 2^{\frac{1}{3}}, 2^{\frac{1}{2}}, \omega 2^{\frac{1}{3}}, \omega^2 2^{\frac{1}{3}}$, $\sqrt{-3}$, and $\sqrt{-6}$ in the present example) and the Galois correspondence.

Note. Although it would be enough to describe the effect of an automorphism on the elements $2^{\frac{1}{6}}$ and ω (or simply on $2^{\frac{1}{6}} + \omega$), a description of the effect of various automorphisms on several elements helps us in setting up the Galois correspondence.

Action of $G = Aut\,(\mathbb{K}/\mathbb{Q}) = \langle \sigma, \eta \rangle$ on $\mathbb{K} = \mathbb{Q}(2^{\frac{1}{6}}, \omega)$

	e	σ	σ^2	σ^3	σ^4	σ^5	η	$\sigma\eta$	$\sigma^2\eta$	$\sigma^3\eta$	$\sigma^4\eta$	$\sigma^5\eta$
$2^{\frac{1}{6}} \longrightarrow$	$2^{\frac{1}{6}}$	$-\omega 2^{\frac{1}{6}}$	$\omega^2 2^{\frac{1}{6}}$	$-2^{\frac{1}{6}}$	$\omega 2^{\frac{1}{6}}$	$-\omega^2 2^{\frac{1}{6}}$	$2^{\frac{1}{6}}$	$-\omega^2 2^{\frac{1}{6}}$	$\omega 2^{\frac{1}{6}}$	$-2^{\frac{1}{6}}$	$\omega 2^{\frac{1}{6}}$	$-\omega 2^{\frac{1}{6}}$
$\omega \longrightarrow$	ω	ω	ω	ω	ω	ω	ω^2	ω^2	ω^2	ω^2	ω^2	ω^2
$\omega 2^{\frac{1}{6}} \longrightarrow$	$\omega 2^{\frac{1}{6}}$	$-\omega^2 2^{\frac{1}{6}}$	$2^{\frac{1}{6}}$	$-\omega 2^{\frac{1}{6}}$	$\omega^2 2^{\frac{1}{6}}$	$-2^{\frac{1}{6}}$	$\omega^2 2^{\frac{1}{6}}$	$-\omega 2^{\frac{1}{6}}$	$2^{\frac{1}{6}}$	$-\omega^2 2^{\frac{1}{6}}$	$\omega 2^{\frac{1}{6}}$	$-2^{\frac{1}{6}}$
$\omega^2 2^{\frac{1}{6}} \longrightarrow$	$\omega^2 2^{\frac{1}{6}}$	$-2^{\frac{1}{6}}$	$\omega 2^{\frac{1}{6}}$	$-\omega^2 2^{\frac{1}{6}}$	$2^{\frac{1}{6}}$	$-\omega 2^{\frac{1}{6}}$	$\omega 2^{\frac{1}{6}}$	$-2^{\frac{1}{6}}$	$\omega^2 2^{\frac{1}{6}}$	$-\omega 2^{\frac{1}{6}}$	$2^{\frac{1}{6}}$	$-\omega^2 2^{\frac{1}{6}}$
$2^{\frac{1}{3}} \longrightarrow$	$2^{\frac{1}{3}}$	$\omega^2 2^{\frac{1}{3}}$	$\omega 2^{\frac{1}{3}}$	$2^{\frac{1}{3}}$	$\omega^2 2^{\frac{1}{3}}$	$\omega 2^{\frac{1}{3}}$	$2^{\frac{1}{3}}$	$\omega 2^{\frac{1}{3}}$	$\omega^2 2^{\frac{1}{3}}$	$2^{\frac{1}{3}}$	$\omega 2^{\frac{1}{3}}$	$\omega^2 2^{\frac{1}{3}}$
$\omega 2^{\frac{1}{3}} \longrightarrow$	$\omega 2^{\frac{1}{3}}$	$2^{\frac{1}{3}}$	$\omega^2 2^{\frac{1}{3}}$	$\omega 2^{\frac{1}{3}}$	$2^{\frac{1}{3}}$	$\omega^2 2^{\frac{1}{3}}$	$\omega^2 2^{\frac{1}{3}}$	$2^{\frac{1}{3}}$	$\omega 2^{\frac{1}{3}}$	$\omega^2 2^{\frac{1}{3}}$	$2^{\frac{1}{3}}$	$\omega 2^{\frac{1}{3}}$
$\omega^2 2^{\frac{1}{3}} \longrightarrow$	$\omega^2 2^{\frac{1}{3}}$	$\omega 2^{\frac{1}{3}}$	$2^{\frac{1}{3}}$	$\omega^2 2^{\frac{1}{3}}$	$\omega 2^{\frac{1}{3}}$	$2^{\frac{1}{3}}$	$\omega 2^{\frac{1}{3}}$	$\omega^2 2^{\frac{1}{3}}$	$2^{\frac{1}{3}}$	$\omega 2^{\frac{1}{3}}$	$\omega^2 2^{\frac{1}{3}}$	$2^{\frac{1}{3}}$
$\sqrt{2} \longrightarrow$	$\sqrt{2}$	$-\sqrt{2}$	$\sqrt{2}$	$-\sqrt{2}$	$\sqrt{2}$	$-\sqrt{2}$	$\sqrt{2}$	$-\sqrt{2}$	$\sqrt{2}$	$-\sqrt{2}$	$\sqrt{2}$	$-\sqrt{2}$
$\sqrt{-3} \longrightarrow$	$\sqrt{-3}$	$\sqrt{-3}$	$\sqrt{-3}$	$\sqrt{-3}$	$\sqrt{-3}$	$\sqrt{-3}$	$-\sqrt{-3}$	$-\sqrt{-3}$	$-\sqrt{-3}$	$-\sqrt{-3}$	$-\sqrt{-3}$	$-\sqrt{-3}$
$\sqrt{-6} \longrightarrow$	$\sqrt{-6}$	$-\sqrt{-6}$	$\sqrt{-6}$	$-\sqrt{-6}$	$\sqrt{-6}$	$-\sqrt{-6}$	$-\sqrt{-6}$	$\sqrt{-6}$	$-\sqrt{-6}$	$\sqrt{-6}$	$-\sqrt{-6}$	$\sqrt{-6}$

Here $\sigma\eta$ means σ followed by η

Subfields of $\mathbb{K} = \mathbb{Q}\left(2^{\frac{1}{6}}, \omega\right)$

Degree over $\mathbb{Q}$	# of Subfields	Subfields
1	1	$\mathbb{Q}$
2	3	$\mathbb{Q}(\omega) = \mathbb{Q}(\sqrt{-3}), \mathbb{Q}(\sqrt{2}), \mathbb{Q}(\sqrt{-6})$
3	3	$\mathbb{Q}(2^{\frac{1}{3}}), \mathbb{Q}(\omega 2^{\frac{1}{3}}), \mathbb{Q}(\omega^2 2^{\frac{1}{3}})$
4	1	$\mathbb{Q}(2^{\frac{1}{2}}, \omega)$
6	7	$\mathbb{Q}(2^{\frac{1}{3}}, \omega), \mathbb{Q}(2^{\frac{1}{6}}), \mathbb{Q}(\omega 2^{\frac{1}{6}}), \mathbb{Q}(\omega^2 2^{\frac{1}{6}})$ $\mathbb{Q}(2^{\frac{1}{3}}, \sqrt{-6}), \mathbb{Q}(\omega 2^{\frac{1}{3}}, \sqrt{-6}), \mathbb{Q}(\omega^2 2^{\frac{1}{3}}, \sqrt{-6})$
12	1	$\mathbb{K} = \mathbb{Q}(2^{\frac{1}{6}}, \omega)$

Subgroups of the Group $G = Aut(\mathbb{K}/\mathbb{Q})$

Order	# of Subgroups	Subgroups
12	1	$G = \langle \sigma, \eta \rangle$
6	3	$\langle \sigma \rangle, \langle \sigma^2, \eta \rangle, \langle \sigma^2, \sigma\eta \rangle$
4	3	$\langle \sigma^3, \eta \rangle, \langle \sigma^3, \sigma\eta \rangle, \langle \sigma^3, \sigma^2\eta \rangle$
3	1	$\langle \sigma^2 \rangle$
2	7	$\langle \sigma^3 \rangle, \langle \eta \rangle, \langle \sigma\eta \rangle, \langle \sigma^2\eta \rangle$ $\langle \sigma^3\eta \rangle, \langle \sigma^4\eta \rangle, \langle \sigma^5\eta \rangle$
1	1	$\{e\}$

Galois Correspondence for $x^6 - 2$

Subgroup	Order	Fixed Field
$G = \langle \sigma, \eta \rangle$	12	$\mathbb{Q}$
$\langle \sigma \rangle$	6	$\mathbb{Q}(\omega)$
$\langle \sigma^2 \eta \rangle$	6	$\mathbb{Q}(\sqrt{2})$
$\langle \sigma^2, \sigma\eta \rangle$	6	$\mathbb{Q}(\sqrt{-6})$
$\langle \sigma^3, \eta \rangle$	4	$\mathbb{Q}(2^{\frac{1}{3}})$
$\langle \sigma^3, \sigma\eta \rangle$	4	$\mathbb{Q}(\omega^2 2^{\frac{1}{3}})$
$\langle \sigma^3, \sigma^2\eta \rangle$	4	$\mathbb{Q}(\omega 2^{\frac{1}{3}})$
$\langle \sigma^2 \rangle$	3	$\mathbb{Q}(2^{\frac{1}{2}}, \omega)$
$\langle \sigma^3 \rangle$	2	$\mathbb{Q}(2^{\frac{1}{3}}, \omega)$
$\langle \eta \rangle$	2	$\mathbb{Q}(2^{\frac{1}{6}})$
$\langle \sigma\eta \rangle$	2	$\mathbb{Q}(\omega^2 2^{\frac{1}{3}}, \sqrt{-6})$
$\langle \sigma^2\eta \rangle$	2	$\mathbb{Q}(\omega^2 2^{\frac{1}{6}})$
$\langle \sigma^3\eta \rangle$	2	$\mathbb{Q}(2^{\frac{1}{3}}, \sqrt{-6})$
$\langle \sigma^4\eta \rangle$	2	$\mathbb{Q}(\omega 2^{\frac{1}{6}})$
$\langle \sigma^5\eta \rangle$	2	$\mathbb{Q}(\omega 2^{\frac{1}{3}}, \sqrt{-6})$
$\{e\}$	1	$\mathbb{K} = \mathbb{Q}(2^{\frac{1}{6}}, \omega)$

6. The Galois Group of $x^7 - 2$

The polynomial $x^7 - 2$ is easily seen to be irreducible over $\mathbb{Q}$. Its seven zeros are $2^{\frac{1}{7}}$, $\varepsilon 2^{\frac{1}{7}}$, $\varepsilon^2 2^{\frac{1}{7}}$, $\varepsilon^3 2^{\frac{1}{7}}$, $\varepsilon^3 2^{\frac{1}{7}}$, $\varepsilon^5 2^{\frac{1}{7}}$, $\varepsilon^6 2^{\frac{1}{7}}$, where $\varepsilon = e^{\frac{2\pi i}{7}} = \cos\frac{2\pi}{7} + i\sin\frac{2\pi}{7}$. The splitting field of the polynomial $x^7 - 2$ is $\mathbb{K} = \mathbb{Q}(2^{\frac{1}{7}}, \varepsilon)$. It has degree 42 over $\mathbb{Q}$. This is evident from the fact that $[\mathbb{Q}(2^{\frac{1}{7}}) : \mathbb{Q}] = 7$ and $[\mathbb{Q}(\varepsilon) : \mathbb{Q}] = 6$. The Galois group $\mathbb{G} = Aut(\mathbb{K}/\mathbb{Q})$ of $x^7 - 2$ is therefore order 42. Let σ and η be the following two automorphism of $\mathbb{K}$.

$$\sigma : \begin{cases} 2^{\frac{1}{7}} & \longrightarrow \quad \varepsilon 2^{\frac{1}{6}} \\ \\ \varepsilon & \longrightarrow \quad \varepsilon \end{cases} \qquad \eta : \begin{cases} 2^{\frac{1}{7}} & \longrightarrow \quad 2^{\frac{1}{7}} \\ \\ \varepsilon & \longrightarrow \quad \varepsilon^3 \end{cases} .$$

It can be easily checked that $\sigma^7 = \eta^6 = e$ (identity), $\eta^{-1}\sigma\eta = \sigma^3$ and $\eta\sigma\eta^{-1} = \sigma^5$ (recall that we have decided to compose the automorphism from left to right). So the group $G = Aut\,(\mathbb{K}/\mathbb{Q}) = \langle \sigma, \eta \rangle$ is a (Frobenius) group of order 42. We have

$$G = \langle \sigma, \eta : \sigma^7 = \eta^6 = e, \eta^{-1}\sigma\eta = \sigma^3 \rangle = \left\{ \sigma^i \eta^j : 0 \leq i \leq 6, 0 \leq j \leq 5 \right\}.$$

Let $x = (1, 2, 3, 4, 5, 6, 7,)$ and $y = (2, 4, 3, 7, 5, 6)$ be two elements of the symmetric group S_7 on seven letters. Then it can be easily seen that the Galois group $G = Aut(\mathbb{K}/\mathbb{Q})$ of $x^7 - 2$ is isomorphic to the subgroup $\langle x, y \rangle$ of S_7 generated by the permutations x and y. In group theory such a group is known as a Frobenius group.

The tables on the following pages show the effect of the elements of G on $2^{\frac{1}{7}}$ and ε, the subgroups of G, the subfields of the splitting field $\mathbb{K}$, and the Galois correspondence.

Action of $G = Aut\,(\mathbb{K}/\mathbb{Q}) = \langle \sigma, \eta \rangle$ on the elements $2^{\frac{1}{7}}$ and $\varepsilon = e^{\frac{2i\pi}{7}}$

	e	σ	σ^2	σ^3	σ^4	σ^5	σ^6
$2^{\frac{1}{7}} \longrightarrow$	$2^{\frac{1}{7}}$	$\varepsilon 2^{\frac{1}{7}}$	$\varepsilon^2 2^{\frac{1}{7}}$	$\varepsilon^3 2^{\frac{1}{7}}$	$\varepsilon^4 2^{\frac{1}{7}}$	$\varepsilon^5 2^{\frac{1}{7}}$	$\varepsilon^6 2^{\frac{1}{7}}$
$\varepsilon \longrightarrow$	ε	ε	ε	ε	ε	ε	ε
	η	$\sigma\eta$	$\sigma^2\eta$	$\sigma^3\eta$	$\sigma^4\eta$	$\sigma^5\eta$	$\sigma^6\eta$
$2^{\frac{1}{7}} \longrightarrow$	$2^{\frac{1}{7}}$	$\varepsilon^3 2^{\frac{1}{7}}$	$\varepsilon^6 2^{\frac{1}{7}}$	$\varepsilon^2 2^{\frac{1}{7}}$	$\varepsilon^5 2^{\frac{1}{7}}$	$\varepsilon 2^{\frac{1}{7}}$	$\varepsilon^4 2^{\frac{1}{7}}$
$\varepsilon \longrightarrow$	ε^3	ε^3	ε^3	ε^3	ε^3	ε^3	ε^3
	η^2	$\sigma\eta^2$	$\sigma^2\eta^2$	$\sigma^3\eta^2$	$\sigma^4\eta^2$	$\sigma^5\eta^2$	$\sigma^6\eta^2$
$2^{\frac{1}{7}} \longrightarrow$	$2^{\frac{1}{7}}$	$\varepsilon^2 2^{\frac{1}{7}}$	$\varepsilon^4 2^{\frac{1}{7}}$	$\varepsilon^6 2^{\frac{1}{7}}$	$\varepsilon 2^{\frac{1}{7}}$	$\varepsilon^3 2^{\frac{1}{7}}$	$\varepsilon^5 2^{\frac{1}{7}}$
$\varepsilon \longrightarrow$	ε^2	ε^2	ε^2	ε^2	ε^2	ε^2	ε^2
	η^3	$\sigma\eta^3$	$\sigma^2\eta^3$	$\sigma^3\eta^3$	$\sigma^4\eta^3$	$\sigma^5\eta^3$	$\sigma^6\eta^3$
$2^{\frac{1}{7}} \longrightarrow$	$2^{\frac{1}{7}}$	$\varepsilon^6 2^{\frac{1}{7}}$	$\varepsilon^5 2^{\frac{1}{7}}$	$\varepsilon^4 2^{\frac{1}{7}}$	$\varepsilon^3 2^{\frac{1}{7}}$	$\varepsilon^2 2^{\frac{1}{7}}$	$\varepsilon 2^{\frac{1}{7}}$
$\varepsilon \longrightarrow$	ε^6	ε^6	ε^6	ε^6	ε^6	ε^6	ε^6
	η^4	$\sigma\eta^4$	$\sigma^2\eta^4$	$\sigma^3\eta^4$	$\sigma^4\eta^4$	$\sigma^5\eta^4$	$\sigma^6\eta^4$
$2^{\frac{1}{7}} \longrightarrow$	$2^{\frac{1}{7}}$	$\varepsilon^4 2^{\frac{1}{7}}$	$\varepsilon 2^{\frac{1}{7}}$	$\varepsilon^5 2^{\frac{1}{7}}$	$\varepsilon^2 2^{\frac{1}{7}}$	$\varepsilon^6 2^{\frac{1}{7}}$	$\varepsilon^3 2^{\frac{1}{7}}$
$\varepsilon \longrightarrow$	ε^4	ε^4	ε^4	ε^4	ε^4	ε^4	ε^4
	η^5	$\sigma\eta^5$	$\sigma^2\eta^5$	$\sigma^3\eta^5$	$\sigma^4\eta^5$	$\sigma^5\eta^5$	$\sigma^6\eta^5$
$2^{\frac{1}{7}} \longrightarrow$	$2^{\frac{1}{7}}$	$\varepsilon^5 2^{\frac{1}{7}}$	$\varepsilon^3 2^{\frac{1}{7}}$	$\varepsilon 2^{\frac{1}{7}}$	$\varepsilon^6 2^{\frac{1}{7}}$	$\varepsilon^4 2^{\frac{1}{7}}$	$\varepsilon^2 2^{\frac{1}{7}}$
$\varepsilon \longrightarrow$	ε^5	ε^5	ε^5	ε^5	ε^5	ε^5	ε^5

Here $\sigma\eta$ means σ followed by η

Subfields of $\mathbb{K} = \mathbb{Q}\left(2^{\frac{1}{7}}, \omega\right)$

Subfields	Degree over $\mathbb{Q}$
$\mathbb{Q}$	1
$\mathbb{Q}(\varepsilon + \varepsilon^2 + \varepsilon^4) = \mathbb{Q}(\sqrt{-7})$	2
$\mathbb{Q}(\varepsilon + \varepsilon^6) = \mathbb{Q}(2\cos\frac{2\pi}{7})$	3
$\mathbb{Q}(\varepsilon)$	6
$\mathbb{Q}(2^{\frac{1}{7}}), \mathbb{Q}(\varepsilon 2^{\frac{1}{7}}), \mathbb{Q}(\varepsilon^2 2^{\frac{1}{7}}), \mathbb{Q}(\varepsilon^3 2^{\frac{1}{7}})$ $\mathbb{Q}(\varepsilon^4 2^{\frac{1}{7}}), \mathbb{Q}(\varepsilon^5 2^{\frac{1}{7}}), \mathbb{Q}(\varepsilon^6 2^{\frac{1}{7}})$	7
$\mathbb{Q}(2^{\frac{1}{7}}, \varepsilon + \varepsilon^2 + \varepsilon^4) = \mathbb{Q}(2^{\frac{1}{7}}, \sqrt{-7})$, $\mathbb{Q}(\varepsilon 2^{\frac{1}{7}}, \sqrt{-7}), \mathbb{Q}(\varepsilon^2 2^{\frac{1}{7}}, \sqrt{-7})$, $\mathbb{Q}(\varepsilon^3 2^{\frac{1}{7}}, \sqrt{-7}), \mathbb{Q}(\varepsilon^4 2^{\frac{1}{7}}, \sqrt{-7})$, $\mathbb{Q}(\varepsilon^5 2^{\frac{1}{7}}, \sqrt{-7}), \mathbb{Q}(\varepsilon^6 2^{\frac{1}{7}}, \sqrt{-7})$	14
$\mathbb{Q}(2^{\frac{1}{7}}, \varepsilon + \varepsilon^6) = \mathbb{Q}(2^{\frac{1}{7}}, \cos\frac{2\pi}{7})$, $\mathbb{Q}(\varepsilon 2^{\frac{1}{7}}, \cos\frac{2\pi}{7}), \mathbb{Q}(\varepsilon^2 2^{\frac{1}{7}}, \cos\frac{2\pi}{7})$, $\mathbb{Q}(\varepsilon^3 2^{\frac{1}{7}}, \cos\frac{2\pi}{7}), \mathbb{Q}(\varepsilon^4 2^{\frac{1}{7}}, \cos\frac{2\pi}{7})$, $\mathbb{Q}(\varepsilon^5 2^{\frac{1}{7}}, \cos\frac{2\pi}{7}), \mathbb{Q}(\varepsilon^6 2^{\frac{1}{7}}, \cos\frac{2\pi}{7})$,	21
$\mathbb{K} = \mathbb{Q}(2^{\frac{1}{7}}, \varepsilon)$	42

Subgroups of the Group $G = Aut(\mathbb{K}/\mathbb{Q})$

Subgroup(s)	Order
$G = \langle \sigma, \eta \rangle$	42
$\langle \sigma, \eta^2 \rangle = \{\sigma^i \eta^{2j} : 0 \leq i \leq 6, 0 \leq j \leq 2\}$	21
$\langle \sigma, \eta^3 \rangle$	14
$\langle \sigma \rangle$	7
$\langle \eta \rangle, \langle \sigma\eta \rangle, \langle \sigma^2\eta \rangle, \langle \sigma^3\eta \rangle,$ $\langle \sigma^4\eta \rangle, \langle \sigma^5\eta \rangle, \langle \sigma^6\eta \rangle$	6
$\langle \eta^2 \rangle, \langle \sigma\eta^2 \rangle, \langle \sigma^2\eta^2 \rangle, \langle \sigma^3\eta^2 \rangle,$ $\langle \sigma^4\eta^2 \rangle, \langle \sigma^5\eta^2 \rangle, \langle \sigma^6\eta^2 \rangle$	3
$\langle \eta^3 \rangle, \langle \sigma^3\eta^3 \rangle, \langle \sigma^6\eta^3 \rangle, \langle \sigma^2\eta^3 \rangle,$ $\langle \sigma^5\eta^3 \rangle, \langle \sigma\eta^3 \rangle, \langle \sigma^4\eta^3 \rangle$	2
$\{e\}$	1

Galois Correspondence for $x^7 - 2$

Subgroup	Order	Fixed Field
$G = \langle \sigma, \eta \rangle$	42	$\mathbb{Q}$
$\langle \sigma, \eta^2 \rangle$	21	$\mathbb{Q}(\varepsilon + \varepsilon^2 + \varepsilon^4) = \mathbb{Q}(\sqrt{-7})$
$\langle \sigma\eta^3 \rangle$	14	$\mathbb{Q}(\varepsilon + \varepsilon^6) = \mathbb{Q}(\cos \frac{2\pi}{7})$
$\langle \sigma \rangle$	7	$Q(\varepsilon)$
$\langle \eta \rangle$	6	$\mathbb{Q}(2^{\frac{1}{7}})$
$\langle \sigma\eta \rangle$	6	$\mathbb{Q}(\varepsilon^2 2^{\frac{1}{7}})$
$\langle \sigma^2\eta \rangle$	6	$\mathbb{Q}(\varepsilon^4 2^{\frac{1}{7}})$
$\langle \sigma^3\eta \rangle$	6	$\mathbb{Q}(\varepsilon^6 2^{\frac{1}{7}})$
$\langle \sigma^4\eta \rangle$	6	$\mathbb{Q}(\varepsilon 2^{\frac{1}{7}})$
$\langle \sigma^5\eta \rangle$	6	$\mathbb{Q}(\varepsilon^3 2^{\frac{1}{7}})$
$\langle \sigma^6\eta \rangle$	6	$\mathbb{Q}(\varepsilon^5 2^{\frac{1}{7}})$
$\langle \eta^2 \rangle$	3	$\mathbb{Q}(2^{\frac{1}{7}}, \sqrt{-7})$
$\langle \sigma\eta^2 \rangle$	3	$\mathbb{Q}(\varepsilon^5 2^{\frac{1}{7}}, \sqrt{-7})$
$\langle \sigma^2\eta^2 \rangle$	3	$\mathbb{Q}(\varepsilon^3 2^{\frac{1}{7}}, \sqrt{-7})$
$\langle \sigma^3\eta^2 \rangle$	3	$\mathbb{Q}(\varepsilon 2^{\frac{1}{7}}, \sqrt{-7})$
$\langle \sigma^4\eta^2 \rangle$	3	$\mathbb{Q}(\varepsilon^6 2^{\frac{1}{7}}, \sqrt{-7})$
$\langle \sigma^5\eta^2 \rangle$	3	$\mathbb{Q}(\varepsilon^4 2^{\frac{1}{7}}, \sqrt{-7})$
$\langle \sigma^6\eta^2 \rangle$	3	$\mathbb{Q}(\varepsilon^2 2^{\frac{1}{7}}, \sqrt{-7})$

Galois Correspondence for $x^7 - 2$. *(continued)*

Subgroup	Order	Fixed Field
$\langle \eta^3 \rangle$	2	$\mathbb{Q}(2^{\frac{1}{7}}, \cos\frac{2\pi}{7})$
$\langle \sigma\eta^3 \rangle$	2	$\mathbb{Q}(\varepsilon^3 2^{\frac{1}{7}}, \cos\frac{2\pi}{7})$
$\langle \sigma^2\eta^3 \rangle$	2	$\mathbb{Q}(\varepsilon^6 2^{\frac{1}{7}}, \cos\frac{2\pi}{7})$
$\langle \sigma^3\eta^3 \rangle$	2	$\mathbb{Q}(\varepsilon^2 2^{\frac{1}{7}}, \cos\frac{2\pi}{7})$
$\langle \sigma^4\eta^3 \rangle$	2	$\mathbb{Q}(\varepsilon^5 2^{\frac{1}{7}}, \cos\frac{2\pi}{7})$
$\langle \sigma^5\eta^3 \rangle$	2	$\mathbb{Q}(\varepsilon 2^{\frac{1}{7}}, \cos\frac{2\pi}{7})$
$\langle \sigma^6\eta^3 \rangle$	2	$\mathbb{Q}(\varepsilon^4 2^{\frac{1}{7}}, \cos\frac{2\pi}{7})$
$\{e\}$	1	$\mathbb{Q}(2^{\frac{1}{7}}, \varepsilon)$

7. The Galois Group of $x^8 - 2$

The polynomial $x^8 - 2$ is easily seen to be irreducible over $\mathbb{Q}$. Its eight zeros are

$$\varepsilon^i 2^{\frac{1}{8}}, \qquad 0 \leq i \leq 7, \qquad \varepsilon = \cos\frac{\pi}{4} + i\sin\frac{\pi}{4} = \frac{1+i}{\sqrt{2}}.$$

So the splitting field of $x^8 - 2$ is $\mathbb{K} = \mathbb{Q}\left(2^{\frac{1}{8}}, \frac{1+i}{\sqrt{2}}\right) = \mathbb{Q}(2^{\frac{1}{8}}, i)$. Evidently $\mathbb{K}$ is of degree 16 over $\mathbb{Q}$. Let σ and η be the following two automorphism of $\mathbb{K}$.

$$\sigma : \begin{cases} 2^{\frac{1}{8}} & \longrightarrow \quad \varepsilon 2^{\frac{1}{8}} \\ i & \longrightarrow \quad i \end{cases} \qquad\qquad \eta : \begin{cases} 2^{\frac{1}{8}} & \longrightarrow \quad 2^{\frac{1}{8}} \\ i & \longrightarrow \quad -i \end{cases} .$$

Then $\sigma^8 = \eta^2 = e$ (identity), $\eta^{-1}\sigma\eta = \sigma^3$. Therefore $G = Aut\,(\mathbb{K}/\mathbb{Q})$ has order 16. We have

$$\begin{aligned} G &= \langle \sigma, \eta, \sigma^8 = \eta^2 = e, \eta^{-1}\sigma\eta = \sigma^3 \rangle \\ &= \left\{e, \sigma, \sigma^2, \sigma^3, \sigma^4, \sigma^5, \sigma^6, \sigma^7, \eta, \sigma\eta, \sigma^2\eta, \sigma^3\eta, \sigma^4\eta, \sigma^5\eta, \sigma^6\eta, \sigma^7\eta\right\}. \end{aligned}$$

The tables on the following pages show the effect of the elements of G on $2^{\frac{1}{8}}$, i, and ε; the subgroups of G; the subfields of the splitting field $\mathbb{K}$; and the Galois correspondence.

Action of $G = Aut\,(\mathbb{K}/\mathbb{Q}) = \langle \sigma, \eta \rangle$ on the Elements $2^{\frac{1}{8}}$, i, and $\varepsilon = e^{\frac{i\pi}{4}}$ of $\mathbb{K}$

	e	σ	σ^2	σ^3	σ^4	σ^5	σ^6	σ^7
$2^{\frac{1}{8}} \longrightarrow$	$2^{\frac{1}{8}}$	$\varepsilon 2^{\frac{1}{8}}$	$\varepsilon^6 2^{\frac{1}{8}}$	$\varepsilon^7 2^{\frac{1}{8}}$	$\varepsilon^4 2^{\frac{1}{8}}$	$\varepsilon^5 2^{\frac{1}{8}}$	$\varepsilon^2 2^{\frac{1}{8}}$	$\varepsilon^3 2^{\frac{1}{8}}$
$i \longrightarrow$	i	i	i	i	i	i	i	i
$\varepsilon = \frac{1+i}{\sqrt{2}} \longrightarrow$	ε	ε^5	ε	ε^5	ε	ε^5	ε	ε^5
	η	$\sigma\eta$	$\sigma^2\eta$	$\sigma^3\eta$	$\sigma^4\eta$	$\sigma^5\eta$	$\sigma^6\eta$	$\sigma^7\eta$
$2^{\frac{1}{8}} \longrightarrow$	$2^{\frac{1}{8}}$	$\varepsilon^7 2^{\frac{1}{8}}$	$\varepsilon^2 2^{\frac{1}{8}}$	$\varepsilon 2^{\frac{1}{8}}$	$\varepsilon^4 2^{\frac{1}{8}}$	$\varepsilon^3 2^{\frac{1}{8}}$	$\varepsilon^6 2^{\frac{1}{8}}$	$\varepsilon^5 2^{\frac{1}{8}}$
$i \longrightarrow$	$-i$	$-i$	$-i$	$-i$	$-i$	$-i$	$-i$	$-i$
$\varepsilon = \frac{1+i}{\sqrt{2}} \longrightarrow$	ε^7	ε^3	ε^7	ε^3	ε^7	ε^3	ε^7	ε^3

Here $\sigma\eta$ means σ followed by η

Subfields of $\mathbb{K} = \mathbb{Q}\left(2^{\frac{1}{8}}, i\right)$

Subfields	Degree over $\mathbb{Q}$
$\mathbb{K} = \mathbb{Q}(2^{\frac{1}{8}}, i) = \mathbb{Q}(2^{\frac{1}{8}}, \varepsilon)$	16
$\mathbb{Q}(2^{\frac{1}{8}})$	8
$\mathbb{Q}(\varepsilon 2^{\frac{1}{8}}) = \mathbb{Q}\left(\frac{1+i}{\sqrt{2}} 2^{\frac{1}{8}}\right) = \mathbb{Q}\left((1+i)2^{\frac{1}{8}}\right)$	8
$\mathbb{Q}(\varepsilon^2 2^{\frac{1}{8}}) = \mathbb{Q}\left(i2^{\frac{1}{8}}\right)$	8
$\mathbb{Q}(\varepsilon^3 2^{\frac{1}{8}}) = \mathbb{Q}\left(\frac{1-i}{\sqrt{2}} 2^{\frac{1}{8}}\right) = \mathbb{Q}\left((1-i)2^{\frac{1}{8}}\right)$	8
$\mathbb{Q}(2^{\frac{1}{4}}, i)$	8
$\mathbb{Q}(2^{\frac{1}{2}}, i)$	4
$\mathbb{Q}(2^{\frac{1}{4}})$	4
$\mathbb{Q}\left(\frac{1+i}{\sqrt{2}}\right) = \mathbb{Q}\left((1+i)2^{\frac{1}{4}}\right)$	4
$\mathbb{Q}(i2^{\frac{1}{4}})$	4
$\mathbb{Q}\left(\frac{1-i}{\sqrt{2}}\right) = \mathbb{Q}\left((1-i)2^{\frac{1}{4}}\right)$	4
$\mathbb{Q}(2^{\frac{1}{2}})$	2
$\mathbb{Q}(i)$	2
$\mathbb{Q}(i2^{\frac{1}{2}})$	2
$\mathbb{Q}$	1

Subgroups of the Group $G = Aut(\mathbb{K}/\mathbb{Q})$

Subgroup(s)	Order
$G = \langle \sigma, \eta \rangle$	16
$\langle \sigma \rangle$	8
$\langle \sigma^2, \eta \rangle$	8^*
$\langle \sigma^2, \sigma\eta \rangle$	$8^\#$
$\langle \sigma^2 \rangle$	4
$\langle \sigma\eta \rangle$	4
$\langle \sigma^3\eta \rangle$	4
$\langle \sigma^4, \eta \rangle$	$4^\%$
$\langle \sigma^4, \sigma^2\eta \rangle$	$4^\%$
$\langle \sigma^4 \rangle$	2
$\langle \eta \rangle$	2
$\langle \sigma^2\eta \rangle$	2
$\langle \sigma^4\eta \rangle$	2
$\langle \sigma^6\eta \rangle$	2
$\{e\}$	1

- $(*)$ This group is dihedral.
- (#) This group is quaternion.
- (%) This group is Klein-Four.

Galois Correspondence for $x^8 - 2$

Subgroup	Order	Fixed Field
$G = \langle \sigma, \eta \rangle$	16	$\mathbb{Q}$
$\langle \sigma \rangle$	8	$\mathbb{Q}(i)$
$\langle \sigma^2, \eta \rangle$	8	$\mathbb{Q}(2^{\frac{1}{2}})$
$\langle \sigma^2, \sigma\eta \rangle$	8	$\mathbb{Q}(i2^{\frac{1}{2}})$
$\langle \sigma^4, \eta \rangle$	4	$\mathbb{Q}(2^{\frac{1}{4}})$
$\langle \sigma^4, \sigma^2\eta \rangle$	4	$\mathbb{Q}(i2^{\frac{1}{4}})$
$\langle \sigma^2 \rangle$	4	$\mathbb{Q}(2^{\frac{1}{2}}, i)$
$\langle \sigma^3\eta \rangle$	4	$\mathbb{Q}\left((1+i)2^{\frac{1}{4}}\right)$
$\langle \sigma\eta \rangle$	4	$\mathbb{Q}\left((1-i)2^{\frac{1}{4}}\right)$
$\langle \sigma^4 \rangle$	2	$\mathbb{Q}(2^{\frac{1}{4}}, i)$
$\langle \eta \rangle$	2	$\mathbb{Q}(2^{\frac{1}{8}})$
$\langle \sigma^2\eta \rangle$	2	$\mathbb{Q}\left((1+i)2^{\frac{1}{8}}\right)$
$\langle \sigma^4\eta \rangle$	2	$\mathbb{Q}(i2^{\frac{1}{8}})$
$\langle \sigma^6\eta \rangle$	2	$\mathbb{Q}\left((1-i)2^{\frac{1}{8}}\right)$
$\{e\}$	1	$\mathbb{K} = \mathbb{Q}(2^{\frac{1}{8}})$

8. The Galois Group of $x^{16}-2$

The polynomial $x^{16}-2$ is easily seen to be irreducible over $\mathbb{Q}$. Its 16 zeros can be written as

$$\alpha^i 2^{\frac{1}{16}}, \qquad 0 \le i \le 15, \qquad \text{where} \qquad \alpha = e^{\frac{2\pi i}{16}} = \cos 22.5^\circ + i \sin 22.5^\circ = \frac{\sqrt{2+\sqrt{2}}}{2} + i\frac{\sqrt{2-\sqrt{2}}}{2}.$$

So the splitting field of $x^{16}-2$ over $\mathbb{Q}$ is $\mathbb{K} = \mathbb{Q}(2^{\frac{1}{16}}, \alpha) = \mathbb{Q}(2^{\frac{1}{16}}, \sqrt{2+\sqrt{2}}, i)$, and has degree 64 over $\mathbb{Q}$. So the Galois group $G = Aut(\mathbb{K}/\mathbb{Q})$ of $x^{16}-2$ has order 64.

Note. If $n \ge 3$ then the Galois group of $x^{2^n}-2$ over $\mathbb{Q}$ has order $\frac{1}{2}\cdot 2^n \cdot \phi(2^n) = 2^{2n-2}$.

Let σ and η be the following two automorphism of $\mathbb{K}$.

$$\sigma : \begin{cases} 2^{\frac{1}{16}} \longrightarrow \alpha 2^{\frac{1}{16}} \\ \alpha \longrightarrow \alpha^{13} \end{cases} \qquad \eta : \begin{cases} 2^{\frac{1}{16}} \longrightarrow \alpha 2^{\frac{1}{16}} \\ \alpha \longrightarrow \alpha^{3} \end{cases}.$$

It can be easily seen that order$(\sigma) = 16$ and order$(\eta) = 8$ and $\sigma^8 = \eta^4$. We also note that

$$\sigma^{-1} : \begin{cases} 2^{\frac{1}{16}} \longrightarrow \alpha^{11} 2^{\frac{1}{16}} \\ \alpha \longrightarrow \alpha^{5} \end{cases} \qquad \eta^{-1} : \begin{cases} 2^{\frac{1}{16}} \longrightarrow \alpha^5 2^{\frac{1}{16}} \\ \alpha \longrightarrow \alpha^{11} \end{cases}.$$

We also have $\eta^{-1}\sigma\eta = \sigma^{15}\eta^2$—remembering that here we are following the convention that

$\sigma\eta$ **means** σ **followed by** η.

So

$$G = \langle \sigma, \eta : \sigma^{16} = \eta^8 = (\sigma\eta)^2 = e, \sigma^8 = \eta^4, \eta^{-1}\sigma\eta = \sigma^{15}\eta^2 \rangle = \{\sigma^i \eta^j : 0 \le i \le 15, 0 \le j \le 3\}.$$

Effect of the elements of $G = Aut(\mathbb{K}/\mathbb{Q})$ on elements of $\mathbb{K}$

The table on the following pages give the effect of elements of G on the elements $2^{\frac{1}{16}}$, $\alpha = e^{\frac{2\pi i}{16}}$, and $\sqrt{2+\sqrt{2}}$. First we note the following:

$$\begin{aligned} \alpha &= \frac{\sqrt{2+\sqrt{2}}}{2} + i\frac{\sqrt{2-\sqrt{2}}}{2}, & \alpha^3 &= \frac{\sqrt{2-\sqrt{2}}}{2} + i\frac{\sqrt{2+\sqrt{2}}}{2}, \\ \alpha^5 &= -\frac{\sqrt{2-\sqrt{2}}}{2} + i\frac{\sqrt{2+\sqrt{2}}}{2}, & \alpha^7 &= -\frac{\sqrt{2+\sqrt{2}}}{2} + i\frac{\sqrt{2-\sqrt{2}}}{2}, \\ \alpha^9 &= -\frac{\sqrt{2+\sqrt{2}}}{2} - i\frac{\sqrt{2-\sqrt{2}}}{2}, & \alpha^{11} &= -\frac{\sqrt{2-\sqrt{2}}}{2} - i\frac{\sqrt{2+\sqrt{2}}}{2}, \\ \alpha^{13} &= -\frac{\sqrt{2-\sqrt{2}}}{2} - i\frac{\sqrt{2+\sqrt{2}}}{2}, & \alpha^{15} &= \frac{\sqrt{2+\sqrt{2}}}{2} - i\frac{\sqrt{2-\sqrt{2}}}{2}, \end{aligned}$$

$$\begin{aligned} \alpha^2 &= \frac{1+i}{\sqrt{2}}, & \alpha^4 &= i \\ \alpha^6 &= \frac{-1+i}{\sqrt{2}}, & \alpha^8 &= -1 \\ \alpha^{10} &= -\frac{1+i}{\sqrt{2}}, & \alpha^{12} &= -i \\ \alpha^{14} &= \frac{1-i}{\sqrt{2}}, & \alpha^{16} &= 1. \end{aligned}$$

We also note that

$$\alpha + \alpha^{15} = \sqrt{2+\sqrt{2}} = -(\alpha^7 + \alpha^9), \qquad \alpha^3 + \alpha^{13} = \sqrt{2-\sqrt{2}} = -(\alpha^5 + \alpha^{11}).$$

In the table we shall denote $\sqrt{2+\sqrt{2}}$ by β. Therefore we have

$$\alpha + \alpha^{15} = \sqrt{2+\sqrt{2}} = \beta \quad \text{and} \quad \alpha^3 + \alpha^{13} = \sqrt{2-\sqrt{2}} = \frac{\sqrt{2}}{\beta}.$$

Automorphism	e	σ	σ^2	σ^3	σ^4	σ^5	σ^6	σ^7
Order	1	16	8	16	4	16	8	16
$2^{\frac{1}{16}} \longrightarrow$	$2^{\frac{1}{16}}$	$\alpha 2^{\frac{1}{16}}$	$\alpha^{14} 2^{\frac{1}{16}}$	$\alpha^7 2^{\frac{1}{16}}$	$\alpha^{12} 2^{\frac{1}{16}}$	$\alpha^{13} 2^{\frac{1}{16}}$	$\alpha^{10} 2^{\frac{1}{16}}$	$\alpha^3 2^{\frac{1}{16}}$
$\alpha \longrightarrow$	α	α^{13}	α^9	α^5	α	α^{13}	α^9	α^5
$\beta \longrightarrow$	β	$\frac{\sqrt{2}}{\beta}$	$-\beta$	$-\frac{\sqrt{2}}{\beta}$	β	$\frac{\sqrt{2}}{\beta}$	$-\beta$	$-\frac{\sqrt{2}}{\beta}$

Automorphism	σ^8	σ^9	σ^{10}	σ^{11}	σ^{12}	σ^{13}	σ^{14}	σ^{15}
Order	2	16	8	16	4	16	8	16
$2^{\frac{1}{16}} \longrightarrow$	$\alpha^8 2^{\frac{1}{16}}$	$\alpha^9 2^{\frac{1}{16}}$	$\alpha^6 2^{\frac{1}{16}}$	$\alpha^{15} 2^{\frac{1}{16}}$	$\alpha^4 2^{\frac{1}{16}}$	$\alpha^5 2^{\frac{1}{16}}$	$\alpha^2 2^{\frac{1}{16}}$	$\alpha^{11} 2^{\frac{1}{16}}$
$\alpha \longrightarrow$	α	α^{13}	α^9	α^5	α	α^{13}	α^9	α^5
$\beta \longrightarrow$	β	$\frac{\sqrt{2}}{\beta}$	$-\beta$	$-\frac{\sqrt{2}}{\beta}$	β	$\frac{\sqrt{2}}{\beta}$	$-\beta$	$-\frac{\sqrt{2}}{\beta}$

Here $\sigma\eta$ means σ followed by η.

$$\alpha = \frac{\sqrt{2+\sqrt{2}}}{2} + i\frac{\sqrt{2-\sqrt{2}}}{2}, \qquad \beta = \sqrt{2+\sqrt{2}}.$$

Automorphism	η	$\sigma\eta$	$\sigma^2\eta$	$\sigma^3\eta$	$\sigma^4\eta$	$\sigma^5\eta$	$\sigma^6\eta$	$\sigma^7\eta$
Order	8	2	8	2	8	2	8	2
$2^{\frac{1}{16}} \longrightarrow$	$\alpha 2^{\frac{1}{16}}$	$\alpha^4 2^{\frac{1}{16}}$	$\alpha^{11} 2^{\frac{1}{16}}$	$\alpha^6 2^{\frac{1}{16}}$	$\alpha^5 2^{\frac{1}{16}}$	$\alpha^8 2^{\frac{1}{16}}$	$\alpha^{15} 2^{\frac{1}{16}}$	$\alpha^{10} 2^{\frac{1}{16}}$
$\alpha \longrightarrow$	α^3	α^7	α^{11}	α^{15}	α^3	α^7	α^{11}	α^{15}
$\beta \longrightarrow$	$\frac{\sqrt{2}}{\beta}$	$-\beta$	$-\frac{\sqrt{2}}{\beta}$	β	$\frac{\sqrt{2}}{\beta}$	$-\beta$	$-\frac{\sqrt{2}}{\beta}$	β

Automorphism	$\sigma^8\eta$	$\sigma^9\eta$	$\sigma^{10}\eta$	$\sigma^{11}\eta$	$\sigma^{12}\eta$	$\sigma^{13}\eta$	$\sigma^{14}\eta$	$\sigma^{15}\eta$
Order	8	2	8	2	8	2	8	2
$2^{\frac{1}{16}} \longrightarrow$	$\alpha^9 2^{\frac{1}{16}}$	$\alpha^{12} 2^{\frac{1}{16}}$	$\alpha^3 2^{\frac{1}{16}}$	$\alpha^{14} 2^{\frac{1}{16}}$	$\alpha^{13} 2^{\frac{1}{16}}$	$2^{\frac{1}{16}}$	$\alpha^7 2^{\frac{1}{16}}$	$\alpha^2 2^{\frac{1}{16}}$
$\alpha \longrightarrow$	α^3	α^7	α^{11}	α^{15}	α^3	α^7	α^{11}	α^{15}
$\beta \longrightarrow$	$\frac{\sqrt{2}}{\beta}$	$-\beta$	$-\frac{\sqrt{2}}{\beta}$	β	$\frac{\sqrt{2}}{\beta}$	$-\beta$	$-\frac{\sqrt{2}}{\beta}$	β

Here $\sigma\eta$ means σ followed by η.

$$\alpha = \frac{\sqrt{2+\sqrt{2}}}{2} + i\frac{\sqrt{2-\sqrt{2}}}{2}, \qquad \beta = \sqrt{2+\sqrt{2}}$$

Automorphism	η^2	$\sigma\eta^2$	$\sigma^2\eta^2$	$\sigma^3\eta^2$	$\sigma^4\eta^2$	$\sigma^5\eta^2$	$\sigma^6\eta^2$	$\sigma^7\eta^2$
Order	4	16	8	16	2	16	8	16
$2^{\frac{1}{16}} \longrightarrow$	$\alpha^4 2^{\frac{1}{16}}$	$\alpha^{13} 2^{\frac{1}{16}}$	$\alpha^2 2^{\frac{1}{16}}$	$\alpha^3 2^{\frac{1}{16}}$	$\alpha 2^{\frac{1}{16}}$	$\alpha^9 2^{\frac{1}{16}}$	$\alpha^{14} 2^{\frac{1}{16}}$	$\alpha^{15} 2^{\frac{1}{16}}$
$\alpha \longrightarrow$	α^9	α^5	α	α^{13}	α^9	α^5	α^1	α^{13}
$\beta \longrightarrow$	$-\beta$	$-\frac{\sqrt{2}}{\beta}$	β	$\frac{\sqrt{2}}{\beta}$	$-\beta$	$-\frac{\sqrt{2}}{\beta}$	β	$\frac{\sqrt{2}}{\beta}$

Automorphism	$\sigma^8\eta^2$	$\sigma^9\eta^2$	$\sigma^{10}\eta^2$	$\sigma^{11}\eta^2$	$\sigma^{12}\eta^2$	$\sigma^{13}\eta^2$	$\sigma^{14}\eta^2$	$\sigma^{15}\eta^2$
Order	4	16	8	16	2	16	8	16
$2^{\frac{1}{16}} \longrightarrow$	$\alpha^{12} 2^{\frac{1}{16}}$	$\alpha^5 2^{\frac{1}{16}}$	$\alpha^{10} 2^{\frac{1}{16}}$	$\alpha 112^{\frac{1}{16}}$	$\alpha^8 2^{\frac{1}{16}}$	$\alpha 2^{\frac{1}{16}}$	$\alpha^6 2^{\frac{1}{16}}$	$\alpha^7 2^{\frac{1}{16}}$
$\alpha \longrightarrow$	α^9	α^5	α	α^{13}	α^9	α^5	α^1	α^{13}
$\beta \longrightarrow$	$-\beta$	$-\frac{\sqrt{2}}{\beta}$	β	$\frac{\sqrt{2}}{\beta}$	$-\beta$	$-\frac{\sqrt{2}}{\beta}$	β	$\frac{\sqrt{2}}{\beta}$

Here $\sigma\eta$ means σ followed by η.

$$\alpha = \frac{\sqrt{2+\sqrt{2}}}{2} + i\frac{\sqrt{2-\sqrt{2}}}{2}, \qquad \beta = \sqrt{2+\sqrt{2}}$$

Automorphism	η^3	$\sigma\eta^3$	$\sigma^2\eta^3$	$\sigma^3\eta^3$	$\sigma^4\eta^3$	$\sigma^5\eta^3$	$\sigma^6\eta^3$	$\sigma^7\eta^3$
Order	8	2	8	2	8	2	8	2
$2^{\frac{1}{16}} \longrightarrow$	$\alpha^{13}2^{\frac{1}{16}}$	$\alpha^{8}2^{\frac{1}{10}}$	$\alpha^{7}2^{\frac{1}{16}}$	$\alpha^{10}2^{\frac{1}{16}}$	$\alpha 2^{\frac{1}{16}}$	$\alpha^{12}2^{\frac{1}{16}}$	$\alpha^{11}2^{\frac{1}{16}}$	$\alpha^{14}2^{\frac{1}{16}}$
$\alpha \longrightarrow$	α^{11}	α^{15}	α^{3}	α^{7}	α^{11}	α^{15}	α^{3}	α^{7}
$\beta \longrightarrow$	$-\frac{\sqrt{2}}{\beta}$	β	$\frac{\sqrt{2}}{\beta}$	$-\beta$	$-\frac{\sqrt{2}}{\beta}$	β	$\frac{\sqrt{2}}{\beta}$	$-\beta$

Automorphism	$\sigma^8\eta^3$	$\sigma^9\eta^3$	$\sigma^{10}\eta^3$	$\sigma^{11}\eta^3$	$\sigma^{12}\eta^3$	$\sigma^{13}\eta^3$	$\sigma^{14}\eta^3$	$\sigma^{15}\eta^3$
Order	8	2	8	2	8	2	8	2
$2^{\frac{1}{16}} \longrightarrow$	$\alpha^{5}2^{\frac{1}{16}}$	$2^{\frac{1}{16}}$	$\alpha^{15}2^{\frac{1}{16}}$	$\alpha^{2}2^{\frac{1}{16}}$	$\alpha^{9}2^{\frac{1}{16}}$	$\alpha^{4}2^{\frac{1}{16}}$	$\alpha^{3}2^{\frac{1}{16}}$	$\alpha^{6}2^{\frac{1}{16}}$
$\alpha \longrightarrow$	α^{11}	α^{15}	α^{3}	α^{7}	α^{11}	α^{15}	α^{3}	α^{7}
$\beta \longrightarrow$	$-\frac{\sqrt{2}}{\beta}$	β	$\frac{\sqrt{2}}{\beta}$	$-\beta$	$-\frac{\sqrt{2}}{\beta}$	β	$\frac{\sqrt{2}}{\beta}$	$-\beta$

Here $\sigma\eta$ means σ followed by η.

$$\alpha = \frac{\sqrt{2+\sqrt{2}}}{2} + i\frac{\sqrt{2-\sqrt{2}}}{2}, \qquad \beta = \sqrt{2+\sqrt{2}}.$$

Now we shall study the properties of this group G and describe its permutation representation on sixteen letters. For this we shall find it more convenient to replace σ by x and η by y. So G is a group of order 64 with the following defining relations:

$$G = \langle x, y, x^{16} = y^8 = (xy)^2 = e, x^8 = y^4, y^{-1}xy = x^{15}y^2 \rangle = \{x^i y^j : 0 \le i \le 15, 0 \le j \le 3\}.$$

The following relations between x and y are easily verifiable consequences of the defining relations for G, namely,

$$x^{16} = y^8 = e, x^8 = y^4, y^{-1}xy = x^{15}y^2.$$

We emphasize this as we shall be using these relations very often. We observe that

$$\begin{aligned} yx &= x^7y^3 & yx^6 &= x^{10}y \\ yx^2 &= x^{14}y & yx^7 &= xy^3 \\ yx^3 &= x^5y^3 & y^2x &= x^9y^2 \\ yx^4 &= x^{12}y & y^3x &= x^9y^2 \\ yx^5 &= x^3y^3 & x^2y^2 &= y^2x^2. \end{aligned}$$

If we let x and y be the following two permutations on sixteen letters

$$x = (1, 2, 3, 4, 5, 6, 7, 8, 9, 10, 11, 12, 13, 14, 15, 16)$$

$$y = (1, 6, 13, 10, 9, 14, 5, 2)(3, 4, 15, 8, 11, 12, 7, 16)$$

then as it can be easily verified that (remember we multiply permutations from left to right)

$$x^{16} = y^8 = e, \qquad x^8 = y^4, \qquad y^{-1}xy = x^{15}y^2$$

and this gives us a permutation representation of G on sixteen letters. The following two tables lists the orders of elements of G. The latter table is sorted on the orders of the elements.

Orders of Elements of $G = \langle x, y, x^{16} = y^8 = (xy)^2 = e, x^8 = y^4, y^{-1}xy = x^{15}y^2 \rangle$.

Elements	Order	Element	Order	Elements	Order	Element	Order
e	1	y	8	y^2	4	y^3	8
x	16	xy	2	xy^2	16	xy^3	2
x^2	8	x^2y	8	x^2y^2	8	x^2y^3	8
x^3	16	x^3y	2	x^3y^2	16	x^3y^3	2
x^4	4	x^4y	8	x^4y^2	2	x^4y^3	8
x^5	16	x^5y	2	x^5y^2	16	x^5y^3	2
x^6	8	x^6y	8	x^6y^2	8	x^6y^3	8
x^7	16	x^7y	2	x^7y^2	16	x^7y^3	2
x^8	2	x^8y	8	x^8y^2	4	x^8y^3	8
x^9	16	x^9y	2	x^9y^2	16	x^9y^3	2
x^{10}	8	$x^{10}y$	8	$x^{10}y^2$	8	$x^{10}y^3$	8
x^{11}	16	$x^{11}y$	2	$x^{11}y^2$	16	$x^{11}y^3$	2
x^{12}	4	$x^{12}y$	8	$x^{12}y^2$	2	$x^{12}y^3$	8
x^{13}	16	$x^{13}y$	2	$x^{13}y^2$	16	$x^{13}y^3$	2
x^{14}	8	$x^{14}y$	8	$x^{14}y^2$	8	$x^{14}y^3$	8
x^{15}	16	$x^{15}y$	2	$x^{15}y^2$	16	$x^{15}y^3$	2

Listing of Elements of $G = \langle x, y, x^{16} = y^8 = (xy)^2 = e, x^8 = y^4, y^{-1}xy = x^{15}y^2 \rangle$.

(Sorted by their orders)

Order	#Elements	Elements
1	1	e
2	19	$x^8, x^4y^2, x^{12}y^2,$ $xy, x^3y, x^5y, x^7y, x^{11}y, x^{13}y, x^{15}y,$ $xy^3, x^3y^3, x^5y^3, x^7y^3, x^9y^3, x^{11}y^3, x^{13}y^3, x^{15}y^3$
4	4	x^4, x^{12}, y^2, x^8y^2
8	24	$x^2, x^6, x^{10}, x^{14}, x^2y^2, x^6y^2, x^{10}y^2, x^{14}y^2,$ $y, x^2y, x^4y, x^6y, x^8y, x^{10}y, x^{12}y, x^{14}y,$ $y^3, x^2y^3, x^4y^3, x^6y^3, x^8y^3, x^{10}y^3, x^{12}y^3, x^{14}y^3$
16	16	$x, x^3, x^5, x^7, x^9, x^{11}, x^{13}, x^{15},$ $xy^2, x^3y^2, x^5y^2, x^7y^2, x^9y^2, x^{11}y^2, x^{13}y^2, x^{15}y^2$

Some useful facts about G

(1) $Z(G) =$ center of $G = \{e, x^8\}$.

(2) Commutator subgroup of $G = G' = \langle x^2y^2 \rangle = \{e, x^2y^2, x^{12}, x^{14}y^2, x^8, x^{10}y^2, x^4, x^6y^2\}$

(3) G has 8 linear characters, 6 characters of degree two and 2 characters of degree four. So G has 16 conjugate classes.*

The two tables in the following pages give a list of centralizers of elements of G and its conjugacy classes

* **We have not discussed the character theory of finite groups in this book. We give the information about the characters of the group G simply to reconfirm that the group G has 16 conjugate classes. It has no implication for the discussion of Galois group and Galois correspondence of the polynomial $x^{16}-2$ and so this may be ignored by the readers not familiar with the theory of group characters.**

Centralizers of Elements of $G = \langle x, y, x^{16} = y^8 = (xy)^2 = e, x^8 = y^4, y^{-1}xy = x^{15}y^2 \rangle$.

Elements	Order	Centralizer	Order Of Centralizer
e	1	G	64
x^8	2	G	64
$x, x^3, x^5, x^7, x^9, x^{11}, x^{13}, x^{15}$	16	$\langle x \rangle$	16
x^2, x^{14}, x^6, x^{10}	8	$\langle x, y^2 \rangle$	32
x^4, x^{12}	4	$\langle x, y^2 \rangle$	32
y, y^3, y^5, y^7	8	$\langle y \rangle$	8
$xy, x^9y, x^5y^3, x^{13}y^3$	2	$\langle x^8, xy, x^4y^2 \rangle$	8
$x^5y, xy^3, x^{13}y, x^9y^3$	2	$\langle x^8, x^5y, x^4y^2 \rangle$	8
$x^3y^3, x^{11}y^3, x^{15}y^3, x^7y$	2	$\langle x^8, x^3y^3, x^4y^2 \rangle$	8
$x^3y, x^{11}y, x^7y^3, x^{15}y^3$	2	$\langle x^8, x^3y, x^4y^2 \rangle$	8
$x^4y^2, x^{12}y^2$	2	$\langle x^2, y^2, xy \rangle$	32
$y^2, x^8y^2 = y^6$	4	$\langle x^2, y \rangle$	32
$x^2y^2, x^6y^2, x^{10}y^2, x^{14}y^2$	8	$\langle x^2, y^2 \rangle$	16
$x^2y, x^{10}y, x^2y^3, x^{10}y^3$	8	$\langle x^2y \rangle$	8
$x^4y, x^4y^3, x^{12}y, x^{12}y^3$	8	$\langle x^4y \rangle$	8
$x^6y^3, x^6y, x^{14}y^3, x^{14}y$	8	$\langle x^6y^3 \rangle$	8
$xy^2, x^3y^2, x^5y^2, x^7y^2, x^9y^2, x^{11}y^2, x^{13}y^2, x^{15}y^2$	16	$\langle xy^2 \rangle$	8

Description of Conjugacy Classes of

$$G = \langle x, y, x^{16} = y^8 = (xy)^2 = e, x^8 = y^4, y^{-1}xy = x^{15}y^2 \rangle.$$

Class Representative	Order	# Elements in the Class	Elements
e	1	1	e
x^8	2	1	x^8
x	16	4	$x, x^9, x^7y^2, x^{15}y^2$
x^2	8	2	x^2, x^{14}
x^3	16	4	$x^3, x^{11}, x^5y^2, x^{13}y^2$
x^4	4	2	x^4, x^{12}
x^5	16	4	$x^5, x^{13}, x^3y^2, x^{11}y^2$
x^6	8	2	x^6, x^{10}
x^7	16	4	$x^7, x^{15}, xy^2, x^9y^2$
y	8	8	$y, x^4y, x^8y, x^{12}y, x^2y^3, x^6y^3, x^{10}y^3, x^{14}y^3$
y^2	4	2	y^2, x^8y^2
y^3	8	8	$y^3, x^4y^3, x^8y^3, x^{12}y^3, x^2y, x^6y, x^{10}y, x^{14}y$
x^4y^2	8	4	$x^2y^2, x^6y^2, x^{10}y^2, x^{14}y^2$
xy	2	8	$xy, x^5y, x^4y, x^{13}y, x^3y^3, x^7y^3, x^{11}y^3, x^{15}y^3$
x^3y	2	8	$x^3y, x^7y, x^{11}y, x^{15}y, xy^3, x^5y^3, x^9y^3, x^{13}y^3$
x^4y^2	2	2	$x^4y^2, x^{12}y^2$

Description of Subgroups of $G = \langle x, y, x^{16} = y^8 = (xy)^2 = e, x^8 = y^4, y^{-1}xy = x^{15}y^2 \rangle$.

#	Subgroup	Order
1	G	64
2	$\langle x, y^2 : y^{-2}xy^2 = x^9 \rangle = \{x^i y^j : 0 \le i \le 15, j = 0, 2\}$	32
3	$\langle x^2, y : (x^2)^8 = e, y^{-1}x^2y = x^{14} \rangle = \langle x^2 \rangle \cdot \langle y \rangle =$ $\{e, x^2, x^4, x^6, x^8, x^{10}, x^{12}, x^{14}, y, x^2y, x^4y, x^6y, x^8y, x^{10}y, x^{12}y, x^{14}y,$ $y^2, x^2y^2, x^4y^2, x^6y^2, x^8y^2, x^{10}y^2, x^{12}y^2, x^{14}y^2, y^3, x^2y^3, x^4y^3, x^6y^3, x^8y^3, x^{10}y^3, x^{12}y^3, x^{14}y^3\}$	32
4	$\langle x^2, xy, y^2 \rangle = \langle x^2 \rangle \cdot \langle xy \rangle \cdot \langle y^2 \rangle =$ $\{e, x^2, x^4, x^6, x^8, x^{10}, x^{12}, x^{14}, y^2, x^2y^2, x^4y^2, x^6y^2, x^8y^2, x^{10}y^2, x^{12}y^2, x^{14}y^2,$ $xy, x^3y, x^5y, x^7y, x^9y, x^{11}y, x^{13}y, x^{15}y, x^9y^3, x^{11}y^3, x^{13}y^3, x^{15}y^3, xy^3, x^3y^3, x^5y^3, x^7y^3\}$	32
5	$\langle x \rangle$	16
6	$\langle xy^2 \rangle =$ $\{e, xy^2, x^2, x^3y^2, x^4, x^5y^2, x^6, x^7y^2, x^8, x^9y^2, x^{10}, x^{11}y^2, x^{12}, x^{13}y^2, x^{14}, x^{15}y^2\}$	16
7	$\langle x^2, y^2 \rangle = \langle x^2 \rangle \cdot \langle y^2 \rangle = \langle x^2 \rangle \times \langle x^4y^2 \rangle \approx C_4 \times C_2$ $= \{e, x^2, x^4, x^6, x^8, x^{10}, x^{12}, x^{14}, x^4y^2, x^6y^2, x^8y^2, x^{10}y^2, x^{12}y^2, x^{14}y^2, y^2, x^2y^2\}$	16
8	$\langle x^2, xy \rangle$ Dihedral $= \{e, x^2, x^4, x^6, x^8, x^{10}, x^{12}, x^{14}, xy, x^3y, x^5y, x^7y, x^9y, x^{11}y, x^{13}y, x^{15}y\}$	16
9	$\langle x^2, xy^3 \rangle$ Dihedral $= \{e, x^2, x^4, x^6, x^8, x^{10}, x^{12}, x^{14}, xy^3, x^3y^3, x^5y^3, x^7y^3, x^9y^3, x^{11}y^3, x^{13}y^3, x^{15}y^3\}$	16
10	$\langle x^2y^2, xy^3 \rangle$ Dihedral $= \{e, x^2y^2, x^{12}, x^{14}y^2, x^8, x^{10}y^2, x^4, x^6y^2, xy^3, x^3y, x^{13}y^3, x^{15}y, x^9y^3, x^{11}y, x^5y^3, x^7y\}$	16
11	$\langle x^2y^2, xy \rangle$ Dihedral $= \{e, x^2y^2, x^{12}, x^{14}y^2, x^8, x^{10}y^2, x^4, x^6y^2, xy, x^{11}y^3, x^{13}y, x^7y^3, x^9y, x^3y^3, x^5y, x^{15}y^3\}$	16

Description of Subgroups of $G = \langle x, y, x^{16} = y^8 = (xy)^2 = e, x^8 = y^4, y^{-1}xy = x^{15}y^2 \rangle$.

(*continued*)

#	Subgroup	Order
12	$\langle y, x^4 : x^{-4}yx^4 = y^5 = x^8y \rangle =$ $\{e, y, y^2, y^3, x^8, x^8y, x^8y^2, x^8y^3, x^4, x^4y, x^4y^2, x^4y^3, x^{12}, x^{12}y, x^{12}y^2, x^{12}y^3\}$	16
13	$\langle x^2y, x^4 : x^{-4}(x^2y)x^4 = x^{10}y = (x^2y)^5 \rangle =$ $\{e, x^2y, y^2, x^2y^3, x^8, x^{10}y, x^8y^2, x^8y^2, x^{10}y^3, x^4, x^6y, x^4y^2, x^6y^3, x^{12}, x^{12}y^2, x^{14}y^3\}$	16
14	$\langle x^4, y^2, xy \rangle$ Extension of $C_4 \times C_2$ by C_2 $= \{e, x^4, x^8, x^{12}, y^2, x^4y^2, x^8y^2, x^{12}y^2, xy, x^5y, x^9y, x^{13}y, x^9y^3, x^{13}y^3, xy^3, x^5y^3\}$	16
15	$\langle x^4, y^2, x^3y \rangle$ Extension of $C_4 \times C_2$ by C_2 $= \{e, x^4, x^8, x^{12}, y^2, x^4y^2, x^8y^2, x^{12}y^2, x^3y, x^7y, x^{11}y, x^{15}y, x^{11}y^3, x^{15}y^3, x^3y^3, x^7y^3\}$	16
16	$\langle x^2 \rangle = \{e, x^2, x^4, x^6, x^8, x^{10}, x^{12}, x^{14}\}$	8
17	$\langle y \rangle = \{e, y, y^2, y^3, x^8, x^8y, x^8y^2, x^8y^3\}$	8
18	$\langle x^2y^2 \rangle = \{e, x^2y^2, x^{12}, x^{14}y^2, x^8, x^{10}y^2, x^4, x^6y^2\}$	8
19	$\langle x^2y \rangle = \{e, x^2y, y^2, x^2y^3, x^8, x^{10}y, x^8y^2, x^{10}y^3\}$	8
20	$\langle x^4y \rangle = \{e, x^4y, y^2, x^4y^3, x^8, x^{12}y, x^8y^2, x^{12}y^3\}$	8
21	$\langle x^6y \rangle = \{e, x^6y, y^2, x^6y^3, x^8, x^{14}y, x^8y^2, x^{14}y^3\}$	8
22	$\langle x^4, x^4y^2 \rangle = \langle x^4 \rangle \times \langle x^4y^2 \rangle = \{e, x^4, x^8, x^{12}, x^4y^2, x^8y^2, x^{12}y^2, y^2\} \approx C_4 \times C_2$	8
23	$\langle xy, x^8, x^4y^2 \rangle = \{e, xy, x^8, x^9y, x^4y^2, x^{12}y^2, x^5y^3, x^{13}y^3\} \approx C_2 \times C_2 \times C_2$	8
24	$\langle x^3y, x^8, x^4y^2 \rangle = \{e, x^3y, x^8, x^{11}y, x^4y^2, x^{12}y^2, x^7y^3, x^{15}y^3\} \approx C_2 \times C_2 \times C_2$	8
25	$\langle x^5y, x^8, x^4y^2 \rangle = \{e, x^5y, x^8, x^{13}y, x^4y^2, xy^3, x^{12}y^2, x^9y^3\} \approx C_2 \times C_2 \times C_2$	8
26	$\langle x^3y^3, x^8, x^4y^2 \rangle = \{e, x^3y^3, x^8, x^{11}y^3, x^4y^2, x^7y, x^{12}y^2, x^{15}y\} \approx C_2 \times C_2 \times C_2$	8

Description of Subgroups of $G = \langle x, y, x^{16} = y^8 = (xy)^2 = e, x^8 = y^4, y^{-1}xy = x^{15}y^2 \rangle$.

(continued)

#	Subgroup	Order
27	$\langle x^4, xy \rangle = \{e, x^4, x^8, x^{12}, xy, x^5y, x^9y, x^{13}y\}$	8
28	$\langle x^4, x^3y \rangle = \{e, x^4, x^8, x^{12}, x^3y, x^7y, x^{11}y, x^{15}y\}$	8
29	$\langle x^4, xy^3 \rangle = \{e, x^4, x^8, x^{12}, xy^3, x^5y^3, x^9y^3, x^{13}y^3\}$	8
30	$\langle x^4, x^3y^3 \rangle = \{e, x^4, x^8, x^{12}, x^3y^3, x^7y^3, x^{11}y^3, x^{15}y^3\}$	8
31	$\langle y^2, xy \rangle = \{e, y^2, x^8, x^8y^2, xy, x^9y^3, x^9y, xy^3\}$	8
32	$\langle y^2, x^3y \rangle = \{e, y^2, x^8, x^8y^2, x^3y, x^{11}y^3, x^{11}y, x^3y^3\}$	8
33	$\langle y^2, x^5y \rangle = \{e, y^2, x^8, x^8y^2, x^5y, x^{13}y^3, x^{13}y, x^5y^3\}$	8
34	$\langle y^2, x^7y \rangle = \{e, y^2, x^8, x^8y^2, x^7y, x^{15}y^3, x^{15}y, x^7y^3\}$	8
	#27–34 are Dihedral groups	
35	$\langle x^4 \rangle = \{e, x^4, x^8, x^{12}\}$	4
36	$\langle y^2 \rangle = \{e, y^2, x^8, x^8y^2\}$	4
37	$\langle x^8, x^4y^2 \rangle = \{e, x^8, x^4y^2, x^{12}y^2\}$ Normal	4
38	$\langle x^8, xy \rangle = \{e, x^8, xy, x^9y\}$	4
39	$\langle x^8, x^3y \rangle = \{e, x^8, x^3y, x^{11}y\}$	4
40	$\langle x^8, x^5y \rangle = \{e, x^8, x^5y, x^{13}y\}$	4
41	$\langle x^8, x^7y \rangle = \{e, x^8, x^7y, x^{15}y\}$	4
42	$\langle x^8, xy^3 \rangle = \{e, x^8, xy^3, x^9y^3\}$	4
43	$\langle x^8, x^3y^3 \rangle = \{e, x^8, x^3y^3, x^{11}y^3\}$	4
44	$\langle x^8, x^5y^3 \rangle = \{e, x^8, x^5y^3, x13y^3\}$	4
45	$\langle x^8, x^7y^3 \rangle = \{e, x^8, x^7y^3, x^{15}y^3\}$	4
46	$\langle x^4y^2, xy \rangle = \{e, x^4y^2, xy, x^{13}y^3\}$	4

Description of Subgroups of $G = \langle x, y, x^{16} = y^8 = (xy)^2 = e, x^8 = y^4, y^{-1}xy = x^{15}y^2\rangle$.

(*continued*)

#	Subgroup	Order
47	$\langle x^{12}y^2, xy\rangle = \{e, x^{12}y^2, xy, x^5y^3\}$	4
48	$\langle x^4y^2, x^5y\rangle = \{e, x^4y^2, x^5y, xy^3\}$	4
49	$\langle x^{12}y^2, x^5y\rangle = \{e, x^{12}y^2, x^5y, x^9y^3\}$	4
50	$\langle x^4y^2, x^9y\rangle = \{e, x^4y^2, x^9y, x^5y^3\}$	4
51	$\langle x^{12}y^2, x^9y\rangle = \{e, x^{12}y^2, x^9y, x^{13}y^3\}$	4
52	$\langle x^4y^2, x^{13}y\rangle = \{e, x^4y^2, x^{13}y, x^9y^3\}$	4
53	$\langle x^{12}y^2, x^{13}y\rangle = \{e, x^{12}y^2, x^{13}y, xy^3\}$	4
54	$\langle x^4y^2, x^3y^3\rangle = \{e, x^4y^2, x^3y^3, x^7y\}$	4
55	$\langle x^{12}y^2, x^3y^3\rangle = \{e, x^{12}y^2, x^3y^3, x^{15}y\}$	4
56	$\langle x^4y^2, x^7y^3\rangle = \{e, x^4y^2, x^7y^3, x^{11}y\}$	4
57	$\langle x^{12}y^2, x^7y^3\rangle = \{e, x^{12}y^2, x^7y^3, x^3y\}$	4
58	$\langle x^4y^2, x^{11}y^3\rangle = \{e, x^4y^2, x^{11}y^3, x^{15}y\}$	4
59	$\langle x^{12}y^2, x^{11}y^3\rangle = \{e, x^{12}y^2, x^{11}y^3, x^7y\}$	4
60	$\langle x^4y^2, x^{15}y^3\rangle = \{e, x^4y^2, x^{15}y^3, x^3y\}$	4
61	$\langle x^{12}y^2, x^{15}y^3\rangle = \{e, x^{12}y^2, x^{15}y^3, x^{11}y\}$	4
62–64	$\langle x^8\rangle, \langle x^4y^2\rangle, \langle x^{12}y^2\rangle$	2
65–72	$\langle xy\rangle, \langle x^3y\rangle, \langle x^5y\rangle, \langle x^7y\rangle, \langle x^9y\rangle, \langle x^{11}y\rangle, \langle x^{13}y\rangle, \langle x^{15}y\rangle$	2
73–80	$\langle xy^3\rangle, \langle x^3y^3\rangle, \langle x^5y^3\rangle, \langle x^7y^3\rangle, \langle x^9y^3\rangle, \langle x^{11}y^3\rangle, \langle x^{13}y^3\rangle, \langle x^{15}y^3\rangle$	2
81	$\{e\}$	1

Description of Subfields of the Splitting Field $\mathbb{K} = \mathbb{Q}(2^{\frac{1}{16}}, \alpha)$.

#	Subfield	Degree
1	$\mathbb{Q}$	1
2	$\mathbb{Q}(\sqrt{2})$	2
3	$\mathbb{Q}(i)$	2
4	$\mathbb{Q}(i\sqrt{2})$	2
5	$\mathbb{Q}(2^{\frac{1}{4}})$	4
6	$\mathbb{Q}(\alpha 2^{\frac{1}{4}}) = Q(\alpha^5 2^{\frac{1}{4}})$	4
7	$\mathbb{Q}(\alpha^2 2^{\frac{1}{4}}) \neq Q(\alpha^6 2^{\frac{1}{4}})$	4
8	$\mathbb{Q}(\alpha^3 2^{\frac{1}{4}}) = Q(\alpha^7 2^{\frac{1}{4}})$	4
9	$\mathbb{Q}(\alpha^6 2^{\frac{1}{4}}) = Q(i\alpha^2 2^{\frac{1}{4}})$	4
10	$\mathbb{Q}(2^{\frac{1}{2}}, i) = Q(\alpha^2)$	4
11	$\mathbb{Q}(\beta) = Q(\sqrt{2+\sqrt{2}})$	4
12	$\mathbb{Q}(i\sqrt{2+\sqrt{2}})$	4
13	$\mathbb{Q}(i2^{\frac{1}{4}})$	4
14	$\mathbb{Q}(i2^{\frac{1}{4}}\sqrt{2+\sqrt{2}})$	4
15	$\mathbb{Q}(2^{\frac{1}{4}}\sqrt{2+\sqrt{2}})$	4
16	$\mathbb{Q}(2^{\frac{1}{4}}\sqrt{2+\sqrt{2}}, i)$	8
17	$\mathbb{Q}(\alpha^7 2^{\frac{1}{8}}) = Q(i\alpha^3 2^{\frac{1}{8}})$	8
18	$\mathbb{Q}(\alpha)$	8
19	$\mathbb{Q}(\alpha 2^{\frac{1}{8}})$	8
20	$\mathbb{Q}(\alpha^3 2^{\frac{1}{8}})$	8

Description of Subfields of the Splitting Field $\mathbb{K} = \mathbb{Q}(2^{\frac{1}{16}}, \alpha)$. (*continued*)

#	Subfield	Degree
21	$\mathbb{Q}(\alpha^5 2^{\frac{1}{8}})$	8
22	$\mathbb{Q}(2^{\frac{1}{4}}, i)$	8
23	$\mathbb{Q}(i2^{\frac{1}{8}})$	8
24	$\mathbb{Q}(\alpha^6 2^{\frac{1}{8}})$	8
25	$\mathbb{Q}(2^{\frac{1}{8}})$	8
26	$\mathbb{Q}(\alpha^2 2^{\frac{1}{8}})$	8
27	$\mathbb{Q}(2^{\frac{1}{4}}, i\sqrt{2+\sqrt{2}})$	8
28	$\mathbb{Q}(i2^{\frac{1}{4}}, \sqrt{2+\sqrt{2}})$	8
29	$\mathbb{Q}(2^{\frac{1}{4}}, \sqrt{2+\sqrt{2}})$	8
30	$\mathbb{Q}(i2^{\frac{1}{4}}, i\sqrt{2+\sqrt{2}})$	8
31	$\mathbb{Q}(2^{\frac{1}{8}}\sqrt{2+\sqrt{2}})$	8
32	$\mathbb{Q}(\alpha^6 2^{\frac{1}{8}}\sqrt{2+\sqrt{2}}) = \mathbb{Q}(i\alpha^2 2^{\frac{1}{8}}\sqrt{2+\sqrt{2}})$	8
33	$\mathbb{Q}(i2^{\frac{1}{8}}\sqrt{2+\sqrt{2}})$	8
34	$\mathbb{Q}(\alpha^2 2^{\frac{1}{8}}\sqrt{2+\sqrt{2}})$	8
35	$\mathbb{Q}(2^{\frac{1}{4}}, i, \sqrt{2+\sqrt{2}})$	16
36	$\mathbb{Q}(2^{\frac{1}{8}}\sqrt{2+\sqrt{2}}, i)$	16
37	$\mathbb{Q}(2^{\frac{1}{8}}, i)$	16
38	$\mathbb{Q}(i2^{\frac{1}{8}}, i\sqrt{2+\sqrt{2}})$	16
39	$\mathbb{Q}(\alpha^6 2^{\frac{1}{8}}, \sqrt{2+\sqrt{2}}) = \mathbb{Q}(i\alpha^2 2^{\frac{1}{8}}, \sqrt{2+\sqrt{2}})$	16
40	$\mathbb{Q}(2^{\frac{1}{8}}, i\sqrt{2+\sqrt{2}})$	16

Description of Subfields of the Splitting Field $\mathbb{K} = \mathbb{Q}(2^{\frac{1}{16}}, \alpha)$. (*continued*)

#	Subfield	Degree
41	$\mathbb{Q}(\alpha^2 2^{\frac{1}{8}}, \sqrt{2+\sqrt{2}})$	16
42	$\mathbb{Q}(2^{\frac{1}{8}}, \sqrt{2+\sqrt{2}})$	16
43	$\mathbb{Q}(\alpha^2 2^{\frac{1}{8}}, i\sqrt{2+\sqrt{2}})$	16
44	$\mathbb{Q}(i2^{\frac{1}{8}}, \sqrt{2+\sqrt{2}})$	16
45	$\mathbb{Q}(\alpha^6 2^{\frac{1}{8}}, i\sqrt{2+\sqrt{2}}) = \mathbb{Q}(i\alpha^2 2^{\frac{1}{8}}, i\sqrt{2+\sqrt{2}})$	16
46	$\mathbb{Q}(\alpha^2 2^{\frac{1}{16}})$	16
47	$\mathbb{Q}(\alpha^4 2^{\frac{1}{16}})$	16
48	$\mathbb{Q}(\alpha^6 2^{\frac{1}{16}})$	16
49	$\mathbb{Q}(2^{\frac{1}{16}})$	16
50	$\mathbb{Q}(\alpha^5 2^{\frac{1}{16}}\sqrt{2+\sqrt{2}}) = \mathbb{Q}(i\alpha 2^{\frac{1}{16}}\sqrt{2+\sqrt{2}})$	16
51	$\mathbb{Q}(\alpha^7 2^{\frac{1}{16}}\sqrt{2+\sqrt{2}}) = \mathbb{Q}(i\alpha^3 2^{\frac{1}{16}}\sqrt{2+\sqrt{2}})$	16
52	$\mathbb{Q}(\alpha^9 2^{\frac{1}{16}}\sqrt{2+\sqrt{2}}) = \mathbb{Q}(\alpha 2^{\frac{1}{16}}\sqrt{2+\sqrt{2}})$	16
53	$\mathbb{Q}(\alpha^{11} 2^{\frac{1}{16}}\sqrt{2+\sqrt{2}}) = \mathbb{Q}(\alpha^3 2^{\frac{1}{16}}\sqrt{2+\sqrt{2}})$	16
54	$\mathbb{Q}(\alpha^6 2^{\frac{1}{16}}\sqrt{2+\sqrt{2}}) = \mathbb{Q}(i\alpha^2 2^{\frac{1}{16}}\sqrt{2+\sqrt{2}})$	16
55	$\mathbb{Q}(2^{\frac{1}{16}}\sqrt{2+\sqrt{2}})$	16
56	$\mathbb{Q}(\alpha^2 2^{\frac{1}{16}}\sqrt{2+\sqrt{2}})$	16
57	$\mathbb{Q}(\alpha^4 2^{\frac{1}{16}}\sqrt{2+\sqrt{2}}) = \mathbb{Q}(i2^{\frac{1}{16}}\sqrt{2+\sqrt{2}})$	16
58	$\mathbb{Q}(\alpha 2^{\frac{1}{16}})$	16
59	$\mathbb{Q}(\alpha^3 2^{\frac{1}{16}})$	16
60	$\mathbb{Q}(\alpha^5 2^{\frac{1}{16}}) = \mathbb{Q}(i\alpha 2^{\frac{1}{16}})$	16

Description of Subfields of the Splitting Field $\mathbb{K} = \mathbb{Q}(2^{\frac{1}{16}}, \alpha)$. (*continued*)

#	Subfield	Degree
61	$\mathbb{Q}(\alpha^7 2^{\frac{1}{16}}) = \mathbb{Q}(i\alpha^3 2^{\frac{1}{16}})$	16
62	$\mathbb{Q}(2^{\frac{1}{8}}, i, \sqrt{2+\sqrt{2}})$	32
63	$\mathbb{Q}(\alpha^2 2^{\frac{1}{16}}, i\sqrt{2+\sqrt{2}})$	32
64	$\mathbb{Q}(\alpha^3 2^{\frac{1}{16}}, \sqrt{2+\sqrt{2}})$	32
65	$\mathbb{Q}(\alpha^4 2^{\frac{1}{16}}, i\sqrt{2+\sqrt{2}}) = \mathbb{Q}(i 2^{\frac{1}{16}}, i\sqrt{2+\sqrt{2}})$	32
66	$\mathbb{Q}(\alpha^5 2^{\frac{1}{16}}, \sqrt{2+\sqrt{2}}) = \mathbb{Q}(i\alpha 2^{\frac{1}{16}}, \sqrt{2+\sqrt{2}})$	32
67	$\mathbb{Q}(\alpha^6 2^{\frac{1}{16}}, i\sqrt{2+\sqrt{2}}) = \mathbb{Q}(i\alpha^2 2^{\frac{1}{16}}, i\sqrt{2+\sqrt{2}})$	32
68	$\mathbb{Q}(\alpha^7 2^{\frac{1}{16}} \sqrt{2+\sqrt{2}}) = \mathbb{Q}(i\alpha^3 2^{\frac{1}{16}} \sqrt{2+\sqrt{2}})$	32
69	$\mathbb{Q}(2^{\frac{1}{16}}, i\sqrt{2+\sqrt{2}})$	32
70	$\mathbb{Q}(\alpha 2^{\frac{1}{16}}, \sqrt{2+\sqrt{2}})$	32
71	$\mathbb{Q}(2^{\frac{1}{16}}, i)$	32
72	$\mathbb{Q}(2^{\frac{1}{8}}, \alpha\sqrt{2+\sqrt{2}})$	32
73	$\mathbb{Q}(\alpha^4 2^{\frac{1}{16}}, \sqrt{2+\sqrt{2}}) = \mathbb{Q}(i 2^{\frac{1}{16}}, \sqrt{2+\sqrt{2}})$	32
74	$\mathbb{Q}(\alpha 2^{\frac{1}{16}}, i\sqrt{2+\sqrt{2}})$	32
75	$\mathbb{Q}(\alpha^6 2^{\frac{1}{16}}, \sqrt{2+\sqrt{2}}) = \mathbb{Q}(i\alpha^2 2^{\frac{1}{16}}, \sqrt{2+\sqrt{2}})$	32
76	$\mathbb{Q}(\alpha^3 2^{\frac{1}{16}}, i\sqrt{2+\sqrt{2}})$	32
77	$\mathbb{Q}(2^{\frac{1}{16}}, \sqrt{2+\sqrt{2}})$	32
78	$\mathbb{Q}(\alpha^5 2^{\frac{1}{16}}, i\sqrt{2+\sqrt{2}}) = \mathbb{Q}(i\alpha 2^{\frac{1}{16}}, \sqrt{2+\sqrt{2}})$	32
79	$\mathbb{Q}(\alpha^2 2^{\frac{1}{16}}, i\sqrt{2+\sqrt{2}})$	32
80	$\mathbb{Q}(\alpha^7 2^{\frac{1}{16}}) = \mathbb{Q}(i\alpha^3 2^{\frac{1}{16}})$	32
81	$\mathbb{K} = \mathbb{Q}(2^{\frac{1}{16}}, \alpha)$	64

Galois Correspondence for $x^{16} - 2$

#	Subgroup	Order	Fixed Subfield	Degree
1	$G = \langle \sigma, \eta \rangle$	64	$\mathbb{Q}$	1
2	$\langle \sigma, \eta^2 \rangle$	32	$\mathbb{Q}(i)$	2
3	$\langle \sigma^2, \eta \rangle$	32	$\mathbb{Q}(i\sqrt{2})$	2
4	$\langle \sigma^2, \sigma\eta, \eta^2 \rangle$	32	$\mathbb{Q}(\sqrt{2})$	2
5	$\langle \sigma \rangle$	16	$\mathbb{Q}(\alpha 2^{\frac{1}{4}})$	4
6	$\langle \sigma\eta^2 \rangle$	16	$\mathbb{Q}(\alpha^3 2^{\frac{1}{4}})$	4
7	$\langle \sigma^2, \eta^2 \rangle$	16	$\mathbb{Q}(2^{\frac{1}{2}}, i)$	4
8	$\langle \sigma^2, \sigma\eta \rangle$	16	$\mathbb{Q}(i2^{\frac{1}{4}}\sqrt{2+\sqrt{2}})$	4
9	$\langle \sigma^2\eta^3 \rangle$	16	$\mathbb{Q}(2^{\frac{1}{4}}\sqrt{2+\sqrt{2}})$	4
10	$\langle \sigma^2\eta^2, \sigma\eta^3 \rangle$	16	$\mathbb{Q}(\sqrt{2+\sqrt{2}})$	4
11	$\langle \sigma^2\eta^2, \sigma\eta \rangle$	16	$\mathbb{Q}(i\sqrt{2+\sqrt{2}})$	4
12	$\langle \eta, \sigma^4 \rangle$	16	$\mathbb{Q}\left((1+i)2^{\frac{1}{4}}\right) = \mathbb{Q}\left(\frac{1-i}{2^{\frac{1}{4}}}\right) = \mathbb{Q}(\alpha^6, 2^{\frac{1}{4}})$	4
13	$\langle \sigma^2\eta, \sigma^4 \rangle$	16	$\mathbb{Q}\left((1-i)2^{\frac{1}{4}}\right) = \mathbb{Q}\left(\frac{1+i}{2^{\frac{1}{4}}}\right) = \mathbb{Q}(\alpha^2, 2^{\frac{1}{4}})$	4
14	$\langle \sigma^4, \eta^2, \sigma\eta \rangle$	16	$\mathbb{Q}(2^{\frac{1}{4}})$	4
15	$\langle \sigma^4, \eta^2, \sigma^3\eta \rangle$	16	$\mathbb{Q}(i2^{\frac{1}{4}})$	4

Galois Correspondence for $x^{16}-2$. (*continued*)

#	Subgroup	Order	Fixed Subfield	Degree
16	$\langle\sigma^2\rangle$	8	$\mathbb{Q}(2^{\frac14}\sqrt{2+\sqrt{2}}, i)$	8
17	$\langle\eta\rangle$	8	$\mathbb{Q}(\alpha^7 2^{\frac18}) = \mathbb{Q}(i\alpha^3 2^{\frac18})$	8
18	$\langle\sigma^2\eta^2\rangle$	8	$\mathbb{Q}(\alpha)$	8
19	$\langle\sigma^2\eta\rangle$	8	$\mathbb{Q}(\alpha 2^{\frac18})$	8
20	$\langle\sigma^4\eta\rangle$	8	$\mathbb{Q}(\alpha^3 2^{\frac18})$	8
21	$\langle\sigma^6\eta\rangle$	8	$\mathbb{Q}(i\alpha 2^{\frac18})$	8
22	$\langle\sigma^4, \sigma^4\eta^2\rangle$	8	$\mathbb{Q}(2^{\frac14}, i)$	8
23	$\langle\sigma^8, \sigma^4\eta^2, \sigma\eta\rangle$	8	$\mathbb{Q}(i2^{\frac18})$	8
24	$\langle\sigma^8, \sigma^4\eta^2, \sigma^3\eta\rangle$	8	$\mathbb{Q}(i\alpha^2 2^{\frac18})$	8
25	$\langle\sigma^8, \sigma^4\eta^2, \sigma^5\eta\rangle$	8	$\mathbb{Q}(2^{\frac18})$	8
26	$\langle\sigma^8, \sigma^4\eta^2, \sigma^3\eta^3\rangle$	8	$\mathbb{Q}(\alpha^2 2^{\frac18})$	8
27	$\langle\sigma^4, \sigma\eta\rangle$	8	$\mathbb{Q}(2^{\frac14}, i\sqrt{2+\sqrt{2}})$	8
28	$\langle\sigma^4, \sigma^3\eta\rangle$	8	$\mathbb{Q}(i2^{\frac14}, \sqrt{2+\sqrt{2}})$	8
29	$\langle\sigma^4, \sigma\eta^3\rangle$	8	$\mathbb{Q}(2^{\frac14}, \sqrt{2+\sqrt{2}})$	8
30	$\langle\sigma^4, \sigma^3\eta^3\rangle$	8	$\mathbb{Q}(i2^{\frac14}, i\sqrt{2+\sqrt{2}})$	8
31	$\langle\eta^2, \sigma\eta\rangle$	8	$\mathbb{Q}(2^{\frac18}\sqrt{2+\sqrt{2}})$	8
32	$\langle\eta^2, \sigma^3\eta\rangle$	8	$\mathbb{Q}(i\alpha^2 2^{\frac18}\sqrt{2+\sqrt{2}})$	8
33	$\langle\eta^2, \sigma^5\eta\rangle$	8	$\mathbb{Q}(i2^{\frac18}\sqrt{2+\sqrt{2}})$	8
34	$\langle\eta^2, \sigma^7\eta\rangle$	8	$\mathbb{Q}(\alpha^2\frac{1}{8}\sqrt{2+\sqrt{2}})$	8

Galois Correspondence for $x^{16}-2$. (*continued*)

#	Subgroup	Order	Fixed Subfield	Degree
35	$\langle\sigma^4\rangle$	4	$\mathbb{Q}(2^{\frac{1}{4}}, i, \sqrt{2+\sqrt{2}})$	16
36	$\langle\eta^2\rangle$	4	$\mathbb{Q}(2^{\frac{1}{8}}\sqrt{2+\sqrt{2}})$	16
37	$\langle\sigma^8, \sigma^4\eta^2\rangle$	4	$\mathbb{Q}(2^{\frac{1}{8}}, i)$	16
38	$\langle\sigma^8\sigma\eta\rangle$	4	$\mathbb{Q}(i2^{\frac{1}{8}}, i\sqrt{2+\sqrt{2}})$	16
39	$\langle\sigma^8, \sigma^3\eta\rangle$	4	$\mathbb{Q}(i\alpha^2 2^{\frac{1}{8}}, \sqrt{2+\sqrt{2}})$	16
40	$\langle\sigma^8, \sigma^5\eta\rangle$	4	$\mathbb{Q}(2^{\frac{1}{8}}, i\sqrt{2+\sqrt{2}})$	16
41	$\langle\sigma^8, \sigma^7\eta\rangle$	4	$\mathbb{Q}(\alpha^2 2^{\frac{1}{8}}, \sqrt{2+\sqrt{2}})$	16
42	$\langle\sigma^8, \sigma\eta^3\rangle$	4	$\mathbb{Q}(2^{\frac{1}{8}}, \sqrt{2+\sqrt{2}})$	16
43	$\langle\sigma^8, \sigma^3\eta^3\rangle$	4	$\mathbb{Q}(\alpha^2 2^{\frac{1}{8}}, i\sqrt{2+\sqrt{2}})$	16
44	$\langle\sigma^8, \sigma^5\eta^3\rangle$	4	$\mathbb{Q}(i2^{\frac{1}{8}}, \sqrt{2+\sqrt{2}})$	16
45	$\langle\sigma^8, \sigma^7\eta^3\rangle$	4	$\mathbb{Q}(i\alpha^2 2^{\frac{1}{8}}, i\sqrt{2+\sqrt{2}})$	16
46	$\langle\sigma^4\eta^2, \sigma\eta\rangle$	4	$\mathbb{Q}(\alpha^2 2^{\frac{1}{16}})$	16
47	$\langle\sigma^{12}\eta^2, \sigma\eta\rangle$	4	$\mathbb{Q}(i\alpha^2 2^{\frac{1}{16}}\sqrt{2+\sqrt{2}})$	16
48	$\langle\sigma^4\eta^2, \sigma^5\eta\rangle$	4	$\mathbb{Q}(i2^{\frac{1}{16}})$	16
49	$\langle\sigma^{12}\eta^2, \sigma^5\eta\rangle$	4	$\mathbb{Q}(2^{\frac{1}{16}}\sqrt{2+\sqrt{2}})$	16
50	$\langle\sigma^4\eta^2, \sigma^9\eta\rangle$	4	$\mathbb{Q}(2^{\frac{1}{16}}\sqrt{2+\sqrt{2}})$	16
51	$\langle\sigma^{12}\eta^2, \sigma^9\eta\rangle$	4	$\mathbb{Q}(i\alpha^2 2^{\frac{1}{16}})$	16
52	$\langle\sigma^4\eta^2, \sigma 13\eta\rangle$	4	$\mathbb{Q}(2^{\frac{1}{16}})$	16

Galois Correspondence for $x^{16}-2$. (*continued*)

#	Subgroup	Order	Fixed Subfield	Degree
53	$\langle \sigma^{12}\eta^2, \sigma^{13}\eta \rangle$	4	$\mathbb{Q}(i2^{\frac{1}{16}}\sqrt{2+\sqrt{2}})$	16
54	$\langle \sigma^{4}\eta^2, \sigma^{3}\eta^3 \rangle$	4	$\mathbb{Q}(i\alpha 2^{\frac{1}{16}}\sqrt{2+\sqrt{2}})$	16
55	$\langle \sigma^{12}\eta^2, \sigma^{3}\eta^3 \rangle$	4	$\mathbb{Q}(\alpha 2^{\frac{1}{16}})$	16
56	$\langle \sigma^{4}\eta^2, \sigma^{7}\eta^3 \rangle$	4	$\mathbb{Q}(i\alpha^3 2^{\frac{1}{16}}\sqrt{2+\sqrt{2}})$	16
57	$\langle \sigma^{12}\eta^2, \sigma^{7}\eta^3 \rangle$	4	$\mathbb{Q}(\alpha^3 2^{\frac{1}{16}})$	16
58	$\langle \sigma^{4}\eta^2, \sigma^{11}\eta^3 \rangle$	4	$\mathbb{Q}(\alpha 2^{\frac{1}{16}}\sqrt{2+\sqrt{2}})$	16
59	$\langle \sigma^{12}\eta^2, \sigma^{11}\eta^3 \rangle$	4	$\mathbb{Q}(i\alpha 2^{\frac{1}{16}})$	16
60	$\langle \sigma^{4}\eta^2, \sigma^{15}\eta^3 \rangle$	4	$\mathbb{Q}(\alpha^3 2^{\frac{1}{16}}\sqrt{2+\sqrt{2}})$	16
61	$\langle \sigma^{12}\eta^2, \sigma^{15}\eta^3 \rangle$	4	$\mathbb{Q}(i\alpha^3 2^{\frac{1}{16}})$	16
62	$\langle \sigma^{8} \rangle$	2	$\mathbb{Q}(2^{\frac{1}{8}}, i\sqrt{2+\sqrt{2}})$	32
63	$\langle \sigma^{4}\eta^2 \rangle$	2	$\mathbb{Q}(2^{\frac{1}{16}}, i)$	32
64	$\langle \sigma^{12}\eta^2 \rangle$	2	$\mathbb{Q}(2^{\frac{1}{8}}, \alpha\sqrt{2+\sqrt{2}})$	32
65	$\langle \sigma\eta \rangle$	2	$\mathbb{Q}(\alpha^2 2^{\frac{1}{16}}, i\sqrt{2+\sqrt{2}})$	32
66	$\langle \sigma^{3}\eta \rangle$	2	$\mathbb{Q}(\alpha^3 2^{\frac{1}{16}}, \sqrt{2+\sqrt{2}})$	32
67	$\langle \sigma^{5}\eta \rangle$	2	$\mathbb{Q}(i2^{\frac{1}{16}}, i\sqrt{2+\sqrt{2}})$	32
68	$\langle \sigma^{7}\eta \rangle$	2	$\mathbb{Q}(i\alpha 2^{\frac{1}{16}}, \sqrt{2+\sqrt{2}})$	32
69	$\langle \sigma^{9}\eta \rangle$	2	$\mathbb{Q}(i\alpha^2 2^{\frac{1}{16}}, i\sqrt{2+\sqrt{2}})$	32
70	$\langle \sigma^{11}\eta \rangle$	2	$\mathbb{Q}(i\alpha^3 2^{\frac{1}{16}}, \sqrt{2+\sqrt{2}})$	32

Galois Correspondence for $x^{16} - 2$. (*continued*)

#	Subgroup	Order	Fixed Subfield	Degree
71	$\langle \sigma^{13}\eta \rangle$	2	$\mathbb{Q}(2^{\frac{1}{16}}, i\sqrt{2+\sqrt{2}})$	32
72	$\langle \sigma^{15}\eta \rangle$	2	$\mathbb{Q}(\alpha 2^{\frac{1}{16}}, \sqrt{2+\sqrt{2}})$	32
73	$\langle \sigma\eta^3 \rangle$	2	$\mathbb{Q}(i2^{\frac{1}{16}}, \sqrt{2+\sqrt{2}})$	32
74	$\langle \sigma^3\eta^3 \rangle$	2	$\mathbb{Q}(\alpha 2^{\frac{1}{16}}, i\sqrt{2+\sqrt{2}})$	32
75	$\langle \sigma^5\eta^3 \rangle$	2	$\mathbb{Q}(i\alpha^2 2^{\frac{1}{16}}, \sqrt{2+\sqrt{2}})$	32
76	$\langle \sigma^7\eta^3 \rangle$	2	$\mathbb{Q}(\alpha^3 2^{\frac{1}{16}}, i\sqrt{2+\sqrt{2}})$	32
77	$\langle \sigma^9\eta^3 \rangle$	2	$\mathbb{Q}(2^{\frac{1}{16}}, \sqrt{2+\sqrt{2}})$	32
78	$\langle \sigma^{11}\eta^3 \rangle$	2	$\mathbb{Q}(i\alpha 2^{\frac{1}{16}}, i\sqrt{2+\sqrt{2}})$	32
79	$\langle \sigma^{13}\eta^3 \rangle$	2	$\mathbb{Q}(\alpha^2 2^{\frac{1}{16}}, \sqrt{2+\sqrt{2}})$	32
80	$\langle \sigma^{15}\eta^3 \rangle$	2	$\mathbb{Q}(i\alpha^3 2^{\frac{1}{16}}, i\sqrt{2+\sqrt{2}})$	32
81	$\{e\}$	1	$\mathbb{K} = \mathbb{Q}(2^{\frac{1}{16}}, \alpha)$	64

CHAPTER 6

Resultant and Discriminant

Let

$$f(x) = ax^n + a_1x^{n-1} + \cdots + a_n \quad \text{and} \quad g(x) = bx^m + b_1x^{m-1} + \cdots + b_m$$

be polynomials of degree $n > 0$ and $m > 0$ respectively (so $ab \neq 0$) with coefficients in a field $\mathbb{F}$. Let α_i, $i \leq i \leq n$ and β_j, $i \leq j \leq m$ be the zeros of $f(x)$ and $g(x)$ in some algebraic closure of the field $\mathbb{F}$. The resultant $Res(f(x), g(x))$ of the polynomials $f(x)$ and $g(x)$ is defined as follows,

$$Res(f(x), g(x)) = a^m b^n \prod_{j=1}^{m} \prod_{i=1}^{n} (\alpha_i - \beta_j)$$

It can be shown that $Res(f(x), g(x))$ is the value of the following $(n+m) \times (n+m)$ determinant (see Kurosh, 1980).

$$Res(f,g) = \begin{vmatrix} a_0 & a_1 & \cdots & & a_{n-1} & & a_n & & \cdots & \\ & a_0 & a_1 & & \cdots & & a_{n-1} & a_n & & \cdots \\ & & \vdots & & & & \vdots & & & \vdots \\ & & \vdots & & & & \vdots & & & \vdots \\ & & & & a_0 & & a_1 \cdots & & a_{n-1} & a_n \\ b_0 & b_1 & \cdots & & b_m & & \cdots\cdots & & & \\ & b_0 & b_1 & \cdots\cdots & & & b_m & & & \\ & & \vdots & & & & \vdots & & & \vdots \\ & & \vdots & & & & \vdots & & & \vdots \\ & & & & & b_0 & b_1 & & \cdots & b_m \end{vmatrix}.$$

In this determinant all the missing entries are zeros. The entries in the first m rows are the coefficients of $f(x)$ and those in the last n rows are the coefficients of $g(x)$. The factor $a^m b^n$ in the definition of the resultant ensures that if the field $\mathbb{F} = \mathbb{Q}$ and $f(x)$ and $g(x)$ have integer coefficients then $Res(f(x), g(x))$ is an integer. The discriminant of the polynomial

$$f(x) = ax^n + a_1x^{n-1} + \cdots + a_n = a\prod_{i=1}^{n}(x - \alpha_i)$$

is defined as $a^{2n-2}\prod_{1\leq i<j\leq n}(\alpha_i - \alpha_j)^2$. We shall prove some properties of the resultant and its relationship with the discriminant of a polynomial. In what follows for notational convenience we may sometime suppress x in $f(x)$ and $g(x)$ and so on.

(1) $Res(f, g) = 0$ if and only if f and g have a common factor.
(2) $Res(f, g) = (-1)^{mn}Res(g, f)$.
(3) $Res(f, g) = a^m\prod_{i=1}^{i=n} g(\alpha_i) = (-1)^{mn}b^n\prod_{j=1}^{j=m} f(\beta_j)$.
(4) $Disc(f) = \frac{1}{a}(-1)^{n(n-1)/2}Res(f, f') = (-1)^{n(n-1)/2}a^{n-2}\prod_{i=1}^{i=n} f'(\alpha_i)$.
(5) If $n = deg(f) \geq deg(g) = m$ and $f(x) = q(x)\cdot g(x) + r(x)$, where $k = deg(r(x)) < deg(g(x))$ then

$$Res(f, g) = (-1)^{mn}\cdot b^{deg(f)-deg(r)}\cdot Res(g, r) = (-1)^{mn}\cdot b^{n-k}\cdot Res(g, r).$$

(6) If $c \in \mathbb{F}$ then $Res(f(x), c) = c^{deg(f)} = c^n$.

Proof 1.

This is an immediate consequence of the definition of the resultant of two polynomials.

Proof 2.

We have

$$Res(f, g) = a^m b^n\prod_{i=1}^{n}\prod_{j=1}^{m}(\alpha_i - \beta_j) = (-1)^{mn}a^m b^n\prod_{i=1}^{n}\prod_{j=1}^{m}(\beta_j - \alpha_i) = (-1)^{mn}Res(g, f).$$

Proof 3.

Since $g(x) = b\prod_{j=1}^{m}(x-\beta_j)$ so $g(\alpha_i) = b\prod_{j=1}^{m}(\alpha_i-\beta_j)$, for $1 \le i \le n$. So we have

$$\prod_{i=1}^{n} g(\alpha_i) = \prod_{i=1}^{n} b \prod_{j=1}^{m}(\alpha_i-\beta_j) = b^n \prod_{i=1}^{n}\prod_{j=1}^{m}(\alpha_i-\beta_j).$$

Therefore

$$a^m \prod_{i=1}^{n} g(\alpha_i) = a^m b^n \prod_{i=1}^{n}\prod_{j=1}^{m}(\alpha_i-\beta_j) = Res(f,g).$$

Likewise as $f(x) = a\prod_{i=1}^{n}(x-\alpha_i)$, we have $f(\beta_j) = a\prod_{i=1}^{n}(\beta_j-\alpha_i) = a(-1)^n\prod_{i=1}^{n}(\alpha_i-\beta_j)$,

Therefore

$$\prod_{j=1}^{m} f(\beta_j) = \prod_{j=1}^{m} a(-1)^n \prod_{i=1}^{n}(\alpha_i-\beta_j) = a^m(-1)^{mn}\prod_{i=1}^{n}\prod_{j=1}^{m}(\alpha_i-\beta_j).$$

Hence

$$(-1)^{mn}b^n \prod_{j=1}^{m} f(\beta_j) = a^m b^n \prod_{i=1}^{n}\prod_{j=1}^{m}(\alpha_i-\beta_j) = Res(f(x),g(x)).$$

Proof 4.

From the last part we have $Res(f,f') = a^{n-1}\prod_{i=1}^{n} f'(\alpha_i)$. Therefore

$$\prod_{i=1}^{n} f'(\alpha_i) = \frac{1}{a^{n-1}}Res(f,f') \quad \text{or} \quad a^{n-1}\prod_{i=1}^{n} f'(\alpha_i) = Res(f,f')$$

We shall take the logarithmic derivative of $f(x) = a\prod_{i=1}^{n}(x-\alpha_i)$. As

$$\log(f(x)) = \log(a) + \sum_{i=1}^{n}\log(x-\alpha_i)$$

so

$$\frac{f'(x)}{f(x)} = \sum_{i=1}^{n}\frac{1}{x-\alpha_i}.$$

Therefore

$$f'(x) = f(x)\sum_{i=1}^{n}\frac{1}{x-\alpha_i} = a\sum_{i=1}^{n}\prod_{j=1,j\neq i}^{j=n}(x-\alpha_j)$$

so

$$f'(\alpha_i) = a \prod_{j=1, j\neq i}^{j=n} (\alpha_i - \alpha_j).$$

Thus

$$\prod_{i=1}^{n} f'(\alpha_i) = \prod_{i=1}^{n} a \prod_{j=1, j\neq i}^{j=n} (\alpha_i - \alpha_j) = a^n \prod_{i\neq j}(\alpha_i - \alpha_j) = a^n(-1)^{\frac{n(n-1)}{2}} \prod_{1\leq i<j\leq n} (\alpha_i - \alpha_j)^2.$$

So

$$a^{n-2}\prod_{i=1}^{n} f'(\alpha_i) = (-1)^{\frac{n(n-1)}{2}} a^{2n-2} \prod_{1\leq i<j\leq n} (\alpha_i - \alpha_j)^2 = (-1)^{\frac{n(n-1)}{2}} Disc(f)$$

Finally we have

$$Disc(f) = (-1)^{\frac{n(n-1)}{2}} a^{n-2} \prod_{i=1}^{n} f'(\alpha_i) = \frac{1}{a}(-1)^{\frac{n(n-1)}{2}} \cdot a^{n-1} \prod_{i=1}^{i=n} f'(\alpha_i) = \frac{1}{a}(-1)^{\frac{n(n-1)}{2}} Res(f, f').$$

Proof 5.

We first note that as $f(x) = q(x)g(x) + r(x)$, $f(\beta_j) = r(\beta_j)$, for $1 \leq j \leq m$. As $g(x) = b\prod_{j=1}^{j=m}(x - \beta_j)$, $Res(g(x), r(x)) = b^k \prod_{j=1}^{j=m} r(\beta_j)$ where k is the degree of the polynomial remainder $r(x)$. So we have

$$\begin{aligned} Res(f(x), g(x)) &= (-1)^{mn} Res(g(x), f(x)) \\ &= (-1)^{mn} b^n \textstyle\prod_{j=1}^{j=m} f(\beta_j) \\ &= (-1)^{mn} b^n \textstyle\prod_{j=1}^{j=m} r(\beta_j) \\ &= (-1)^{mn} b^n \cdot \frac{1}{b^k} \cdot Res(g(x), r(x)) \\ &= (-1)^{mn} b^{n-k} Res(g(x), r(x)). \end{aligned}$$

Proof 6.

If $c = 0$ we have nothing to prove. We consider a nonzero $c \in \mathbb{F}$ as a polynomial of degree zero. So by (3) we have $Res(f(x), c) = c^{deg(f)} = c^n$.

We shall now give a few illustrations of the calculation of discriminant of certain polynomials using the methods we have just discussed.

1. Discriminant of ax^2+bx+c

First Method

Let $f(x)=ax^2+bx+c$, so $f^{'}(x)=2ax+b$. By (4) we have

$$Disc(ax^2+bx+c)=-a^{-1}Res(ax^2+bx+c,2ax+b).$$

Using the definition of the resultant of two polynomials in the form of a determinant we find that

$$Disc(ax^2+bx+c)=-a^{-1}Res(ax^2+bx+c,2ax+b)=-a^{-1}\times det\begin{pmatrix} a & b & c \\ 2a & b & 0 \\ 0 & 2a & b \end{pmatrix}=b^2-4ac$$

Second Method

We can find discriminant of ax^2+bx+c by property (4) above, which says that $Disc(f)=\frac{1}{a}(-1)^{n(n-1)/2}Res(f,f^{'})$. If α and β are the roots of $ax^2+bx+c=0$, then the discriminant is given by

$$\begin{aligned}-a^{-1}\times a\times(2a)^2\times\left[\left(\alpha+\frac{b}{2a}\right)\left(\beta+\frac{b}{2a}\right)\right] &= -4a^2\left[\alpha\beta+\frac{b}{2a}(\alpha+\beta)+\frac{b^2}{4a^2}\right]\\ &= -4a^2\left[\frac{c}{a}-\frac{b^2}{2a^2}+\frac{b^2}{4a^2}\right]\\ &= -4a^2\left[\frac{c}{a}-\frac{b^2}{4a^2}\right]=b^2-4ac\end{aligned}$$

2. Discriminant of x^3+px+q

First Method

$$Disc(x^3+px+q)=(-1)^3\times Res(x^3+px+q,3x^2+p).$$

If we consider the resultant in the form of a determinant then this discriminant is equal to

$$(-1)^3 \times det \begin{pmatrix} 1 & 0 & p & q & 0 \\ 0 & 1 & 0 & p & q \\ 3 & 0 & p & 0 & 0 \\ 0 & 3 & 0 & p & 0 \\ 0 & 0 & 3 & 0 & p \end{pmatrix}$$

Subtracting three times the first row from the third row this reduces to

$$(-1) \times det \begin{pmatrix} 1 & 0 & p & q & 0 \\ 0 & 1 & 0 & p & q \\ 0 & 0 & -2p & -3q & 0 \\ 0 & 3 & 0 & p & 0 \\ 0 & 0 & 3 & 0 & p \end{pmatrix} = (-1) \times det \begin{pmatrix} 1 & 0 & p & q \\ 0 & -2p & -3q & 0 \\ 3 & 0 & p & 0 \\ 0 & 3 & 0 & p \end{pmatrix}$$

Again subtracting three times the first row from the third row we have

$$(-1) \times det \begin{pmatrix} 1 & 0 & p & q \\ 0 & -2p & -3q & 0 \\ 0 & 0 & -2p & -3q \\ 0 & 3 & 0 & p \end{pmatrix} = (-1) \times det \begin{pmatrix} -2p & -3q & 0 \\ 0 & -2p & -3q \\ 3 & 0 & p \end{pmatrix}.$$

Finally

$$(-1) \times det \begin{pmatrix} -2p & -3q & 0 \\ 0 & -2p & -3q \\ 3 & 0 & p \end{pmatrix} = -\left[-2p(-2p^2) + 3(9q^2)\right] = -(4p^3 + 27q^2).$$

Second Method

We write $f(x)=x^3+x+q$ and so $f'(x)=3x^2+p$. By the division algorithm for polynomials over a field we have

$$x^3+px+q=(3x^2+p)\frac{x}{3}+\left(\frac{2p}{3}x+q\right).$$

Using the properties of the resultants we have

$$\begin{aligned} Disc(x^3+px+q) &= (-1)\cdot Res(x^3+px+q,3x^2+p), && \text{by property (4)}\\ &= (-1)3^{3-1}\cdot Res(3x^2+p,\tfrac{2}{3}px+q), && \text{by property (5)}\\ &= -9\left(\frac{2}{3}p\right)^2\left[3\left(-\frac{3q}{2p}\right)^2+p\right], && \text{by property (3)}\\ &= -4p^2\left(\frac{27q^2}{4p^2}+p\right)=-27q^2-4p^3. \end{aligned}$$

Another Method

Another interesting way to compute this discriminant is to find a (monic) cubic polynomial whose roots are $(\alpha-\beta)^2$, $(\beta-\gamma)^2$, and $(\gamma-\alpha)^2$ where α, β, and γ are roots of $x^3+px+q=0$. Then the discriminant of x^3+px+q would be $(-1)\times$ the constant term of this polynomial. This can be done as follows. Let $y=(\alpha-\beta)^2$. Then we have

$$\begin{aligned} y &= \alpha^2+\beta^2-2\alpha\beta\\ &= (\alpha+\beta+\gamma)^2-2(\alpha\beta+\beta\gamma+\gamma\alpha)-\frac{2\alpha\beta\gamma}{\gamma}-\gamma^2\\ &= -2p+\frac{2q}{\gamma}-\gamma^2\\ \gamma y &= 2p\gamma+2q-\gamma^3. \end{aligned}$$

This gives $\gamma^3+\gamma(2p+y)-2q=0$. Which together with the fact that $\gamma^3+p\gamma+q=0$ gives $\gamma=\dfrac{3q}{p+y}$. Since γ is a root of the equation $x^3+px+q=0$ we have

$$\left(\frac{3q}{p+y}\right)^3+\frac{3q}{p+y}\cdot p+q=0$$

or (assuming $q\neq 0$)

$$27q^2+3p(p+y)^2+(p+y)^3=0$$

which is cubic in y with roots $(\alpha-\beta)^2$, $(\beta-\gamma)^2$, and $(\gamma-\alpha)^2$. Its constant term is equal to $4p^3+27q^2$. Therefore the discriminant of the cubic equation $x^3+px+q=0$ is

$$(\alpha-\beta)^2(\beta-\gamma)^2(\gamma-\alpha)^2=-(4p^3+27q^2).$$

3. Discriminant of Trinomial x^n+ax+b

Using the methods discussed above we shall show that the discriminant of the polynomial x^n+ax+b is given by the expression

$$(-1)^{\frac{n(n-1)}{2}}\left[n^nb^{n-1}+(-1)^{n-1}a^n(n-1)^{n-1}\right]$$

Let $f(x)=x^n+ax+b$, so $f'(x)=nx^{n-1}+a$. By the division algorithm we have

$$x^n+ax+b=(nx^{n-1}+a)\frac{1}{n}x+\frac{a(n-1)}{n}x+b.$$

We have

$$\begin{aligned}
Disc(x^n+ax+b)&=(-1)^{\frac{n(n-1)}{2}}Res(x^n+ax+b,nx^{n-1}+a)\\
&=(-1)^{\frac{n(n-1)}{2}}(-1)^{n(n-1)}\cdot n^{n-1}Res(nx^{n-1}+a,\frac{a(n-1)}{n}x+b)\\
&=(-1)^{\frac{n(n-1)}{2}}\cdot n^{n-1}\cdot(-1)^{n-1}\cdot\left(\frac{a(n-1)}{n}\right)^{n-1}\cdot\left[n\left(\frac{-bn}{a(n-1)}\right)^{n-1}+a\right]\\
&=(-1)^{\frac{n(n-1)}{2}}\cdot(-1)^{n-1}a^{n-1}(n-1)^{n-1}\left[(-1)^{n-1}\frac{n^nb^{n-1}}{a^{n-1}(n-1)^{n-1}}+a\right]\\
&=(-1)^{\frac{n(n-1)}{2}}\left[n^nb^{n-1}+(-1)^{n-1}a^n(n-1)^{n-1}\right].
\end{aligned}$$

We can check that for $n=2$ the above expression reduces to a^2-4b and for $n=3$ it simplifies to $-(4a^3+27b^2)$.

The concept of resultant has some useful applications. Its equivalent definition (as a certain determinant) ensures its quick computation with the help of some symbolic computation software such as Mathematica or Maple. If f and g are polynomials in two variables x and y then we can determine their resultant with respect to any of the variable. This provides a method of finding the common solution of the two equations in two unknowns. Our present interest is in the application of the resultant in finding a polynomial whose roots are sums of the roots of the given polynomial taken k at a times for small values of k such as $k = 2, 3, 4$. For this it will be convenient to begin with a monic polynomial. Let

$$f(x) = x^n + a_1 x^{n-1} + \cdots + a_n \in \mathbb{F}[x]$$

be a polynomial of degree n with coefficients in a field $\mathbb{F}$, with zeros $\alpha_1, \alpha_2, \ldots, \alpha_n$ in some extension of $\mathbb{F}$. The polynomial $(-1)^n f(x-y)$ can be regarded as a monic polynomial of degree n in y with coefficient in the polynomial ring $\mathbb{F}[x]$. As a polynomial in y its n zeros are $x - \alpha_i$, $1 \le i \le n$ since

$$(-1)^n f(x-y) = (-1)^n \prod_{1 \le i \le n} (x - y - \alpha_i) = \prod_{1 \le i \le n} (-x + y + \alpha_i) = \prod_{1 \le i \le n} [y - (x - \alpha_i)].$$

This observation (and similar ones) will be used repeatedly in what follows.

Let $f_k(x)$ denote the monic polynomial of degree ${}_nC_k$ with zeros $\alpha_{i_1} + \alpha_{i_2} + \cdots + \alpha_{i_k}$ where $1 \le i_1 < i_2 < i_3 \cdots < i_k \le n$. The following examples will illustrate how the concept of the resultant can be used to find $f_k(x)$ for small values of k.

Note: We caution the readers that in the examples that follow the polynomials $R(x), A(x), B(x), C(x)$, and so on, are locally defined and do not have the same meaning for different examples. For instance in Example 1, the polynomial $R(x)$ has degree n^2 where as in Example 2 the polynomial has degree $\dfrac{n^2(n-1)}{2}$.

4. The Polynomials $f_k(x)$, $2 \le k \le 5$

Example 1. Computation of $f_2(x)$

Let

$$R(x) = Res_y\left((-1)^n f(x-y), f(y)\right) = \prod_{j=1}^{n} \prod_{i=1}^{n} (x - \alpha_i - \alpha_j)$$

Then $R(x)$ is a polynomial of degree n^2 with zeros $\alpha_i + \alpha_j$. So the n^2 zeros of $R(x)$ are $2\alpha_i$, $1 \le i \le n$ and $\alpha_i + \alpha_j$, $i < j$, each appearing twice. Let

$$A(x) = 2^n f\left(\frac{x}{2}\right)$$

Then $A(x)$ is a monic polynomial of degree n with zeros $2\alpha_i$, $1 \le i \le n$. Therefore, $R(x)/A(x)$ is a polynomial of degree $n^2 - n$, with zeros $\alpha_i + \alpha_j$, $i < j$, each zero appearing twice. Therefore,

$$f_2(x) = (R(x)/A(x))^{\frac{1}{2}}$$

is a polynomial of degree $\frac{n^2-n}{2} = {}_nC_2$, with zeros $\alpha_i + \alpha_j$, $1 \leq i < j \leq n$.

Example 2. Computation of $f_3(x)$

We first note that as a polynomial in y, $f_2(x-y)$ has degree $\frac{n(n-1)}{2}$ and has roots $x - \alpha_i - \alpha_j$ for $1 \leq i < j \leq n$. In other words we can write

$$f_2(x-y) = \prod_{1\leq i<j\leq n} (x - y - \alpha_i - \alpha_j) = (-1)^{\frac{n(n-1)}{2}} \prod_{1\leq i<j\leq n} (y - x + \alpha_i + \alpha_j).$$

Let

$$R(x) = Res_y\left((-1)^{\frac{n(n-1)}{2}} f_2(x-y), f(y)\right) = \prod_{1\leq i<j\leq n}\prod_{1\leq k\leq n} (x - \alpha_i - \alpha_j - \alpha_k).$$

Then $R(x)$ is a polynomial of degree ${}_nC_2 \cdot n = \dfrac{n^2(n-1)}{2}$ with the zeros of following types:

(1) $\alpha_i + \alpha_j + \alpha_k$, $\quad i < j < k$. (Each appearing three times.)

(2) $2\alpha_i + \alpha_j$, $\quad i \neq j$.

We check that the total number adds up to the right degree, namely

$$3 \cdot {}_nC_3 + 2 \cdot {}_nC_2 = \frac{n(n-1)(n-2)}{2} + n(n-1) = \frac{n^2(n-1)}{2}.$$

We shall now find a polynomial with zeros $2\alpha_i + \alpha_j$, $i \neq j$. Let

$$A(x) = Res_y\left((-1)^n f\left(\frac{x-y}{2}\right)2^n, f(y)\right).$$

Here as before we have multiplied by $(-1)^n$ to ensure that the first polynomial in the argument of Res_y is monic (as a polynomial in y).

We note that as a polynomial in y the roots of $f\left(\frac{x-y}{2}\right)$ are $x - 2\alpha_i$, for $1 \leq i \leq n$. Also $A(x)$ is a polynomial of degree n^2. In fact

$$A(x) = \prod_{j=1}^{n}\prod_{i=1}^{n} (x - 2\alpha_i - \alpha_j).$$

So the zeros of $A(x)$ are $3\alpha_i$, $1 \leq i \leq n$, and $2\alpha_i + \alpha_j$, $i \neq j$. Let

$$B(x) = f\left(\frac{x}{3}\right)3^n, \qquad C(x) = \frac{A(x)}{B(x)}.$$

So $B(x)$ is a monic polynomial with zeros $3\alpha_i$, $1 \le i \le n$ and so $C(x)$ is a polynomial of degree $n^2 - n$ with zeros, $2\alpha_i + \alpha_j$, $i \ne j$, $1 \le i, j \le n$. We also note that

$$\frac{R(x)}{C(x)} = \frac{R(x) \cdot B(x)}{A(x)}$$

is a polynomial of degree

$$\frac{n^2(n-1)}{2} - (n^2 - n) = \frac{n(n-1)(n-2)}{2}$$

with zeros $\alpha_i + \alpha_j + \alpha_k$, $\quad i < j < k$, each repeated three times. Therefore

$$f_3(x) = \left(\frac{R(x)}{C(x)}\right)^{\frac{1}{3}}$$

is the required polynomial of degree ${}_nC_3$.

Example 3. Computation of $f_4(x)$

We shall carry out this computation in several steps. We first note that

$$f_3(x - y) = \prod_{1 \le i < j < k \le n} (x - y - \alpha_i - \alpha_j - \alpha_k)$$

is a polynomial of degree ${}_nC_3$. Therefore the polynomial $(-1)^{{}_nC_3} \cdot f_3(x - y)$ is monic in y of degree ${}_nC_3$ and as a polynomial in y its roots are $x - \alpha_i - \alpha_j - \alpha_k$, $1 \le i < i < j \le n$.

(i) Let

$$R(x) = Res_y\left((-1)^{{}_nC_3} \cdot f_3(x - y), f(y)\right) \prod_{1 \le i < j < k \le n} \prod_{l=1}^{l=n} (x - \alpha_i - \alpha_j - \alpha_k - \alpha_l)\,.$$

The degree of $R(x) = {}_nC_3 \cdot n = \dfrac{n^2(n-1)(n-2)}{6}$. Its zeros are $\alpha_i + \alpha_j + \alpha_k + \alpha_l$, $1 \le i < j < k \le n$, and $1 \le l \le n$. We divide these zeros in two groups:

(1) $\alpha_i + \alpha_j + \alpha_k + \alpha_l$, $\quad i < j < k < l$, each appearing four times, and so $4 \cdot {}_nC_4$ in all.

(2) $2\alpha_i + \alpha_j + \alpha_k$, $\quad i, j, k$ distinct. These are $3 \cdot {}_nC_3$ in all.

We note that

$$4 \cdot {}_nC_4 + 3 \cdot {}_nC_3 = n \cdot {}_nC_3$$

We note that $(-1)^n \cdot 2^n \cdot f\left(\frac{x-y}{2}\right)$ is a monic polynomial in y of degree n with roots $x - 2\alpha_i$, $\quad 1 \le i \le n$.

(ii) Let

$$A(x) = Res_y\left(2^n \cdot (-1)^n f\left(\frac{x-y}{2}\right), f_2(y)\right) = \prod_{i=1}^{i=n} \prod_{1\le j<k\le n} (x - 2\alpha_i - \alpha_j - \alpha_k).$$

Degree of $A(x) = n \cdot {}_nC_2 = \dfrac{n^2(n-1)}{2}$. Its zeros are $2\alpha_i + \alpha_j + \alpha_k, \quad j<k.$

These can be divided into two types:

(1) $3\alpha_i + \alpha_j, \quad i \neq j.$

(2) $2\alpha_i + \alpha_j + \alpha_k, \quad i, j, k$ distinct.

(iii) Let

$$B(x) = Res_y\left(3^n(-1)^n f\left(\frac{x-y}{3}\right), f(y)\right) = \prod_{i=1}^{i=n} \prod_{j=1}^{j=n} (x - 3\alpha_i - \alpha_j).$$

Degree of $B(x) = n^2$. Its zeros are $3\alpha_i + \alpha_j$. These can be divided into two types:

(1) $4\alpha_i \quad 1 \le i \le n.$

(2) $3\alpha_i + \alpha_j, \quad i \neq j.$

(iv) Let

$$C(x) = 4^n f\left(\frac{x}{4}\right).$$

Degree of $C(x) = n$. Its zeros are $4\alpha_i, \quad 1 \le i \le n.$

(v) Let

$$D(x) = \frac{B(x)}{C(x)}.$$

Degree of $D(x) = n^2 - n$. Its zeros are $3\alpha_i + \alpha_j, \quad 1 \le i, j \le n, \quad i \neq j.$

(vi) Let

$$E(x) = \frac{A(x)}{D(x)} = \frac{A(x) \cdot C(x)}{B(x)}.$$

Degree of $E(x) = n \cdot {}_nC_2 - n(n-1) = \dfrac{n(n-1)(n-2)}{2}$. Its zeros are $2\alpha_i + \alpha_j + \alpha_k$, i, j, k distinct.

(vii) Let

$$F(x) = \frac{R(x)}{E(x)} = \frac{R(x) \cdot B(x)}{A(x) \cdot C(x)}.$$

Degree of $F(x) = n \cdot {}_nC_3 - \dfrac{n(n-1)(n-2)}{2} = {}_nC_3 \cdot (n-3) = \dfrac{n(n-1)(n-2)(n-3)}{6}$.
Its zeros are $\alpha_i + \alpha_j + \alpha_k + \alpha_l, \quad i<j<k<l$, each repeated four times.

(viii) Let

$$f_4(x) = (F(x))^{\frac{1}{4}}$$

The degree of $f_4(x)$ is ${}_nC_4$. Its zeros are $1 \le \alpha_i + \alpha_j + \alpha_k + \alpha_l \le n, \quad i < j < k < l.$

Example 4. Computation of $f_5(x)$

As with the computation of $f_4(x)$ we will carry out the computation of $f_5(x)$ in several steps. First we note that

$$f_4(x-y) = \prod_{1 \le i<j<k<l \le n} (x - y - \alpha_i - \alpha_j - \alpha_k - \alpha_l)$$

is a polynomial of degree ${}_nC_4$. So the polynomial $(-1)^{{}_nC_4} f_4(x-y)$ can be regarded as a monic in y of degree ${}_nC_4$ with roots $x - \alpha_i - \alpha_j - \alpha_k - \alpha_l$, $1 \le i < j < k < l \le n$.

(i) Let

$$R(x) = Res_y \left((-1)^{{}_nC_4} f_4(x-y), f(y) \right) = \prod_{1 \le i<j<k<l \le n} \prod_{m=1}^{m=n} (x - \alpha_i - \alpha_j - \alpha_k - \alpha_l - \alpha_m).$$

Degree $R(x) = n \cdot {}_nC_4 = \dfrac{n^2(n-1)(n-2)(n-3)}{24}$. Its zeros are of the following types:

(1) $\alpha_i + \alpha_j + \alpha_k + \alpha_l + \alpha_m$, $\quad 1 \le i < j < k < l < m \le n$, each of these zeros is repeated five times.

(2) $2\alpha_i + \alpha_j + \alpha_k + \alpha_l$, i, j, k, l all distinct.

We note that

$$5 \times {}_nC_5 + 4 \times {}_nC_4 = n \times {}_nC_4 = \frac{n^2(n-1)(n-2)(n-3)(n-4)}{24}.$$

(ii) Let

$$A(x) = Res_y \left((-1)^n \cdot 2^n f\left(\frac{x-y}{2}\right), f_3(y) \right).$$

Degree of $A(x) = n \cdot {}_nC_3 = \dfrac{n^2(n-1)(n-2)}{6}$. Its zeros are of the following types:

(1) $2\alpha_i + \alpha_j + \alpha_k + \alpha_l$, $\quad i, j, k, l$ are all distinct.

(2) $3\alpha_i + \alpha_j + \alpha_k$, $\quad i, j, k$ are all distinct.

(iii) Let

$$B(x) = Res_y\left((-1)^n \cdot 3^n f\left(\frac{x-y}{3}\right), f_2(y)\right) = \prod_{i=1}^{n} \prod_{1\le j<k\le n} (x - 3\alpha_i - \alpha_j - \alpha_k).$$

Degree of $B(x) = n \cdot {}_nC_2 = \dfrac{n^2(n-1)}{2}$. Its zeros are of the following types:

(1) $4\alpha_i + \alpha_j, \quad i \neq j.$

(2) $3\alpha_i + \alpha_j + \alpha_k, \quad i, j, k$ are all distinct.

(iv) Let

$$C(x) = Res_y\left((-1)^n \cdot 4^n f\left(\frac{x-y}{4}\right), f(y)\right) = \prod_{i=1}^{n} \prod_{j=1}^{n} (x - 4\alpha_i - \alpha_j).$$

Degree of $C(x) = n^2$. Its zeros are of the following types:

(1) $5\alpha_i, \quad i \le i \le n.$

(2) $4\alpha_i + \alpha_j \quad i \neq j.$

(v) Let

$$D(x) = 5^n \cdot f\left(\frac{x}{5}\right).$$

Degree of $D(x) = n$. Its zeros are $5\alpha_i, \quad 1 \le i \le n.$

(vi) Let

$$E(x) = \frac{C(x)}{D(x)}.$$

Degree of $E(x) = n^2 - n$. Its zeros are $4\alpha_i + \alpha_j, \quad 1 \neq j.$

(vii) Let

$$F(x) = \frac{B(x)}{E(x)} = \frac{B(x) \cdot D(x)}{C(x)}.$$

Degree of $F(x) = \dfrac{n(n-1)(n-2)}{2}$. Its zeros are $3\alpha_i + \alpha_j + \alpha_k, \quad i, j, k$ distinct.

(viii) Let

$$G(x) = \frac{A(x)}{F(x)} = \frac{A(x) \cdot C(x)}{B(x) \cdot D(x)}.$$

Degree of $G(x) = 4 \cdot {}_nC_4$. Its zeros are $2\alpha_i + \alpha_j + \alpha_k + \alpha_l, \quad i, j, k, l$ distinct.

(ix) Let

$$H(x) = \frac{R(x)}{G(x)} = \frac{A(x) \cdot B(x) \cdot D(x)}{A(x) \cdot C(x)}.$$

Degree of $H(x) = 5 \cdot {}_nC_5$. Its zeros are $\alpha_i + \alpha_j + \alpha_k + \alpha_l + \alpha_m, \quad i, j, k, l, m$ distinct.

Each zero is repeated five times.

(x) Now

$$f_5(x) = (H(x))^{\frac{1}{5}}.$$

Degree of $f_5(x) = {}_nC_5$. Its zeros are $\alpha_i+\alpha_j+\alpha_k+\alpha_l+\alpha_m,\quad 1 \le i < j < k < l < m \le n.$

5. The Galois Group of the Polynomial $x^7 - 154x + 99$

We shall apply the techniques developed in the previous section to determine the order of the Galois group of the polynomial $x^7 - 154x + 99$ over $\mathbb{Q}$. The Galois group of this polynomial has been discussed in the paper "Polynomials with $PSL(2,7)$ as a Galois group," by Erbach, Fisher, and McKay (1979) however our method of proof is slightly different. We will use the following result from the their paper:

Lemma 1.

Let $f(x) \in \mathbb{Q}[x]$ be a polynomial of degree $n > 1$ with distinct roots $\alpha_1, \alpha_2, \cdots, \alpha_n$. Let $1 \le r < n$ and $k = \binom{n}{r} = \dfrac{n!}{r!(n-r)!}$. Let $\beta_1, \beta_2, \cdots \beta_k$ be the sums of $\alpha_i's$ taken r at a time then

$$\mathbb{Q}(\alpha_1, \alpha_2, \cdots, \alpha_n) = \mathbb{Q}(\beta_1, \beta_2, \cdots, \beta_k).$$

Proof.

Evidently

$$\mathbb{Q}(\beta_1, \beta_2, \cdots, \beta_k) \subseteq \mathbb{Q}(\alpha_1, \alpha_2, \cdots, \alpha_n).$$

So we only need to show that

$$\mathbb{Q}(\alpha_1, \alpha_2, \cdots, \alpha_n) \subseteq \mathbb{Q}(\beta_1, \beta_2, \cdots, \beta_k).$$

We will do so by showing that for $1 \le i \le n \quad \alpha_i \in \mathbb{Q}(\beta_1, \beta_2, \cdots, \beta_k)$.

Let δ_i denote the sum of those $\beta's$ which contain α_i as one of their summands, then

$$\delta_i = \binom{n-1}{r-1}\alpha_i + \binom{n-2}{r-2}\sum_{j \ne i} \alpha_j,$$

since every $\alpha_i \ne \alpha_j$ can appear in $\binom{n-2}{r-2}$, $\beta's$.

We have

$$\delta_i = \left\{\binom{n-1}{r-1} - \binom{n-2}{r-2}\right\}\alpha_i + \binom{n-2}{r-2}\sum_{j=1}^{j=n} \alpha_j$$

or

$$\delta_i = \binom{n-2}{r-1}\alpha_i + \binom{n-2}{r-2}\sum_{j=1}^{j=n} \alpha_j.$$

This shows that each α_i can be expressed as a rational function of $\delta's$ and the sum of all $\alpha_i's$. Therefore

$$\alpha_i \in \mathbb{Q}(\beta_1, \beta_2, \cdots, \beta_k).$$

Hence

$$\mathbb{Q}(\alpha_1, \alpha_2, \cdots, \alpha_n) = \mathbb{Q}(\beta_1, \beta_2, \cdots, \beta_k).$$

This completes the proof of the Lemma.

We shall now give an illustration of the above lemma by a specific example.
Let $f(x) = x^7 - 154x + 99$. The polynomial $f(x)$ is easily seen to be irreducible over the field $\mathbb{Q}$ using Eisenstein Criterion for prime $p = 11$. Let $\alpha_1, \alpha_2, \alpha_3, \alpha_4, \alpha_5, \alpha_6, \alpha_7 \in \mathbb{C}$ be its roots. Let $f_2(x)$ denote the polynomial of degree $\binom{7}{2} = 21$ whose roots are sums of two distinct roots of $f(x)$, namely

$\beta_1 = \alpha_1 + \alpha_2$	$\beta_2 = \alpha_1 + \alpha_3$	$\beta_3 = \alpha_1 + \alpha_4$
$\beta_4 = \alpha_1 + \alpha_5$	$\beta_5 = \alpha_1 + \alpha_6$	$\beta_6 = \alpha_1 + \alpha_7$
$\beta_7 = \alpha_2 + \alpha_3$	$\beta_8 = \alpha_2 + \alpha_4$	$\beta_9 = \alpha_2 + \alpha_5$
$\beta_{10} = \alpha_2 + \alpha_6$	$\beta_{11} = \alpha_2 + \alpha_7$	$\beta_{12} = \alpha_3 + \alpha_4$
$\beta_{13} = \alpha_3 + \alpha_5$	$\beta_{14} = \alpha_3 + \alpha_6$	$\beta_{15} = \alpha_3 + \alpha_7$
$\beta_{16} = \alpha_4 + \alpha_5$	$\beta_{17} = \alpha_4 + \alpha_6$	$\beta_{18} = \alpha_4 + \alpha_7$
$\beta_{19} = \alpha_5 + \alpha_6$	$\beta_{20} = \alpha_5 + \alpha_7$	$\beta_{21} = \alpha_6 + \alpha_7$

We note that $\beta_1, \beta_2, \beta_3, \beta_4, \beta_5, \beta_6$ are the only $\beta's$ which contain α_1 as one their summands. So

$$\beta_1 + \beta_2 + \beta_3 + \beta_4 + \beta_5 + \beta_6 = 6\alpha_1 + \alpha_2 + \alpha_3 + \alpha_4 + \alpha_5 + \alpha_6 + \alpha_7$$

or

$$\beta_1 + \beta_2 + \beta_3 + \beta_4 + \beta_5 + \beta_6 = 5\alpha_1 + (\alpha_1 + \alpha_2 + \alpha_3 + \alpha_4 + \alpha_5 + \alpha_6 + \alpha_7) = 5\alpha_1.$$

Therefore

$$\alpha_1 \in \mathbb{Q}(\beta_1, \beta_2, \cdots, \beta_{21}).$$

Likewise

$$\alpha_i \in \mathbb{Q}(\beta_1, \beta_2, \cdots, \beta_{21}), \qquad 2 \leq i \leq 7.$$

Hence

$$\mathbb{Q}(\alpha_1, \alpha_2, \cdots, \alpha_7) = \mathbb{Q}(\beta_1, \beta_2, \cdots, \beta_{21}).$$

Let $f_3(x)$ denote the polynomial of degree $\binom{7}{3} = 35$ whose roots are sums of three distinct roots of $f(x)$, namely

$$\gamma_1 = \alpha_1 + \alpha_2 + \alpha_3 \quad \gamma_2 = \alpha_1 + \alpha_2 + \alpha_4 \quad \gamma_3 = \alpha_1 + \alpha_2 + \alpha_5 \quad \gamma_4 = \alpha_1 + \alpha_2 + \alpha_6 \quad \gamma_5 = \alpha_1 + \alpha_2 + \alpha_7$$

$$\gamma_6 = \alpha_1 + \alpha_3 + \alpha_4 \quad \gamma_7 = \alpha_1 + \alpha_3 + \alpha_5 \quad \gamma_8 = \alpha_1 + \alpha_3 + \alpha_6 \quad \gamma_9 = \alpha_1 + \alpha_3 + \alpha_7 \quad \gamma_{10} = \alpha_1 + \alpha_4 + \alpha_5$$

$$\gamma_{11} = \alpha_1 + \alpha_4 + \alpha_6 \quad \gamma_{12} = \alpha_1 + \alpha_4 + \alpha_7 \quad \gamma_{13} = \alpha_1 + \alpha_5 + \alpha_6 \quad \gamma_{14} = \alpha_1 + \alpha_5 + \alpha_7 \quad \gamma_{15} = \alpha_1 + \alpha_6 + \alpha_7$$

$$\gamma_{16} = \alpha_2 + \alpha_3 + \alpha_4 \quad \gamma_{17} = \alpha_2 + \alpha_3 + \alpha_5 \quad \gamma_{18} = \alpha_2 + \alpha_3 + \alpha_6 \quad \gamma_{19} = \alpha_2 + \alpha_3 + \alpha_7 \quad \gamma_{20} = \alpha_2 + \alpha_4 + \alpha_5$$

$$\gamma_{21} = \alpha_2 + \alpha_4 + \alpha_6 \quad \gamma_{22} = \alpha_2 + \alpha_4 + \alpha_7 \quad \gamma_{23} = \alpha_2 + \alpha_5 + \alpha_6 \quad \gamma_{24} = \alpha_2 + \alpha_5 + \alpha_7 \quad \gamma_{25} = \alpha_2 + \alpha_6 + \alpha_7$$

$$\gamma_{26} = \alpha_3 + \alpha_4 + \alpha_5 \quad \gamma_{27} = \alpha_3 + \alpha_4 + \alpha_6 \quad \gamma_{28} = \alpha_3 + \alpha_4 + \alpha_7 \quad \gamma_{98} = \alpha_3 + \alpha_5 + \alpha_6 \quad \gamma_{30} = \alpha_3 + \alpha_5 + \alpha_7$$

$$\gamma_{31} = \alpha_3 + \alpha_6 + \alpha_7 \quad \gamma_{32} = \alpha_4 + \alpha_5 + \alpha_6 \quad \gamma_{33} = \alpha_4 + \alpha_5 + \alpha_7 \quad \gamma_{34} = \alpha_4 + \alpha_6 + \alpha_7 \quad \gamma_{35} = \alpha_5 + \alpha_6 + \alpha_7$$

We note that γ_i, $1 \leq i \leq 15$ are the only roots of $f_3(x)$ which contain α_1 as one of their summand. We have

$$\sum_{i=1}^{i=15} \gamma_i = 15\alpha_1 + 5\left(\sum_{j=2}^{j=7} \alpha_j\right)$$

or

$$\sum_{i=1}^{i=15} \gamma_i = 10\alpha_1 + 5(\sum_{j=1}^{j=7} \alpha_j = 10\alpha_1.$$

Therefore

$$\alpha_1 \in \mathbb{Q}(\gamma_1, \gamma_2, \cdots, \gamma_{35}).$$

Similarly we have

$$\alpha_i \in \mathbb{Q}(\gamma_1, \gamma_2, \cdots, \gamma_{35}) \quad 2 \leq i \leq 7.$$

Hence

$$\mathbb{Q}(\alpha_1, \alpha_2, \cdots, \alpha_7) = \mathbb{Q}(\gamma_1, \gamma_2, \cdots, \gamma_{35})$$

Lemma 2.

The Galois group of $f(x) \in \mathbb{Q}[x]$ acts transitively on its roots if and only if $f(x)$ is a power of an irreducible polynomial.

Proof.

This is an immediate consequence of the definition of the Galois group and the fact that the an element of the Galois group cannot map a root of an irreducible polynomial to the root of a different irreducible polynomial.

As a preparation for the computation of the order of the Galois group of the polynomial $x^7 - 154x + 99$ we shall discuss the calculation of the polynomials $f_2(x)$ and $f_3(x)$ for the polynomial $f(x) = x^7 - 154x + 99$ using Mathematica. We recall that $f_2(x)$ and $f_3(x)$ are respectively the polynomials whose roots are the sum of roots of $f(x)$ taken two (respectively three) at a time.

Computation of $f_2(x)$ for $f(x) = x^7 - 154x + 99$

For $f(x) = x^7 - 154x + 99$, we have

$$R_2(x) = Res_y\left(-(x-y)^7 + 154(x-y) - 99, y^7 - 154y + 99\right)$$

$$R_2(x) = 11930436453209472 + 324772992337368960x + 3637457514178532352x^2$$
$$+2121850216604143872 0x^3 + 66013117849906698240x^4$$
$$+92418364989869377536x^5 - 95910920922794710887x^7$$
$$-90971293541772243 6x^8 - 36741835622247635 88x^9$$
$$-33338966249383945 92x^{10} + 44012218145615929 44x^{11}$$
$$-24472455656338166 40x^{12} + 24530080275985173 12x^{13}$$
$$+1015231139552834 82x^{14} - 1137246015925154 52x^{15}$$
$$+1117678552672096 80x^{16} - 892443255679926 72x^{17}$$
$$+470640038142415 68x^{18} - 233032894729909 12x^{19}$$
$$+413128425060351x^{21} - 659274289698024x^{22} + 498287847840552x^{23}$$
$$-297698449134240x^{24} + 95786693462768x^{25} + 363702995964x^{28}$$
$$-823980498264x^{29} + 593047323792x^{30} - 130794878368x^{31}$$
$$-116837721x^{35} + 166272876x^{36} - 63582596x^{37} + 1386x^{42} - 2156x^{43} + x^{49}$$

The (monic) polynomial $A_2(x)$ whose roots are $2\alpha_i$, $1 \le i \le 7$ is:

$$\mathbf{A_2(x) = 128\left((x/2)^7 - 154(x/2) + 99\right) = x^7 - 9856x + 12672}$$

The polynomial $R_2(x)/A_2(x)$ is:

$$941480149401 + 263614441832 28x + 307550182137660x^2$$

$$+1913645577745440x^3 + 6697759522109040x^4$$

$$+12502484441270208x^5 + 9724154565432384x^6 - 5496722488422x^7$$

$$-760665241246 68x^8 - 349132163366448x^9 - 534789841592160x^{10}$$

$$-691575618444448x^{11} - 247898120234688x^{12} + 8012043140607x^{14}$$

$$-2736887458536x^{15} + 6718925555496x^{16} - 1774605859488x^{17}$$

$$+2339223960304x^{18} + 31969411452x^{21} - 26944987608x^{22}$$

$$+17834526864x^{23} - 9481277344x^{24} + 26178471x^{28} - 42536340x^{29}$$

$$+12308604x^{30} - 11286x^{35} + 7700x^{36} + x^{42}$$

The above polynomial factors as:

$$(970299+13584186x+63392868x^2+98611128x^3-2832489x^7+457380x^8-1256948x^9-5643x^{14}+3850x^{15}+x^{21})^2$$

Therefore the polynomial $f_2(x)$ of degree 21 whose roots are $\alpha_i + \alpha_j$, where $1 \leq i < j \leq 7$ is:

$$970299+13584186x+63392868x^2+98611128x^3-2832489x^7+457380x^8-1256948x^9-5643x^{14}+3850x^{15}+x^{21}$$

Likewise using the method developed in the last section we calculate the polynomial $f_3(x)$ of degree 35. The monic polynomial $f_3(x)$ of degree 35 with roots γ_j, $1 \le j \le 35$ is:

$$\begin{array}{rrrr}
x^{35} & -6160x^{29} & +29898x^{28} & -38277624x^{23} \\
-41255676x^{22} & +37518228x^{21} & +18524283008x^{17} & +6522421752x^{16} \\
+27295157736x^{15} & +35173338750x^{14} & -2894923232432x^{11} & +489571380144x^{10} \\
-4925879415072x^{9} & +3933790086996x^{8} & -702099623709x^{7} & +149674336745472x^{5} \\
-96219216479232x^{4} & -25773004414080x^{3} & +21354775085952x^{2} & +946763427456x \\
-1217267263872 & & &
\end{array}$$

The polynomial $f_3(x)$ factors as a product of two irreducible polynomials (over $\mathbb{Q}$) $g(x)$ and $h(x)$. We have $f_3(x) = g(x) \cdot h(x)$ where

$$g(x) = x^7 - 231x^3 - 462x^2 + 77x + 66$$

and

$$\begin{array}{rrrr}
h(x) = x^{28} & +231x^{24} & +462x^{23} & -6237x^{22} \\
+29832x^{21} & +53361x^{20} & +213444x^{19} & -1245090x^{18} \\
+3958878x^{17} & -11719092x^{16} & +30817248x^{15} & -157564143x^{14} \\
+319312224x^{13} & -796323990x^{12} & +1481906118x^{11} & -2994381313x^{10} \\
+5889443406x^{9} & +965064177x^{8} & -4595839182x^{7} & +33180883659x^{6} \\
-84492127566x^{5} & +181691003340x^{4} & -382065796728x^{3} & +152613801648x^{2} \\
+35862251040x & -18443443392 & &
\end{array}$$

The Galois Group of the polynomial $f(x) = x^7 - 154x + 99$

Theorem 5.1. *The Galois group of the polynomial* $f(x) = x^7 - 154x + 99$ *over the field* $\mathbb{Q}$ *of rational numbers is a subgroup of the alternating group* A_7 *and has order* 168.

Proof. Using Mathematica we have

$$\text{Discriminant}(f(x)) = (-1)^{21}\ \text{Resultant}(f(x), f'(x)) =$$
$$95,065,918,010,223,428,961 = 3^6 \cdot 7^8 \cdot 11^6 \cdot 113^2.$$

Note.

This discriminant can also be calculated by using the formula of the discriminant of the trinomial $x^n + ax + b$ discussed earlier. As we saw the discriminant of the polynomial $x^n + ax + b$ is given by the expression:

$$(-1)^{\frac{n(n-1)}{2}} \left[n^n b^{n-1} + (-1)^{n+1}(n-1)^{n-1}a^n\right]$$

If we substitute $n = 7$, $a = -154$, and $b = 99$ in the above expression we get the same value for the discriminant of $f(x)$ as given above.

Now we shall proceed to calculate the order of the Galois group of the polynomial $f(x) = x^7 - 154x + 99$. $\mathbb{Q}$. We first enumerate the observations (and its consequences) we have made about the polynomial $f(x)$ sofar.

(1) $f(x)$ is irreducible over $\mathbb{Q}$.

(2) Let $\alpha_i \in \mathbb{C}$, $1 \leq i \leq 7$ be the roots of $f(x)$ and $\mathbb{K} = \mathbb{Q}(\alpha_1, \alpha_2, \cdots, \alpha_7)$ be its splitting field. Let $G = Gal(f) = Aut(\mathbb{K}/\mathbb{Q})$. Then G is transitive on the set $\{\alpha_1, \alpha_2, \cdots, \alpha_7\}$.

(3) As $Disc(x^7 - 154 + 99) = 3^6 \cdot 7^8 \cdot 11^6 \cdot 113^2$ is a square of a rational number $G = Gal(f)$ is a subgroup of the alternating group A_7.

(4) The polynomial $f_2(x) = x^{21} + 3850x^{15} - 5643x^{14} - 125694x^9 + 457380x^8 - 2832489x^7 + 9811128x^3 + 63392868x^2 + 13584186x + 970299$ is irreducible over $\mathbb{Q}$ and so the group G is transitive on 21 letters and therefore its order $|G|$ is a multiple of 21.

(5) Since $f_3(x)$ is a product of two different irreducible polynomials one of degree 28 and other of degree 7, the group G *is not* 3−transitive on seven letters and so $G \neq A_7$.

So G is a proper subgroup of A_7. Since the roots of the irreducible polynomial $h(x)$ are algebraic of degree 28 $[\mathbb{K} : \mathbb{Q}]$ is divisible by 28. It follows that the order of the group G is a multiple of 28 and therefore $|G|$ a multiple of $LCM(21, 28) = 84$.

Let $P \subset G$ be a subgroup of order seven, the Sylow-7 subgroup of G, and therefore of the alternating group A_7. We know that the normalizer of P in A_7 has order 21. So the order of normalizer of P in G has order 21 or 7. But latter is not possible as by Burnside's N/C Theorem it would imply that G would have a normal 7-complement which is not possible (as a normal subgroup of a primitive group must be transitive). Hence $N_G(P)$ (the normalizer of the P in G) has order 21. Therefore it has index 120 in A_7. The only way 120 factors as a product of two numbers, none of which is equal to 1, and each congruent to 1 (modulo 7) is $120 = 8 \times 15$. So the order of G is either $21 \times 8 = 168$ or $21 \times 15 = 315$. Since the order of G is a multiple of 84 we conclude that its order must be 168. This completes the verification of the fact that the Galois group of the polynomial $f(x) = x^7 - 154x + 99$ has order 168. □

Note 1. In fact it can be shown that this group of order 168 is a simple group and is isomorphic to $GL(3, 2)$—the group of 3×3 nonsingular matrices over the field $\mathbb{F}_2$.

Note 2. The Galois group of the polynomial $f = x^7 - 7x + 3$ (known as Trink polynomial) can be likewise shown to be of order 168 and isomorphic to $GL(3, 2)$. If we begin with the Trink polynomial $f = x^7 - 7x + 3$ and carry out the construction for f_2 and f_3 as we did for the polynomial $x^7 - 154 + 99$ then we get the following values of f_2 and f_3.

$$f_2 = x^{21} + 175x^{15} - 171x^{14} - 2597x^9 + 630x^8 - 2601x^7 + 9261x^3 + 3969x^2 + 567x + 27.$$

$f_3 = (x^7 + 14x^4 - 42x^2 - 21x + 9) \times (x^{28} - 14x^{25} + 42x^{23} - 63x^{22} + 897x^{21} - 1176x^{20} + 588x^{19} - 10668x^{18} + 14700x^{17} - 51345x^{16} + 62538x^{15} - 179421x^{14} + 293706x^{13} - 487452x^{12} + 760116x^{11} - 956137x^{10} + 1472415x^9 - 1497363x^8 + 1674299x^7 - 1568784x^6 + 1154832x^5 - 876624x^4 - 7056x^3 + 110880x^2 - 4032x - 3456)$.

CHAPTER 7

More Galois Groups

1. Polynomials with Cyclic Galois Group over the $\mathbb{Q}$

Let $n > 1$ be an integer and $\alpha = e^{\frac{2\pi i}{n}} = \cos\frac{2\pi}{n} + i\sin\frac{2\pi}{n}$ be a primitive n-th root of unity. Let $\mathbb{K} = \mathbb{Q}(\alpha)$. Then $\mathbb{K}$ is the splitting field of the polynomial $x^n - 1$. Let $G = Aut(\mathbb{K}/\mathbb{Q})$. Then G has order $\phi(n)$ and is isomorphic to the (multiplicative) group $\mathbb{Z}_n^*$ of the residue classes modulo n corresponding to the integers relatively prime to n. It is known that $\mathbb{Z}_n^*$ is cyclic only when $n = 1, 2, 4, p^k, 2p^k$ where p is an odd prime. In particular, for these value of n the Galois groups of the cyclotomic polynomial Φ_n is cyclic of order $\phi(n)$. This raises the question of determining polynomials whose Galois group is cyclic of a given arbitrary order. This determination involves what are known as Gaussian sums first discussed by Gauss. We shall describe a method of determining irreducible polynomials of degree $m > 1$ whose Galois group is cyclic of order m.

Let $m > 1$ be an integer. Let p be a prime such that $p \equiv 1$ (modulo m). The existence of such a prime is ensured by Dirichlet's Theorem on Primes in Arithmetical Progression. Let $d = \frac{p-1}{m}$. If $d = 1$ then $m = p - 1$, and

$$f(x) = \frac{x^p - 1}{x - 1} = x^{p-1} + x^{p-2} + \cdots + x + 1 = x^m + x^{m-1} + \cdots x + 1$$

is an irreducible polynomial with cyclic Galois group of order $m = p - 1$. We now assume that $d > 1$. Let $\alpha = e^{\frac{2\pi i}{p}}$ be a primitive p-th root of unity. Let $w \in \{1, 2, \cdots, p-1\}$ be of order d (modulo p) and $r \in \{1, 2, \cdots, p-1\}$ be a primitive root modulo p (i.e., r has order $p - 1$ modulo p). Consider the polynomial

$$x^{p-1} + x^{p-2} + \cdots + x + 1$$

Its Galois group over $\mathbb{Q}$ is cyclic or order $p-1$ and can be generated by the automorphism

$$\sigma : \mathbb{Q}[\alpha] \longrightarrow \mathbb{Q}[\alpha]$$

such that $\sigma(\alpha) = \alpha^r$. Let

$$\beta = \beta_0 = \alpha + \alpha^w + \alpha^{w^2} + \cdots\cdots + \alpha^{w^{d-1}}$$

Then β is algebraic of degree precisely m. Its m conjugate being $\beta_i = \sigma^i(\beta), \quad 0 \le i \le m-1$. Then the polynomial

$$g(x) = \prod_{i=0}^{m-1} (x - \beta_i) \in \mathbb{Z}[x]$$

is an irreducible polynomial over $\mathbb{Q}$ of degree m whose Galois group is cyclic or order m. We have the following algorithmic description of the above procedure:

ALGORITHM

Given an integer $m > 1$, this algorithm determines the coefficients a_i, $1 \le i \le m$ of an irreducible polynomial of degree m

$$x^m + a_1 x^{m-1} + a_2 x^{m-2} + \cdots + a_{m-1} x + a_m$$

whose Galois group over $\mathbb{Q}$ is cyclic or order m.

- **Step 1.** Find a prime p, such that $p \equiv 1$ (modulo m).
- **Step 2.** If $p = m+1$, then $a_i = 1, \quad 1 \le i \le m$. Stop. Otherwise go to next step.
- **Step 3.** Let $d = \frac{p-1}{m}$ and $\alpha = e^{\frac{2\pi i}{p}}$.
- **Step 4.** Let r be the least primitive root modulo p.
- **Step 5.** Choose least $w \in \{1, 2, \cdots, p-1\}$ such that its has order d (modulo p).
- **Step 6.** Let $C = [c_{ij}]$ be an $m \times d$ matrix with entries in $\mathbb{F}_p^* = GF(p)^*$ where

$$c_{ij} \equiv r^i w^j \text{ (modulo } p), \quad 0 \le i \le m-1, \quad 0 \le j \le d-1.$$

- **Step 7.** Let

$$\beta_i = \sum_{j=0}^{d-1} \alpha^{c_{ij}}, \qquad 0 \le i \le m-1.$$

Then

$$g(x) = \prod_{i=0}^{m-1} (x - \beta_i)$$

is the required polynomial of degree m with cyclic Galois group (of order m).

We shall now give some examples implementing the above algorithm for various values of m.

- **Example 1.** $m = 7$

In this case we have $p = 29$, $d = 4$, $r = 2$, $w = 12$, and $\alpha = e^{\frac{2\pi i}{29}}$. We have $C = [c_{ij}]$ as the following 7×4 matrix. (Here $0 \le i \le 6$ and $0 \le j \le 3$.)

$$C = \begin{bmatrix} 1 & 12 & 28 & 17 \\ 2 & 24 & 27 & 5 \\ 4 & 19 & 25 & 10 \\ 8 & 9 & 21 & 20 \\ 16 & 18 & 13 & 11 \\ 3 & 7 & 26 & 22 \\ 6 & 14 & 23 & 15 \end{bmatrix}.$$

This gives the following values of zeros β_i, $0 \le i \le 6$, of the degree seven irreducible polynomial that we wish to find.

$$\begin{aligned}
\beta_0 &= \alpha + \alpha^{12} + \alpha^{28} + \alpha^{17} \\
\beta_1 &= \alpha^2 + \alpha^{24} + \alpha^{27} + \alpha^5 \\
\beta_2 &= \alpha^4 + \alpha^{19} + \alpha^{25} + \alpha^{10} \\
\beta_3 &= \alpha^8 + \alpha^9 + \alpha^{21} + \alpha^{20} \\
\beta_4 &= \alpha^{16} + \alpha^{18} + \alpha^{13} + \alpha^{11} \\
\beta_5 &= \alpha^3 + \alpha^7 + \alpha^{26} + \alpha^{22} \\
\beta_6 &= \alpha^6 + \alpha^{14} + \alpha^{23} + \alpha^{15}.
\end{aligned}$$

So

$$g(x) = \prod_{i=0}^{6}(x - \beta_i) = x^7 + x^6 - 12x^5 - 7x^4 + 28x^3 + 14x^2 - 9x + 1$$

has cyclic group of order 7 as its Galois group over $\mathbb{Q}$.

- **Example 2.** $m = 13$

In this case we have $p = 53$, $d = 4$, $r = 2$, $w = 23$, and $\alpha = e^{\frac{2\pi i}{53}}$. We have C as the following 13×4 matrix:

$$C = \begin{bmatrix} 1 & 23 & 52 & 30 \\ 2 & 46 & 51 & 7 \\ 4 & 39 & 49 & 14 \\ 8 & 25 & 45 & 28 \\ 16 & 50 & 37 & 3 \\ 32 & 47 & 21 & 6 \\ 11 & 41 & 42 & 12 \\ 22 & 29 & 31 & 24 \\ 44 & 5 & 9 & 48 \\ 35 & 10 & 18 & 43 \\ 17 & 20 & 36 & 33 \\ 34 & 40 & 19 & 13 \\ 15 & 27 & 38 & 36 \end{bmatrix}.$$

This gives the following values of zeros β_i, $0 \leq i \leq 12$. of the degree thirteen irreducible polynomial that we wish to find.

$$\begin{aligned}
\beta_0 &= \alpha + \alpha^{23} + \alpha^{52} + \alpha^{30} \\
\beta_1 &= \alpha^2 + \alpha^{46} + \alpha^{51} + \alpha^7 \\
\beta_2 &= \alpha^4 + \alpha^{39} + \alpha^{49} + \alpha^{14} \\
\beta_3 &= \alpha^8 + \alpha^{25} + \alpha^{45} + \alpha^{28} \\
\beta_4 &= \alpha^{16} + \alpha^{50} + \alpha^{37} + \alpha^3 \\
\beta_5 &= \alpha^{32} + \alpha^{47} + \alpha^{21} + \alpha^6 \\
\beta_6 &= \alpha^{11} + \alpha^{41} + \alpha^{42} + \alpha^{12} \\
\beta_7 &= \alpha^{22} + \alpha^{29} + \alpha^{31} + \alpha^{24} \\
\beta_8 &= \alpha^{44} + \alpha^5 + \alpha^9 + \alpha^{48} \\
\beta_9 &= \alpha^{35} + \alpha^{10} + \alpha^{18} + \alpha^{43} \\
\beta_{10} &= \alpha^{17} + \alpha^{20} + \alpha^{36} + \alpha^{33} \\
\beta_{11} &= \alpha^{34} + \alpha^{40} + \alpha^{19} + \alpha^{13} \\
\beta_{12} &= \alpha^{15} + \alpha^{27} + \alpha^{38} + \alpha^{26}
\end{aligned}$$

So

$$g(x) = \prod_{i=0}^{12}(x - \beta_i) = x^{13} + x^{12} - 24x^{11} - 19x^{10} + 190x^9 + 116x^8 - 601x^7 - 246x^6 + 738x^5 + 215x^4 - 291x^3 - 68x^2 + 10x + 1$$

has cyclic group of order 13 as its Galois group over $\mathbb{Q}$. In the table on the following pages we give for each m, $1 < m \leq 20$, such that $m + 1$ is not a prime number, the polynomial of degree m as given by the preceding algorithm. As there are infinitely many irreducible polynomials of degree m with cyclic Galois group of order m, we need to emphasize that the polynomials given in our table are the ones that are derived with the help of the algorithm that we have just described by choosing the *least prime* $p \equiv 1$ (modulo m).

Polynomials with Cyclic Galois Group over $\mathbb{Q}$

Degree	Polynomial
3	$x^3 + x^2 - 2x - 1$
5	$x^5 + x^4 - 4x^3 - 3x^2 + 3x + 1$
7	$x^7 + x^6 - 12x^5 - 7x^4 + 28x^3 + 14x^2 - 9x + 1$
8	$x^8 + x^7 - 7x^6 - 6x^5 + 15x^4 + 10x^3 - 10x^2 - 4x + 1$
9	$x^9 + x^8 - 8x^7 - 7x^6 + 21x^5 + 15x^4 - 20x^3 - 10x^2 + 5x + 1$
11	$x^{11} + x^{10} - 10x^9 - 9x^8 + 36x^7 + 28x^6 - 56x^5 - 35x^4 + 35x^3 + 15x^2 - 6x - 1$
13	$x^{13} + x^{12} - 24x^{11} - 19x^{10} + 190x^9 + 116x^8 - 601x^7 - 246x^6 + 738x^5 + 215x^4 - 291x^3 - 68x^2 + 10x + 1$
14	$x^{14} + x^{13} - 13x^{12} - 12x^{11} + 66x^{10} + 55x^9 - 165x^8 - 120x^7 + 212x^6 + 126x^5 - 126x^4 - 56x^3 + 28x^2 + 7x - 1$
15	$x^{15} + x^{14} - 14x^{13} - 13x^{12} + 78x^{11} + 66x^{10} - 220x^9 - 165x^8 + 330x^7 + 210x^6 - 252x^5 - 126x^4 + 84x^3 + 28x^2 - 8x - 1$
17	$x^{17} + x^{16} - 48x^{15} - 105x^{14} + 763x^{13} + 2579x^{12} - 3653x^{11} - 2331x^{10} - 11031x^9 + 74838x^8 + 107759x^7 - 50288x^6 - 198615x^5 - 102976x^4 + 58507x^3 + 75722x^2 + 25763x + 2837$
19	$x^{19} + x^{18} - 90x^{17} - 57x^{16} + 3044x^{15} + 1124x^{14} - 51184x^{13} - 4822x^{12} + 474003x^{11} - 91110x^{10} - 2465084x^9 + 1153239x^8 + 6854098x^7 - 5023125x^6 - 8711114x^5 + 8950277x^4 + 2600136x^3 - 5125792x^2 + 1553447x - 117649$
20	$x^{20} + x^{19} - 19x^{18} - 18x^{17} + 153x^{16} + 136x^{15} - 680x^{14} - 560x^{13} + 180x^{12} + 1365x^{11} - 3003x^{10} - 2002x^9 + 3003x^8 + 1716x^7 - 1716x^6 - 792x^5 + 495x^4 + 165x^3 - 55x^3 - 55x^2 - 10x + 1$

2. Polynomials with the Symmetric Group S_n as a Galois Group

We begin this section by stating an elementary but useful result from the theory of permutation group. For proof we refer the readers to a book by Wielandt (1964).

> Let G be a subgroup of S_n, the symmetric group on n letters. If G is 2−transitive and contains a transposition then $G = S_n$.

In fact the above result is true under a weaker condition namely under the assumption that if G is primitive (rather than 2−transitive) and has a transposition then $G = S_n$. It is known that a transitive permutation group of prime degree is primitive. So if G is a transitive group of prime degree p and contains a transposition, then $G = S_p$. Therefore if $f(x) \in \mathbb{Q}[x]$ is an irreducible polynomial of degree p (a prime number) having exactly two nonreal roots then its Galois group is isomorphic to S_p, the symmetric group on p letters. We shall use these (and some more) facts about permutation groups to find polynomials with Galois group S_n. We again recall the following useful result from the book *Modern Algebra* by Van Der Waerden (1953).

> Let $f(x) \in \mathbb{Z}[x]$ be a monic polynomial of degree n and p be a prime number which does not divide the discriminant of $f(x)$. Let $\alpha_1, \alpha_2, \cdots, \alpha_n$ denote the zeros of $f(x)$ in some extension of the field $\mathbb{Q}$. Let $\bar{f}(x) \in \mathbb{Z}_p[x]$ denote the polynomial obtained by reducing the coefficients of $f(x)$ modulo p. Suppose that in $\mathbb{Z}_p[x]$, $\bar{f}(x)$ factors as a product of irreducible polynomials of degree $n_1, n_2, \cdots n_k$, where $n_1 + n_2 + \cdots + n_k = n$. Let $G = Gal(f)$ be the Galois group of $f(x)$ over $\mathbb{Q}$. Then the group G, considered as a permutation group on the set $\{\alpha_1, \alpha_2, \cdots, \alpha_n\}$ has an element which when expressed as a product of disjoint cycles is of type $n_1 \cdot n_2 \cdot n_3 \cdot \cdot n_k$ (that is it is a product of disjoint cycles of lengths $n_1, n_2, n_3, \cdots, n_k$).

The above result is attributed to Dedekind. We shall give a proof of this due to Tate in one of the appendix.

Suppose that we are able to find a monic polynomial $f(x) \in \mathbb{Z}[x]$ satisfying the following conditions

(i) $f(x)$ is irreducible modulo 2. That is it is irreducible considered as a polynomial in $\mathbb{Z}_2[x]$.

(ii) Modulo 3 the polynomial $f(x)$ factors as a product of a linear polynomials and an irreducible polynomial of degree $n-1$.

(iii) Modulo 5 the polynomial $f(x)$ factors a product of an irreducible polynomial of degree two and each of the rest of the irreducible factors are of odd degree.

Let $\mathbb{K}$ be the splitting field of such a polynomial $f(x) \in \mathbb{Z}[x]$ and $G = Gal(f) = Aut(\mathbb{K}/\mathbb{Q})$. The condition (i) above implies that $f(x)$ is irreducible (over $\mathbb{Q}$) and G is transitive on the set $\{\alpha_1, \alpha_2, \cdots, \alpha_n\}$ of the zeros of $f(x)$. The condition (ii) then implies that G is not only transitive but 2−transitive on $\{\alpha_1, \alpha_2, \cdots, \alpha_n\}$. The condition (iii) ensures that the group G contains a transposition. Therefore by preceding discussion G is isomorphic to the symmetric group S_n.

Example 1. In this example we shall construct a polynomial of degree four with Galois group isomorphic to S_4—the symmetric group on four letters. Let

$g_1(x) = x^4 + x + 1$

$g_2(x) = (x-1)(x^3 + x^2 + 2) = x^4 - x^2 + 2x - 2 \equiv x^4 + 2x^2 + 2x + 1$ (modulo 3)

$g_3(x) = (x^2 - 2)(x-1)(x-2) \equiv x^4 - 3x^3 + 6x - 4 \equiv x^4 + 2x^3 + x + 1$ (modulo 5)

We note that $g_1(x)$ is irreducible (modulo 2), $g_2(x)$ is a product of a linear factor and an irreducible polynomial of degree three (modulo 3), and $g_3(x)$ is a product of an irreducible quadratic polynomials and two linear polynomials (modulo 5). Let

$$f(x) = -15g_1(x) + 10g_2(x) + 6g_3(x) = x^4 + 12x^3 + 20x^2 + 11x + 1$$

Since $f(x) \equiv g_1(x)$ (modulo 2), $f(x) \equiv g_2(x)$ (modulo 3), and $f(x) \equiv g_3(x)$ (modulo 3), the polynomial $f(x) \in \mathbb{Q}[x]$ is irreducible and its Galois group G is transitive of degree four containing a transposition and a three cycle. So its order is divisible by 12 but it is not contained in the alternating group A_4 (as G contains a transposition), and is therefore isomorphic to the symmetric group S_4.

Example 2. In this example we shall construct a polynomial of degree five with Galois group isomorphic to S_5—the symmetric group on five letters. Let

$g_1(x) = x^5 + x^2 + 1$

$g_2(x) = (x^2 + 1)(x^3 + x^2 + 2) \equiv x^5 + x^4 + x^3 + 3x^2 + 2 \equiv x^5 + x^4 + x^3 + 2$ (modulo 3)

We note that $g_1(x)$ is irreducible (modulo 2) and $g_2(x)$ is a product of an irreducible degree two polynomial and an irreducible degree three polynomial (modulo 3).

$$f(x) = 3g_1(x) - 2g_2(x) = x^5 - 2x^4 - 2x^3 + 3x^2 - 1$$

Since $f(x) \equiv g_1(x)$ (modulo 2) and $f(x) \equiv g_2(x)$ (modulo 3) the Galois group G of $f(x)$ is transitive of prime degree five and contains a transposition. It is therfore a primitive group of prime degree containing a transposition and is therefore isomorphic to the symmetric group S_5.

If $p > 3$ is a prime number then there is another interesting method to construct polynomials with Galois group S_p, the symmetric group on p letters. This is due to Richard Brauer and our description of it is essentially the same as given by Jacobson (2014).

We shall show that for any odd integer $N > 1$ there exist monic irreducible polynomial $f(x) \in \mathbb{Z}[x]$ of degree N having exactly two nonreal zeros—that is just one pair of complex conjugate roots.

For $N = 3$, the polynomial $x^3 - 2$ has desired property so we assume that $N > 3$. Let $k = \frac{N-3}{2}$ and $a_i = 2i$ for $1 \leq i \leq k$. and

$$g(x) = x\prod_{i=1}^{k}(x \pm a_i) = x(x^2 - a_1^2)(x^2 - a_2^2)\cdots(x^2 - a_k^2)$$

be a polynomial of degree $N - 2$ having $N - 2$ real zeros, namely $x = 0$ and $x = \pm a_i, 1 \leq i \leq k$. Let $h(x) = (x^2 + m) \cdot g(x)$, where m is a positive even integer to be specified later. The polynomial $h(x)$ has $N - 2$ real zeros. We observe that between two successive zeros of the function $h(x)$ the maximum value of $|h(x)|$ exceeds 2. As $f(-a_k) = 2$ and $f(-\infty) = -\infty$, the polynomial $f(x)$ has one (real) root in $(-\infty, -a_k)$, it follows that the polynomial $f(x) = h(x) + 2$ has at least $N - 2$ real zeros. We shall show that for sufficiently large m the polynomial $f(x)$ has just $N - 2$ real zeros (and therefore just two nonreal zeros). Suppose that the N zeros of the polynomial $f(x)$ are $\alpha_i, 1 \leq i \leq N$. Then we have

$$f(x) = h(x) + 2 = x(x^2 + m)\prod_{i=1}^{k}(x^2 - a_i^2) + 2 = \prod_{i=1}^{N}(x - \alpha_i).$$

Since the coefficients of the even powers of x in $f(x)$ (except for the constant term) are zero, and the coefficient of x^{N-2} is $m - \sum_{i=1}^{k} a_i^2$, we have

$$\sum_{i=1}^{i=N} \alpha_i = 0, \qquad \text{and} \qquad \sum_{i \neq j} \alpha_i \alpha_j = m - \sum_{i=1}^{i=k} a_i^2.$$

Therefore

$$\sum_{i=1}^{k} \alpha_i^2 = \left(\sum_{i=1}^{k} \alpha_i\right)^2 - 2\sum_{i \neq j} \alpha_i \alpha_j = -2\left(m - \sum_{i=1}^{k} a_i^2\right) = -2m + 2\sum_{i=1}^{k} a_i^2 < 0$$

for sufficiently large m. Therefore not all $\alpha_i' s$ can be real. But we know that at least $N-2$ $\alpha_i' s$ are real so the polynomial $f(x)$ has exactly $N-2$ real zeros and two nonreal zeros. As we have chosen m to be even the polynomial $f(x)$ is irreducible by Eisenstein's Criterion as applied for prime 2. Therefore $f(x)$ is an irreducible polynomial of degree N having exactly two nonreal roots.

In all the cases that we have tried we have found that $m = 2$ gives the polynomial with desired properties. Perhaps it is true that $m = 2$ works in all cases.

For $N = 5$ and $m = 2$ the polynomials $f(x)$ is $x^5 - 2x^3 - 8x + 2$ and for $N = 7$ and $m = 2$ the polynomial $f(x)$ is $x^7 - 18x^5 + 24x^3 + 128x + 2$. The polynomial $x^5 - 2x^3 - 8x + 2$ has three real zeros and two nonreal zeros. So its Galois group is transitive of degree 5 and has a transposition and so is isomorphic to S_5—the symmetric group on five letters. Likewise the Galois group of the polynomial $x^7 - 18x^5 + 24x^3 + 128x + 2$ is isomorphic to the S_7—the symmetric group on seven letters. So a given any odd prime $p > 3$, the above procedure can be used to construct polynomials in $\mathbb{Z}[x]$ of degree p having S_p—the symmetric group on p letters as its Galois group.

Remark. It is known that the Galois group of the polynomial $x^n - x - 1$ over $\mathbb{Q}$ is the symmetric group S_n for all $n \geq 2$. See the paper by Osada (1987) for details.

CHAPTER 8

Transcendence of e and π

In this chapter we shall show that e and π are transcendental numbers. In other words, we shall prove that there is no nonzero polynomial with rational coefficients having e or π as one of its zeros. Our treatment of these topics is similar (with slight modification) to the one given in the text *Field Theory and Its Classical Problems* by Hadlock (1978) and *Galois Theory* by Stewart (1989).

1. Transcendence of e

Theorem 1.1. *The number $e = lim_{n\longrightarrow\infty}(1+\frac{1}{n})^n$, the base of natural logarithm, is transcendental.*

Lemma 1. $\lim_{n\to\infty} \dfrac{x^n}{n!} = 0.$

Proof. Without the loss of generality we may assume that $x > 0$. If $0 < x \leq 1$ then the lemma is obvious. So we may assume that $x > 1$. Let N be a integer such that $N \leq x < N+1$. Suppose $n > N$ and let $n = N+k$. Then we have

$$\lim_{n\to\infty} \frac{x^n}{n!} = \lim_{k\to\infty} \frac{x^{N+k}}{(N+k)!} = \frac{x^N}{N!} \cdot \lim_{k\to\infty} \frac{x^k}{(N+1)(N+2)\cdots(N+k)}$$

Since $N \leq x < N+1$, for $1 \leq i \leq k$ we have $\dfrac{x}{N+i} < \dfrac{x}{N}$. Therefore

$$\frac{x^k}{(N+1)(N+2)\cdots(N+k)} < \left(\frac{x}{N}\right)^k$$

and since $0 < \dfrac{x}{N} < 1$ we have

$$\lim_{k\to\infty} \frac{x^k}{(N+1)(N+2)\cdots(N+k)} \leq \lim_{k\to\infty} \left(\frac{x}{N}\right)^k = 0.$$

Hence $\lim_{n\to\infty} \dfrac{x^n}{n!} = 0$, and the proof of the lemma is complete.

Lemma 2. Let m and p be positive integers and $p > 1$. Let

$$f(x) = \frac{1}{(p-1)!} \cdot x^{p-1}(x-1)^p(x-2)^p \cdots (x-m)^p = \frac{1}{(p-1)!} x^{p-1}\left[(x-1)(x-2)\cdots(x-m)\right]^p$$

be a polynomial of degree $mp + p - 1$. Then for an integer j, $0 \le j \le m$, we have

$$f^{(n)}(j) = \begin{cases} (-1)^{mp}(m!)^p & \text{if } n = p-1, \text{ and } \quad j = 0 \\ \text{a multiple of } p \text{ otherwise} \end{cases}$$

where $f^{(n)}(j)$, denotes the nth derivative of $f(x)$ evaluated at j.

Note: In what follows we shall require this lemma only when p is a large prime number. However the proof of the lemma does not require the primality of p.

Proof.

Case 1. $j \neq 0$.

In this case we write $f(x) = \dfrac{1}{(p-1)!}(x-j)^p \cdot g(x)$, where

$$g(x) = \frac{x^{p-1}(x-1)^p \cdots (x-j)^p \cdots (x-m)^p}{(x-j)^p} = x^{p-1}(x-1)^p \cdots \widehat{(x-j)^p} \cdots (x-m)^p$$

where $\widehat{(x-j)^p}$ indicates missing term in the product. Using the Leibnitz's rule for calculating the n-th derivative of the product of two functions we have

$$f^{(n)}(x) = \frac{1}{(p-1)!} \sum_{r=0}^{n} {}_nC_r \frac{d^r}{dx^r}(x-j)^p \cdot \frac{d^{n-r}}{dx^{n-r}}(g(x)).$$

We need to find the value at $x = j$ of the following

$$\frac{d^r}{dx^r}(x-j)^p \cdot \frac{d^{n-r}}{dx^{n-r}}(g(x)).$$

It is clear that if $n < p$ then $r < p$ and therefore $f^{(n)}(j) = 0$. Now suppose $n \ge p$. The above product will take nonzero value only when $r = p$. For, if $r \neq p$ the r-th derivative $\dfrac{d^r}{dx^r}(x-j)^p$ is either zero or has $(x-j)$ as a factor. If $r = p$, we need to evaluate the value of $\dfrac{d^p}{dx^p}(x-j)^p \cdot \dfrac{d^{n-p}}{dx^{n-p}}(g(x))$ at $x = j$. This value is

$$p!\left[\frac{d^{n-p}}{dx^{n-p}}(g(x))\right]_{x=j} = p!\, t, \qquad \text{for some } t \in \mathbb{Z}$$

Therefore for $j \neq 0$ we have

$$f^{(n)}(j) = \frac{1}{(p-1)!} {}_nC_p \cdot p! \cdot t = {}_nC_p \cdot p\, t \equiv 0 \text{ (modulo } p).$$

This proves that if $j \neq 0$, then $f^{(n)}(j)$ is a multiple of p.

Case 2. $j = 0$. In this case we write

$$f(x) = \frac{1}{(p-1)!} x^{p-1} \cdot h(x)$$

where $h(x) = (x-1)^p (x-2)^p \cdots (x-m)^p$. Again by using Leibnitz rule we have

$$f^{(n)}(x) = \frac{1}{(p-1)!} \sum_{r=0}^{n} {}_nC_r \frac{d^r}{dx^r}(x^{p-1}) \cdot \frac{d^{n-r}}{dx^{n-r}}(h(x))$$

Evidently at $x = 0$ the r-th derivative $\dfrac{d^r}{dx^r}(x^{p-1})$ takes the value zero unless $r = p-1$, when its value is $(p-1)!$. In particular, if $n < p-1$ then $f^{(n)}(0) = 0$. So for $n \geq p-1$ we have

$$f^{(n)}(0) = \frac{1}{(p-1)!}\ {}_nC_{p-1} \cdot (p-1)! \left(\frac{d^{n-p+1}}{dx^{n-p+1}} h(x) \right)_{x=0} = {}_nC_{p-1} \left(\frac{d^{n-p+1}}{dx^{n-p+1}} h(x) \right)_{x=0}$$

Since $h(x) = (x-1)^p (x-2)^p \cdots (x-m)^p$ we have

$$\left(\frac{d^{n-p+1}}{dx^{n-p+1}} h(x) \right)_{x=0} = \begin{cases} \equiv 0 \text{ (modulo } p) & \text{if} \quad n > p-1 \\ (-1)^{mp} (m!)^p & \text{if} \quad n = p-1. \end{cases}$$

Therefore

$$f^{(n)}(0) = \begin{cases} \equiv 0 \text{ (modulo } p) & \text{if} \quad n > p-1 \\ (-1)^{mp} (m!)^p & \text{if} \quad n = p-1 \end{cases}$$

Combining the conclusions of the two cases above we have

$$f^{(n)}(j) = \begin{cases} \equiv 0 \text{ (modulo } p) & \text{if} \quad j \neq 0 \quad \text{or} \quad n \neq p-1 \\ (-1)^{mp} (m!)^p & \text{if} \quad j = 0 \quad \text{and} \quad n = p-1 \end{cases}$$

This completes the proof of the Lemma 2. Now we proceed to prove the main theorem

Proof. Suppose that e not transcendental. Then there is a polynomial

$$a_0 + a_1x + a_2x^2 + \cdots + a_mx^m \in \mathbb{Z}[x], \qquad a_0 \neq 0$$

of degree $m \geq 1$ having e as one of its zeros. So

$$a_0 + a_1e + a_2e^2 + \cdots + a_me^m = 0.$$

As in Lemma 2, we let

$$f(x) = \frac{1}{(p-1)!} \cdot x^{p-1}(x-1)^p(x-2)^p \cdots (x-m)^p$$

where now we assume that p is a (large) prime number to be specified later. Clearly, degree$(f(x)) = mp + p - 1$. By lemma 2 we have

$$f^{(i)}(j) = \begin{cases} (-1)^{mp}(m!)^p & \text{if} \quad i = p-1 \quad \text{and} \quad j = 0 \\ \equiv 0 \text{ (modulo } p) & \text{otherwise} \end{cases}$$

Let

$$F(x) = f(x) + f^{'}(x) + f^{''}(x) + \cdots + f^{(mp+p-1)}(x).$$

Then

$$F^{'}(x) = f^{'}(x) + f^{''}(x) + \cdots + f^{(mp+p-1)}(x).$$

Since, evidently, $f^{(mp+p)}(x) = 0$, as degree of $f(x) = mp + p - 1$. Therefore $F(x) - F^{'}(x) = f(x)$. We have

$$\frac{d}{dx}\left(e^{-x}F(x)\right) = e^{-x}F^{'}(x) - e^{-x}F(x) = e^{-x}\left(F^{'}(x) - F(x)\right) = -e^{-x}f(x).$$

Therefore

$$\int_0^j -e^{-x}f(x) = \left[e^{-x}F(x)\right]_0^j = e^{-j}F(j) - F(0)$$

$$\int_0^j e^{-x}f(x) = F(0) - e^{-j}F(j)$$

or

$$e^j \cdot \int_0^j e^{-x}f(x) = e^jF(0) - F(j).$$

We multiply the above equality by a_j and sum it for $j = 0$ to $j = m$. Hence we get

$$\sum_{j=0}^{m} a_j e^j \int_0^j e^{-x} f(x) = F(0) \sum_{j=0}^{m} a_j e^j - \sum_{j=0}^{m} a_j F(j) = -\sum_{j=0}^{m} a_j \sum_{i=0}^{mp+p-1} f^{(i)}(j)$$

because $\sum_{j=0}^{m} a_j e^j = 0$ and

$$F(j) = f(j) + f^{'}(j) + \cdots + f^{(mp+p-1)}(j).$$

By Lemma 2 we have

$$\sum_{i=0}^{mp+p-1} f^{(i)}(j) = \begin{cases} \text{a multiple of } p & \text{if} \quad j \neq 0 \\ (-1)^{mp}(m!)^p + tp & \text{if} \quad j = 0 \end{cases}$$

for some integer t. We note that, by Lemma 2, the above sum gives a contribution which is not a multiple of p only when $i = p - 1$ and $j = 0$. So we have

$$\sum_{j=0}^{m} a_j \sum_{i=0}^{mp+p-1} f^{(i)}(j) \equiv a_0 \sum_{i=0}^{mp+p-1} f^{(i)}(0) \equiv a_0(-1)^{mp}(m!)^p \text{ (modulo } p).$$

Thus we can write

$$\sum_{j=0}^{m} a_j e^j \int_0^j e^{-x} f(x) = -\sum_{j=0}^{m} a_j \sum_{i=0}^{mp+p-1} f^{(i)}(j) = kp - a_0(-1)^{mp}(m!)^p.$$

for some integer k. If p is a large prime then the right hand side of the above equality is a nonzero integer. We shall now make an estimate of the left hand side of the preceding equality. Since

$$f(x) = \frac{1}{(p-1)!} \cdot x^{p-1}(x-1)^p(x-2)^p \cdots (x-m)^p$$

for $0 \leq x \leq m$ we have

$$|f(x)| \leq \frac{m^{mp+p-1}}{(p-1)!} \quad \text{and} \quad e^{-x} \leq 1.$$

Therefore for $0 \leq x \leq m$

$$|e^{-x}f(x)| \leq \frac{m^{mp+p-1}}{(p-1)!}.$$

Thus we have

$$\left|\sum_{j=0}^{m} a_j e^j \int_0^j e^{-x} f(x)\ dx\right| \leq \sum_{j=0}^{m} |a_j e^j| \int_0^j |e^{-x} f(x)|\ dx \leq \sum_{j=0}^{m} |a_j e^j| \cdot j \cdot \frac{m^{mp+p-1}}{(p-1)!}.$$

As $m^{mp+p-1} = m^m \cdot (m^{m+1})^{p-1}$, we have by Lemma 1

$$\lim_{p\to\infty} \frac{m^{mp+p-1}}{(p-1)!} = m^m \lim_{p\to\infty} \frac{(m^{m+1})^{p-1}}{(p-1)!} = 0.$$

So by choosing p as a suitably large prime the finite sum

$$\sum_{j=0}^{m} |a_j e^j| \cdot j \cdot \frac{m^{mp+p-1}}{(p-1)!}$$

can be made arbitrarily small. But this contradicts the fact that the sum

$$\sum_{j=0}^{m} a_j e^j \int_0^j e^{-x} f(x) dx = k \cdot p - a_0(-1)^{mp}(m!)^p$$

is a nonzero integer if p is chosen suitably large (say $p >$ Max $\{|a_0|, m!\}$).

So we have a contradiction. As this contradiction arises from our assumption that e is a zero of the polynomial

$$a_0 + a_1 x + \cdots + a_m x^m \in \mathbb{Z}[x]$$

we conclude that there is no such polynomial having e as one of its zeros. Hence e is transcendental. This completes the proof. □

Comment. If we simply wish to show that e is an irrational then there is a very easy way to prove it. Suppose that $e = \dfrac{p}{q}$ where $gcd(p, q) = 1$ and $q > 1$. Since $e = \lim_{n\to\infty}(1 + \frac{1}{n})^n = \sum_{n=0}^{\infty} \frac{1}{n!}$, we have $2 < e < 3$. So

$$\frac{p}{q} = 1 + \frac{1}{1!} + \frac{1}{2!} + \cdots + \frac{1}{q!} + \frac{1}{(q+1)!} + \cdots$$

Multiplying both sides by $q!$ we get

$$p \cdot (q-1)! = q! \left(1 + \frac{1}{1!} + \frac{1}{2!} + \cdots + \frac{1}{q!}\right) + \frac{1}{(q+1)} + \frac{1}{(q+1)(q+2)} + \cdots$$

So

$$p \cdot (q-1)! - q! \left(1 + \frac{1}{1!} + \frac{1}{2!} + \cdots + \frac{1}{q!}\right) = \frac{1}{(q+1)} + \frac{1}{(q+1)(q+2)} + \cdots$$

Evidently the left hand side is an integer. Now as

$$0 < \frac{1}{(q+1)} + \frac{1}{(q+1)(q+2)} + \cdots < \frac{1}{q+1} \cdot \left(1 + \frac{1}{q+1} + \frac{1}{(q+1)^2} + \cdots\right)$$

$$= \frac{1}{q+1} \left(\frac{1}{1 - \frac{1}{q+1}}\right) = \frac{1}{q}$$

the right hand side is a fraction in the open interval $(0, \frac{1}{q})$. This contradiction proves that e is irrational.

2. Transcendence of π

The proof of the transcendence of π is somewhat more involved. For this proof we shall need the **Fundamental Theorem of Symmetric Polynomials** which we state below without proof. We recommend to the readers the proof given in *Higher Algebra* by Kurosh (1980). A nonzero polynomial $f(x_1, x_2, \cdots, x_n) \in \mathbb{F}[x_1, x_2, \cdots, x_n]$ in n unknowns $x_1, x_2, \cdots x_n$ with coefficients in the field $\mathbb{F}$ is said to be a *symmetric polynomial* (in the variables $x_1, x_2, \cdots x_n$) if it remains unchanged for any rearrangement of these variables. In other words the polynomial $f(x_1, x_2, \cdots, x_n)$ is said to be symmetric if it remains unchanged by application of any permutation in the symmetric group S_n to its variables $x_1, x_2, \cdots x_n$. Since the symmetric group S_n is generated by all transpositions (permutations which interchanges two letters and keeps others unchanged) we can also define a symmetric polynomial as a polynomial $f(x_1, x_2, \cdots, x_n) \in \mathbb{F}[x_1, x_2, \cdots, x_n]$ which remains unchanged by all exchanges of any two of its variables. The following n symmetric polynomials are known as *Elementary Symmetric Polynomials*

$$\begin{aligned}
s_1 &= x_1 + x_2 + x_3 + \cdots + x_n \\
s_2 &= x_1x_2 + x_1x_2 + \cdots + x_{n-1}x_n \\
\vdots &= \vdots \qquad\qquad \vdots \\
s_n &= x_1x_2x_3\cdots x_n
\end{aligned}$$

We recall that with the help of Vieta's formulas we can express the elementary symmetric functions of the roots of a polynomials in terms of the coefficients of a polynomial. That is if

$$f(x) = a_0x^n + a_1x^{n-1} + \cdots\cdots + a_{n-1}x + a_n = a_0\prod_{i=1}^{i=n}(x-\alpha_i)$$

then

$$\begin{aligned}
\alpha_1 + \alpha_2 + \alpha_3 + \cdots + \alpha_n &= -\frac{a_1}{a_0} \\
\alpha_1\alpha_2 + \alpha_1\alpha_2 + \cdots + \alpha_{n-1}\alpha_n &= \frac{a_2}{a_0} \\
\vdots &= \vdots \qquad \vdots \\
\alpha_1\alpha_2\alpha_3\cdots\alpha_i + \cdots\cdots + \alpha_{n-i+1}\cdots\alpha_{n-1}\alpha_n &= (-1)^i\frac{a_i}{a_0} \\
\vdots &= \vdots \qquad \vdots \\
\alpha_1\alpha_2\alpha_3\cdots\alpha_n &= (-1)^n\frac{a_n}{a_0}
\end{aligned}$$

Theorem 2.1. (Fundamental Theorem of Symmetric Polynomials)
Let $f(x_1, x_2, \cdots, x_n) \in \mathbb{F}[x_1, x_2, \cdots, x_n]$ be a nonzero symmetric polynomial of degree m in n variables $x_1, x_2, \cdots x_n$ with coefficients in the field $\mathbb{F}$. Then

$$f(x_1, x_2, \cdots, x_n) = g(s_1, s_2, \cdots, s_n) \in \mathbb{F}[s_1, s_2, \cdots, s_n]$$

where s_i, $1 \leq i \leq n$ are elementary symmetric polynomials in $x_1, x_2, \cdots x_n$, namely

$$\begin{aligned} s_1 &= x_1 + x_2 + x_3 + \cdots + x_n \\ s_2 &= x_1x_2 + x_1x_2 + \cdots + x_{n-1}x_n \\ \vdots &= \vdots \qquad\quad \vdots \\ s_n &= x_1x_2x_3 \cdots x_n \end{aligned}$$

i.e., s_k is the sum of the products of k $x_i^{'s}$ (so it is the sum of ${}_nC_k$ terms).

Proof. For a proof see Kurosh (1980). □

Here are some examples of expressing a symmetric polynomials in $\mathbb{Z}[x_1, x_2, \cdots, x_n]$ as a polynomial in $\mathbb{Z}[s_1, s_2, \cdots, s_n]$

$$\begin{aligned} x_1^2 + x_2^2 &= s_1^2 - 2s_2 \\ x_1^3 + x_2^3 + x_3^3 &= s_1^3 - 3s_1s_2 + 3s_1 \\ x_1^2x_2 + x_1^2x_3 + x_2^2x_1 + x_2^2x_3 + x_3^2x_1 + x_3^2x_2 &= s_1s_2 - 3s_3 \\ x_1^3x_2 + x_2^3x_3 + x_3^3x_1 + x_1^3x_3 + x_2^3x_1 + x_3^3x_2 &= s_1^2s_2 - 2s_2^2 - s_1s_3 \\ x_1^2x_2x_3x_4 + x_2^2x_1x_3x_4 + x_3^2x_1x_2x_4 + x_4^2x_1x_2x_3 &= s_1s_4 \end{aligned}$$

The **weight** of the monomial $s_1^{k_1}s_2^{k_2}\cdots s_n^{k_n} \in \mathbb{F}[s_1, s_2, \cdots, s_n]$ is defined to be $k_1 + 2k_2 + 3k_3 + \cdots + nk_n$. Also the weight of $g(s_1, s_2, \cdots, s_n) \in \mathbb{F}[s_1, s_2, \cdots, s_n]$ is defined to be the maximum of the weights of the monomials appearing in it. In

$$f(x_1, x_2, \cdots, x_n) = g(s_1, s_2, \cdots, s_n) \in \mathbb{F}[s_1, s_2, \cdots, s_n]$$

the degree of $f(x_1, x_2, \cdots, x_n)$ is equal to weight of $g(s_1, s_2, \cdots, s_n)$. Since the degree of $g(s_1, s_2, \cdots, s_n)$ is at most equal to its weight (generally less) it follows that in

$$\text{degree } (f(x_1, x_2, \cdots, x_n)) \geq \text{degree } (g(s_1, s_2, \cdots, s_n))$$

Theorem 2.2. *π is transcendental.*

Proof. Suppose that π is algebraic. Then so is πi. Assume that πi is algebraic number of degree $n > 1$. We know that $e^{\pi i} = -1$. Let $\alpha_1 = \pi i, \alpha_2, \alpha_3, \cdots, \alpha_n$ be the algebraic conjugates of πi, that is the zeros of the minimal polynomial of πi over $\mathbb{Q}$, the field of rational numbers. Evidently

$$(e^{\alpha_1}+1)\cdot(e^{\alpha_2}+1)\cdots(e^{\alpha_n}+1)=0$$

as the first factor on the left hand side is zero. When we multiply out the left hand side we get.

$$1+\sum_{1\le i\le n} e^{\alpha_i}+\sum_{1\le i<j<\le n} e^{\alpha_i+\alpha_j}+\sum_{1\le i<j<k\le n} e^{\alpha_i+\alpha_j+\alpha_k}+\cdots+e^{\sum_{1\le i\le n}\alpha_i}=0.$$

For instance if $n = 3$, then we get

$$e^{\alpha_1+\alpha_2+\alpha_3}+e^{\alpha_1+\alpha_2}+e^{\alpha_2+\alpha_3}+e^{\alpha_3+\alpha_1}+e^{\alpha_1}+e^{\alpha_2}+e^{\alpha_3}+1=0.$$

Now it is possible that a certain sum of some of the algebraic numbers $\alpha_1, \alpha_2, \cdots, \alpha_n$ is zero. Taking into consideration this possibility we can write

$$(e^{\alpha_1}+1)\cdot(e^{\alpha_2}+1)\cdots(e^{\alpha_n}+1)=0$$

as

$$e^{q_1}+e^{q_2}+\cdots+e^{q_m}+k=0$$

where $q_1, q_2, \cdots, q_m$ are sums of certain $\alpha_i's$ (including a single α_i). So $q_i's$ are (nonzero) algebraic numbers from among the following ${}_nC_1 + {}_nC_2 + \cdots + {}_nC_n = 2^n - 1$ numbers

$$\alpha_1, \alpha_2, \cdots, \alpha_n, \alpha_1+\alpha_2, \alpha_1+\alpha_3, \cdots, \alpha_{n-1}+\alpha_n, \cdots, \alpha_1+\alpha_2+\cdots+\alpha_n.$$

As mentioned before, we cannot rule out the possibility that some of these may be zero. Let $\Theta_j(x)$ denote a polynomial of degree ${}_nC_j$ whose roots are sums of j, $\alpha_i's$ from the n algebraic conjugates of $\alpha_1 = \pi i$, namely $\alpha_1, \alpha_2, \cdots, \alpha_n$. The following table gives roots and degree of the polynomials $\Theta_j(x)$ for $1 \leq j \leq n$.

Polynomial	Degree	Zeros
$\Theta_1(x)$	${}_nC_1$	$\alpha_1, \alpha_2, \cdots, \alpha_n$
$\Theta_2(x)$	${}_nC_2$	$\alpha_1+\alpha_2, \alpha_1+\alpha_3, \cdots, \alpha_{n-1}+\alpha_n$
$\vdots$	$\vdots$	$\vdots$
$\Theta_i(x)$	${}_nC_i$	$\alpha_1+\alpha_2+\cdots+\alpha_i, \cdots$
$\vdots$	$\vdots$	$\vdots$
$\Theta_n(x)$	${}_nC_n = 1$	$\alpha_1+\alpha_2+\cdots+\alpha_n$

The polynomial

$$\Theta(x) = \Theta_1(x) \cdot \Theta_2(x) \cdot \Theta_3(x) \cdots \Theta_n(x)$$

has degree $2^n - 1$ and its zeros are the (nonzero) exponents of e in the product

$$(e^{\alpha_1}+1)(e^{\alpha_2}+1)\cdots(e^{\alpha_n}+1).$$

As the elementary symmetric polynomials in ${}_nC_r$ sums

$$\alpha_{i_1}+\alpha_{i_2}+\cdots+\alpha_{i_r}$$

are symmetric in $\alpha_i^{'}s$, the polynomial $\Theta(x)$ has rational coefficients. As we are not disregarding the possibility that a few of the above sums of some of the $\alpha_i^{'}s$ may be zero the polynomial $\Theta(x)$ may have certain number of zero roots. If we divide the polynomial $\Theta(x)$ by a suitable power of x (to avoid zero roots) and multiply by a suitable integer we get a polynomial $g(x) \in \mathbb{Z}[x]$ whose roots are the nonzero exponents of e in the product

$$(e^{\alpha_1}+1)(e^{\alpha_2}+1)\cdots(e^{\alpha_n}+1) = e^{q_1}+e^{q_2}+\cdots+e^{q_m}+k.$$

So if q_1, q_2, $\cdots$ q_m denote these nonzero exponents then the polynomial $g(x) \in \mathbb{Z}[x]$ has q_1, q_2, $\cdots$ q_m as its zeros. Let

$$g(x) = a_0x^m + a_1x^{m-1} + \cdots + a_{m-1}x + a_m, \qquad a_0 \neq 0.$$

As we shall be using the subscripts of a_0 quite often we let $a_0 = a$. Now we let

$$f(x) = \frac{a^{mp-1}x^{p-1}[g(x)]^p}{(p-1)!}$$

where p is a prime number to be specified later. Evidently, $f(x)$ has degree $mp+p-1$. Let $mp-1=s$. So degree of $f(x)$ is $s+p$ and we can write

$$f(x)=\frac{a^s x^{p-1}[g(x)]^p}{(p-1)!}.$$

We may regard the polynomial $f(x)$ as a function of a complex variable x. Let

$$F(x)=f(x)+f^{'}(x)+f^{''}(x)+\cdots+f^{(s+p)}(x).$$

Since degree $f(x)=s+p,\quad f^{(s+p+1)}(x)=0$. So

$$F^{'}(x)=f^{'}(x)+f^{''}(x)+\cdots+f^{(s+p)}(x).$$

Therefore

$$F(x)-F^{'}(x)=f(x).$$

Now

$$\frac{d}{dx}(e^{-x}F(x))=e^{-x}F^{'}(x)-e^{-x}F(x)=-e^{-x}f(x).$$

Let x be a (fixed) complex number and ϕ_1 and ϕ_2 be the following maps

$$\mathbb{R}\xrightarrow{\phi_1}\mathbb{C}\xrightarrow{\phi_2}\mathbb{C}$$

where $\phi_1(u)=xu$ and $\phi_2(z)=e^{-z}F(z)$. Let $\phi=\phi_2\circ\phi_1$. So ϕ is a map from $\mathbb{R}$ to $\mathbb{C}$ such that

$$\phi(u)=\phi_2\circ\phi_1(u)=\phi_2(xu)=e^{-xu}F(xu).$$

Thus ϕ is a complex-valued function of a real variable u and $\phi(u)=e^{-xu}F(xu)$, where x is a fixed complex number. As ϕ depends upon the complex number x we should replace ϕ by ϕ_x, but as long as no confusion is likely we shall continue to use ϕ instead of ϕ_x. Differentiating $\phi(u)=e^{-xu}F(xu)$, we have

$$\phi^{'}(u)=e^{-xu}(-x)F(xu)+e^{-xu}F^{'}(xu)x=-xe^{xu}(F(xu)-F^{'}(xu))=-xe^{-xu}f(xu).$$

By the Fundamental Theorem of Calculus—as applied to a complex valued function of a real variable, we have

$$\int_0^1\phi^{'}(u)\ du=\phi(1)-\phi(0)$$

i.e.

$$\int_0^1 -xe^{-xu} f(xu)\ du = e^{-x}F(x) - F(0)$$

or

$$F(x) - F(0)e^x = \int_0^1 -xe^{(1-u)x} f(xu)\ du.$$

Substituting $x = q_1, q_2, \cdots q_m$ and adding we get

$$\sum_{j=1}^{m} F(q_j) - F(0)\left(e^{q_1} + e^{q_2} + \cdots + e^{q_m}\right) = -\sum_{j=1}^{m} \int_0^1 q_j e^{(1-u)q_j} f(uq_j)\ du.$$

Since $e^{q_1} + e^{q_2} + \cdots + e^{q_m} + k = 0$, we can write the previous equation as

$$\sum_{j=1}^{m} F(q_j) + kF(0) = -\sum_{j=1}^{m} \int_0^1 q_j e^{(1-u)q_j} f(uq_j)\ du \qquad (A)$$

We shall arrive at a (desired) contradiction by showing the following

(1) $\sum_{j=1}^{m} F(q_j)$ is an integer divisible by p.

(2) For sufficiently large primes p, $kF(0)$ is an integer *not* divisible by p.

(3) Right hand side of (A) approaches zero as $p \longrightarrow \infty$.

Proof 1. We recall that

$$F(x) = \sum_{r=0}^{s+p} f^{(r)}(x), \qquad \left(f^{(0)}(x) = f(x)\right).$$

Thus

$$\sum_{j=1}^{m} F(q_j) = \sum_{j=1}^{m} \sum_{r=0}^{s+p} f^{(r)}(q_j) = \sum_{r=0}^{s+p} \sum_{j=1}^{m} f^{(r)}(q_j).$$

Since

$$f(x) = \frac{a^s x^{p-1}[g(x)]^p}{(p-1)!}, \qquad \text{degree}(f(x)) = mp + p - 1 = s + p$$

and the polynomial

$$g(x) = a_0 x^m + a_1 x^{m-1} + \cdots + a_m, \qquad a_0 = a$$

has zeros $q_1, q_2, \cdots, q_m$, if we differentiate the product $x^{p-1}[g(x)]^p$, r-times, where $0 \leq r < p$, by Leibnitz's rule we see that each term of the final expression of the derivative will have a factor $g(x)$ in it. Therefore $f^{(r)}(q_j) = 0$, for $0 \leq r < p$, $1 \leq j \leq m$. So

$$\sum_{j=1}^{m} F(q_j) = \sum_{r=p}^{s+p} \sum_{j=1}^{m} f^{(r)}(q_j)$$

as the remaining summands are equal to zero. As observed before $x^{p-1}[g(x)]^p$ is a polynomial of degree $mp + p - 1 = s + p$, with integer coefficients, so for $p \leq r \leq s + p$, each nonzero term of its rth derivative has a coefficient which is a multiple of r consecutive integers and is therefore divisible by $r!$ (an immediate consequence of the fact that ${}_{k+r}C_r$ is an integer). In particular, each of these coefficients is divisible by $(p-1)!$ as $r \geq p$. It follows that each of the coefficient of $f^{(r)}(x)$ $(r \geq p)$ is divisible by $p \cdot a^s$. Now consider the sum

$$\sum_{j=1}^{m} \frac{1}{p \cdot a^s} f^{(r)}(y_j), \qquad p \leq r \leq s + p.$$

This is a polynomial in $y_1, y_2, \cdots, y_m$ with integral coefficients (as we have seen that each of the coefficients of $f^{(r)}(x)$, for $r \geq p$ is divisible by $p \cdot a^s$). This polynomial is evidently symmetric in $y_1, y_2, \cdots, y_m$. By the Fundamental Theorem of Symmetric Functions it can be expressed as a polynomial in elementary symmetric function with integer coefficients. Substituting $q_1, q_2, \cdots, q_m$ for $y_1, y_2, \cdots, y_m$ we see that the sum

$$\sum_{j=1}^{m} \frac{1}{p \cdot a^s} f^{(r)}(q_j)$$

is expressible as a polynomial (with integer coefficients) of degree at most max $\left(\text{degree}(f^{(r)}(q_j)\right) = mp + p - 1 - r = s + p - r \leq s$ (since $r \geq p$), in $\sigma_1, \sigma_2, \cdots, \sigma_m$ where

$$\begin{aligned}
\sigma_1 &= q_1 + q_2 + \cdots + q_m \\
\sigma_2 &= q_1q_2 + q_1q_3 + \cdots + q_{n-1}q_n \\
\sigma_3 &= q_1q_2q_3 + \cdots + q_{n-2}q_{n-1}q_n \\
\vdots \quad & \vdots \quad \vdots \\
\vdots \quad & \vdots \quad \vdots \\
\sigma_m &= q_1q_2 \cdots q_m
\end{aligned}$$

are elementary symmetric polynomials in $q_1, q_2, \cdots q_m$. Since $q_1, q_2, \cdots q_m$ are the zeros of the polynomial

$$g(x) = a_0 x^m + a_1 x^{m-1} + \cdots + a_m, \qquad a_0 = a$$

we have $\sigma_i = \frac{a_i}{a_0}(-1)^i = \frac{a_i}{a}(-1)^i$, $1 \le i \le m$. We now write (for $p \le r \le s+p$)

$$\sum_{j=1}^{m} f^{(r)}(q_j) = p \cdot a^s \sum_{j=1}^{m} \frac{1}{p \cdot a^s} f^{(r)}(q_j).$$

Since

$$\sum_{j=1}^{m} \frac{1}{p \cdot a^s} f^{(r)}(q_j)$$

is a polynomial in $\sigma_1, \sigma_2, \cdots, \sigma_m$ of degree at most s with integer coefficients and as $\sigma_i = \frac{a_i}{a}(-1)^i$, $1 \le i \le m$, each term of this sum (when expressed as a polynomial in $\sigma_1, \sigma_2, \cdots, \sigma_m$ which in turn are replaced by their values in terms of coefficients of $g(x)$) has a denominator dividing a^s. It follows that

$$\sum_{j=1}^{m} f^{(r)}(q_j) = p \cdot a^s \sum_{j=1}^{m} \frac{1}{p \cdot a^s} f^{(r)}(q_j)$$

is an integer which is a multiple of p. This completes the proof of # 1.

Proof 2. We now show that for large prime p, $kF(0)$ is an integer *not* divisible by p. We have

$$F(x) = \sum_{r=0}^{s+p} f^{(r)}(x) = \sum_{r=0}^{p-2} f^{(r)}(x) + f^{(p-1)}(x) + \sum_{r=p}^{s+p} f^{(r)}(x)$$

Therefore

$$F(0) = \sum_{r=0}^{p-2} f^{(r)}(0) + f^{(p-1)}(0) + \sum_{r=p}^{s+p} f^{(r)}(0)$$

As shown before for $p \le r \le s+p$, $f^{(r)}(x)$ is a polynomial with integral coefficients each divisible by p. Therefore its constant term namely $f^{(r)}(0)$ is also divisible by p. This proves that $\sum_{r=p}^{s+p} f^{(r)}(0)$ is a multiple of p. Since

$$f(x) = \frac{a^s \cdot x^{p-1}[g(x)]^p}{(p-1)!}$$

it is clear that when we differentiate it r times, for $r \leq p-2$ it still retains a factor which is a positive power of x. Therefore

$$\sum_{r=0}^{p-2} f^{(r)}(0) = 0$$

Finally, writing

$$f(x) = \frac{a^s}{(p-1)!} x^{p-1} \cdot [g(x)]^p$$

using Leibnitz's rule for the n-th derivative of product of two functions, we see that in $f^{(p-1)}(x)$ only term which is not a product of a nonzero power of x comes when we differentiate x^{p-1}, $p-1$ times and $[g(x)]^p$ zero times. Therefore

$$f^{(p-1)}(0) = \frac{a^s}{(p-1)!} \ (p-1)! \ [g(0)]^p = a^s \cdot a_m^p$$

Thus $k \cdot f^{(p-1)}(0) = k \cdot a^s \cdot a_m^p$. So if we choose prime $p > |k \cdot a \cdot a_m|$ then

$$k \cdot F(0) = k \cdot \sum_{r=0}^{p-2} f^{(r)}(0) + k \cdot a^s \cdot a_m^p + k \sum_{r=p}^{s+p} f^{(r)}(0) = k \cdot a^s \cdot a_m^p + \text{a multiple of } \ p \not\equiv 0 \ (\text{modulo } p)$$

for sufficiently large prime p. This proves # 2.

Proof 3. Now we shall prove that the right hand side of (A), namely

$$-\sum_{j=1}^{\infty} \int_0^1 q_j e^{(1-u)q_j} f(uq_j) du$$

approaches zero as $p \longrightarrow \infty$. We recall that

$$f(x) = \frac{a^s \cdot x^{p-1}[g(x)]^p}{(p-1)!}, \qquad a_0 = a$$

and $g(x) = a_0 x^m + a_1 x^{m-1} + \cdots + a_m$. We note that $a^s = a^{mp-1} = a^{mp-m+m-1} = (a^{p-1})^m \cdot a^{m-1}$. So we have

$$\left| -\sum_{j=1}^{m} \int_0^1 q_j e^{(1-u)q_j} f(uq_j) du \right| = \left| \sum_{j=1}^{m} \int_0^1 q_j e^{(1-u)q_j} \ \frac{a^s (uq_j)^{p-1} [g(uq_j)]^p}{(p-1)!} du \right|$$

$$\leq \int_0^1 \sum_{j=1}^{m} \left| \frac{[a^m uq_j g(uq_j)]^{p-1}}{(p-1)!} q_j e^{(1-u)q_j} a^{m-1} g(uq_j) \right| du$$

For $1 \leq j \leq m$ the part $a^m uq_j g(uq_j)$, of the integrand can be regarded as a continuous function of u in the closed interval $[0,1]$. Let B denote a bound for its absolute value. In other words we have

$$| a^m uq_j g(uq_j) | \leq B \quad \text{for} \quad 1 \leq j \leq m \quad \text{and} \quad 0 \leq u \leq 1.$$

Also the expression

$$\sum_{j=1}^{m} \left| q_j e^{(1-u)q_j} a^{m-1} g(uq_j) \right| .$$

represents a continuous function for $u \in [0,1]$, and so is bounded in the closed interval $[0,1]$. Let C denote its bound. So the integrand in the last expression (in the above inequality) is bounded by $\frac{B^{p-1}}{(p-1)!} \cdot C$. Therefore

$$\left| -\sum_{j=1}^{m} \int_0^1 q_j e^{(1-u)q_j} f(uq_j) du \right| \leq \frac{B^{p-1} \cdot C}{(p-1)!} \int_0^1 du = \frac{B^{p-1} \cdot C}{(p-1)!} \longrightarrow 0 \quad \text{as} \quad p \longrightarrow \infty.$$

So in the equation (A)

$$\sum_{j=1}^{m} F(q_j) + k \cdot F(0) = -\sum_{j=1}^{m} \int_0^1 q_j e^{(1-u)q_j} f(uq_j) du$$

for sufficiently large prime p, the left hand side is a nonzero integer whereas the right hand side tends to zero as $p \longrightarrow \infty$. As this contradiction arises from our assumption that π is algebraic, we conclude that our assumption is wrong and so π is transcendental. □

Just in case of e, there is a short proof of irrationality of π (in fact for π^2). As this proof is precursor to the proof of transcendence of e and π it is of historical interest. We shall now discuss this proof irrationality of π^2 first given by Legendre in 1794. One can notice some similarity between this proof and the proofs of transcendence of e and π.

The Proof that π^2 is Not Rational

Suppose that $\pi^2 = \frac{a}{b}$, where a and b are positive integers. Let n be a positive integer and let

$$f(x) = \frac{x^n(1-x)^n}{n!} = \frac{1}{n!} \sum_{i=n}^{2n} c_i x^i, c_i \in \mathbb{Z}$$

be a polynomial in x of degree $2n$. Observe that the term of least degree in $f(x)$ is $\frac{x^n}{n!}$. It can be easily seen that

(i) For $0 < x < 1$, $0 < f(x) < \frac{1}{n!}$

(ii) $f(x) = f(1-x)$, $f^{(r)}(x) = f^{(r)}(1-x)\cdot(-1)^r$

(iii) $f^{(r)}(1) = (-1)^r f^{(r)}(0)$, and for $r < n$ or $r > 2n$, $f^{(r)}(0) = 0$.

(iv) If $n \le r \le 2n$, then $f^{(r)}(0) = \frac{1}{n!}\cdot r!\cdot c_r \in \mathbb{Z}$ and $f^{(r)}(1) \in \mathbb{Z}$.

So $f(x)$ and all of its derivatives take integral values at $x = 0$ and at $x = 1$. Let

$$\begin{aligned} F(x) &= a^n f(x) - a^{n-1}bf^{(2)}(x) + a^{n-2}b^2 f^{(4)}(x) - \cdots + (-1)^n b^n f^{(2n)} b^n f^{(2n)}(x) \\ &= b^n \left[\frac{a^n}{b^n} f(x) - \frac{a^{n-1}}{b^{n-1}} f^{(2)}(x) + \cdots + (-1)^n f^{(2n)}(x)\right] \\ &= b^n \left[\pi^{2n} f(x) - \pi^{2(n-1)} f^{(2)}(x) + \cdots + (-1)^n f^{(2n)}(x)\right] \end{aligned}$$

where we have used the fact that $\pi^2 = \frac{a}{b}$. As $f(x)$ is a polynomial of degree $2n$, its derivatives of order higher than $2n$ vanish. Now

$$F^{(2)}(x) = b^n \left[\pi^{2n} f^{(2)}(x) - \pi^{2(n-1)} f^{(4)}(x) + \cdots + (-1)^{n-1}\pi^2 f^{(2n)}(x)\right]$$

and

$$\pi^2 F(x) = b^n \left[\pi^{2n+2} f(x) - \pi^{2n)} f^{(2)}(x) + \cdots + (-1)^{(n-1)}\pi^2 f^{(2n-2)}(x) + (-1)^n \pi^2 f^{(2n)}(x)\right]$$

Therefore

$$F^{(2)}(x) + \pi^2 F(x) = b^n \pi^{2n+2} f(x) = b^n \pi^{2n}\pi^2 f(x) = a^n \pi^2 f(x)$$

Now

$$\begin{aligned} \frac{d}{dx}\left[F^{'}(x)\sin(\pi x) - \pi F(x)\cos(\pi x)\right] &= F^{(2)}(x)\sin(\pi x) + \pi F^{'}(x)\cos(\pi x) \\ &\quad - \pi F^{'}(x)\cos(\pi x) + \pi^2 F(x)\sin(\pi x) \\ &= \left[F^{(2)}(x) + \pi^2 F(x)\right]\cdot\sin(\pi x) = a^n\pi^2\cdot f(x)\cdot\sin(\pi x). \end{aligned}$$

That is

$$\frac{d}{dx}\left[\frac{F^{'}(x)\sin(\pi x)}{\pi} - F(x)\cos(\pi x)\right] = a^n \pi f(x)\sin(\pi x).$$

Using the Fundamental Theorem of Calculus we have

$$\int_0^1 a^n \pi f(x) \sin(\pi x)\ dx = \left[\frac{F'(x)\sin(\pi x)}{\pi} - F(x)\cos(\pi x)\right]_0^1 = F(1) + F(0).$$

As we have seen before $F(1) + F(0)$ is an integer. We shall arrive at a contradiction by showing that the integral on the left hand side has value strictly between 0 and 1 for large values of n. For $0 < x < 1$, $\sin(\pi x) < 1$, and $0 < f(x) < \frac{1}{n!}$. Therefore

$$0 < \int_0^1 a^n \pi f(x) \sin(\pi x)\ dx < \frac{a^n \pi}{n!} < 1, \qquad \text{for large } n$$

This contradiction completes the proof that π^2 is not rational.

CHAPTER 9

A Radical Expression for $\cos\frac{2\pi}{11}$

In an impressive paper Vandermonde (1771), comes very close to introducing the concept of cyclotomy and Galois Theory—some forty years before the birth of Galois. Discussing the work of Vandermonde, Jean-Pierre Tignol (2001) writes:

> $\cdots$ The existence of a cyclic permutation which does preserve the relations (among the roots) is very remarkable; and quite mysterious property of cyclotomic equations, which should have awaken Vandermonde's curiosity. If he had investigated this property he could have developed the theory of cyclotomy about 30 years before Gauss.

The fact that Vandermonde did his investigations without the benefit of the knowledge of the work of Gauss (as Galois did) is indeed very noteworthy. His observations about the existence of cyclic permutations which preserve the relations among the roots is a forerunner to the introduction of the concept of automorphism in later times.

In the last article of the aforementioned paper Vandermonde talks about applying the theory developed earlier to solve (that is to find radical expressions for the roots of) the equation $z^{11} - 1 = 0$, having disposed of the similar problems for $z^n - 1 = 0$, for smaller values of n in the earlier articles of the same paper. In particular, he gives radical expressions for $2\cos\frac{2k\pi}{11}$ for $1 \leq k \leq 5$ (actually for some unknown reasons Vandermonde prefers to work with the equation whose roots are $-2\cos\frac{2k\pi}{11}$ for $1 \leq k \leq 5$). Vandermonde does not provide the details of his calculations—he only points out the method of doing so and gives the final result. Tignol (2001) has reproduced the radical expressions for $-2\cos\frac{2k\pi}{11}$ for $1 \leq k \leq 5$ as these appear in Vandermonde's paper—still not providing any of the details which led him (or Vandermonde) to the radical expressions for $-2\cos\frac{2k\pi}{11}$ for $1 \leq k \leq 5$. These radical expressions for $-2\cos\frac{2k\pi}{11}$ for $1 \leq k \leq 5$, given by Vandermonde and, reproduced from there by Tignol are:

Let

$$a = -2\cos\frac{2\pi}{11}, \quad b = -2\cos\frac{4\pi}{11}, \quad c = -2\cos\frac{6\pi}{11}, \quad d = -2\cos\frac{8\pi}{11}, \quad e = -2\cos\frac{10\pi}{11}$$

then

$$a, b, c, d, e = \frac{1}{5}\left[1 + \Delta' + \Delta'' + \Delta''' + \Delta^{iv}\right]$$

where

$$\Delta' = \sqrt[5]{\frac{11}{4}\left(89 + 25\sqrt{5} - 5\sqrt{-5 + 2\sqrt{5}} + 45\sqrt{-5 - 2\sqrt{5}}\right)}$$

$$\Delta'' = \sqrt[5]{\frac{11}{4}\left(89 + 25\sqrt{5} + 5\sqrt{-5 + 2\sqrt{5}} - 45\sqrt{-5 - 2\sqrt{5}}\right)}$$

$$\Delta''' = \sqrt[5]{\frac{11}{4}\left(89 - 25\sqrt{5} - 5\sqrt{-5 + 2\sqrt{5}} - 45\sqrt{-5 - 2\sqrt{5}}\right)}$$

$$\Delta^{iv} = \sqrt[5]{\frac{11}{4}\left(89 - 25\sqrt{5} + 5\sqrt{-5 + 2\sqrt{5}} + 45\sqrt{-5 - 2\sqrt{5}}\right)}.$$

Despite our repeated efforts we failed to verify the correctness of above expressions. We decided to evaluate the expressions for $\Delta', \Delta'', \Delta'''$, and Δ^{iv} first by hand calculations (using the reductions procedures indicated in Vandermonde's paper) and later by the help of MATHEMATICA. We claim that the correct expressions for $\Delta', \Delta'', \Delta'''$, and Δ^{iv} are:

$$\Delta_c' = \sqrt[5]{\frac{11}{4}\left(89 + 25\sqrt{5} + \mathbf{20}\sqrt{\mathbf{-10 - 2\sqrt{5}}} - \mathbf{25}\sqrt{\mathbf{-10 + 2\sqrt{5}}}\right)}$$

$$\Delta_c'' = \sqrt[5]{\frac{11}{4}\left(89 + 25\sqrt{5} - \mathbf{20}\sqrt{\mathbf{-10 - 2\sqrt{5}}} + \mathbf{25}\sqrt{\mathbf{-10 + 2\sqrt{5}}}\right)}$$

$$\Delta_c''' = \sqrt[5]{\frac{11}{4}\left(89 - 25\sqrt{5} - \mathbf{25}\sqrt{\mathbf{-10 - 2\sqrt{5}}} - \mathbf{20}\sqrt{\mathbf{-10 + 2\sqrt{5}}}\right)}$$

$$\Delta_c^{iv} = \sqrt[5]{\frac{11}{4}\left(89 - 25\sqrt{5} + \mathbf{20}\sqrt{\mathbf{-10 - 2\sqrt{5}}} + \mathbf{25}\sqrt{\mathbf{-10 - 2\sqrt{5}}}\right)}.$$

In the above the corrections have been indicated by the boldface parts of the expressions. The contribution of Vandermonde to the solutions of equations predates the work of Lagrange on the same by some twenty years. The work of Vandermonde and the opinion of Lagrange about it have also been mentioned in the article by Robin Rider Hamburg (1976). Specifically, while discussing cyclotomic equations Lagrange (1869) makes the following statement about solving the equation

$$u^5 + u^4 - 4u^3 - 3u^2 + 3u + 1 = 0.$$

> En prenant u négativement, ce qui change les signes de tous les termes pairs, on a l'equation résolve par Vandermonde. Cet Auteur ne donne l'éxpression dont s'agit que comme un résultat de sa methode génerale, sans indiquer en détail les opérations par lesquelles il y est parvenu, et personne, aprés lui, ne s'est occupé, que je sache, à vérifer ce résultat, que paraît même éste ignoré.

which among other things essentially says that by replacing u by $(-u)$ one obtains the equation solved by Vandermonde, namely $u^5 - u^4 - 4u^3 + 3u^2 + 3u - 1 = 0$, and he does not know of anyone who has verified the results obtained by Vandermonde. Lagrange did point out the needed correction but does not provide the actual computation which led him to do that correction. In this chapter we will do just that.

In the explanation that follows we shall substantially adhere to the notations used by Vandermonde and Tignol. However, at some places we shall use the notions from elementary Galois theory for the justification for some of the steps in the calculations. We begin by considering the equation

$$z^{11} - 1 = 0. \tag{1}$$

Dividing the above by $z - 1$ we get the equation

$$z^{10} + z^9 + z^8 + z^7 + z^6 + z^5 + z^4 + z^3 + z^2 + z + 1 = 0 \tag{2}$$

whose roots are

$$e^{\frac{2\pi ik}{11}} = \cos\frac{2\pi k}{11} + i\sin\frac{2\pi k}{11}, \qquad 1 \le k \le 10.$$

Dividing this equation by z^5 and making the substitution $z + \dfrac{1}{z} = y$ we get

$$y^5 + y^4 - 4y^3 - 3y^2 + 3y + 1 = 0 \tag{3}$$

whose five roots are:

$$e^{\frac{2\pi ik}{11}} + e^{-\frac{2\pi ik}{11}} = 2\cos\frac{2\pi k}{11}, \qquad \text{for} \quad 1 \leq k \leq 5.$$

We would have preferred to work with Equation (3). However, as indicated earlier, for some reasons Vandermonde replaces y by $(-x)$ in (3). In order to provide a convenient comparison with Vandermonde's calculations we shall do likewise. So by replacing y by $(-x)$ in (3) we get the equation

$$x^5 - x^4 - 4x^3 + 3x^2 + 3x - 1 = 0 \tag{4}$$

whose five roots are: $-2\cos\frac{2\pi k}{11}$, for $1 \leq k \leq 5$. As 2 is a primitive root modulo 11, if we let $\alpha = e^{\frac{2\pi i}{11}}$, the mapping $\alpha \to \alpha^2$, which permutes the 10 roots of Equation (2) as

$$\alpha \to \alpha^2 \to \alpha^4 \to \alpha^8 \to \alpha^5 \to \alpha^{10} \to \alpha^9 \to \alpha^7 \to \alpha^3 \to \alpha^6 \to \alpha$$

gives the generating automorphism of the field $\mathbb{Q}(\alpha)$. The restriction of this mapping to the subfield $\mathbb{Q}(\cos\frac{2\pi}{11})$ (its maximum real subfield) induces an automorphism of $\mathbb{Q}(\cos\frac{2\pi}{11})$. To see the effect of this automorphism on $\mathbb{Q}(\cos\frac{2\pi}{11})$ we write:

$$a = -2\cos\frac{2\pi}{11} = -\left(\alpha + \frac{1}{\alpha}\right) = -(\alpha + \alpha^{10})$$

$$b = -2\cos\frac{4\pi}{11} = -\left(\alpha^2 + \frac{1}{\alpha^2}\right) = -(\alpha^2 + \alpha^9)$$

$$c = -2\cos\frac{6\pi}{11} = -\left(\alpha^3 + \frac{1}{\alpha^3}\right) = -(\alpha^3 + \alpha^8)$$

$$d = -2\cos\frac{8\pi}{11} = -\left(\alpha^4 + \frac{1}{\alpha^4}\right) = -(\alpha^4 + \alpha^7)$$

$$e = -2\cos\frac{10\pi}{11} = -\left(\alpha^5 + \frac{1}{\alpha^5}\right) = -(\alpha^5 + \alpha^6).$$

Consider the tower of fields:

$$\begin{array}{c} \mathbb{Q}(\alpha) \\ \Big| \; 2 \\ \mathbb{Q}(\cos\frac{2\pi}{11}) \\ \Big| \; 5 \\ \mathbb{Q} \end{array}$$

The automorphism of the field $\mathbb{Q}(\alpha)$ given by $\alpha \to \alpha^2$ induces an automorphism of its subfield $\mathbb{Q}(\cos\frac{2\pi}{11})$ which cyclically permutes the five roots of (3) as follows:

$$\alpha+\alpha^{-1} \to \alpha^2+\alpha^{-2} \to \alpha^4+\alpha^{-4} \to \alpha^8+\alpha^{-8} \to \alpha^5+\alpha^{-5} \to \alpha^{10}+\alpha^{-10} = \alpha+\alpha^{-1}$$

or

$$2\cos\frac{2\pi}{11} \to 2\cos\frac{4\pi}{11} \to 2\cos\frac{8\pi}{11} \to 2\cos\frac{6\pi}{11} \to 2\cos\frac{10\pi}{11}, \to 2\cos\frac{2\pi}{11}$$

which is the same as saying

$$a \to b \to d \to c \to e \to a$$

In other words the cyclic permutation (a, b, d, c, e) preserves the (rational) relations among the roots of Equation (3). In the words of Tignol:

> Vandermonde's brilliant (but not quite explicit) observation is that the permutation $a \to b \to d \to c \to e \to a$ preserves the relations between the roots.

As $(\alpha+\alpha^{-1})^2 = \alpha^2+\alpha^{-2}+2$, we have $a^2 = -b+2$ (which is same as $\cos\frac{4\pi}{11} = 2\cos^2\frac{2\pi}{11} - 1$). Applying the permutation (a, b, d, c, e) to the relation

$$a^2 = -b+2$$

we get

$$b^2 = -d+2$$

(which is same as $\cos\frac{8\pi}{11} = 2\cos^2\frac{4\pi}{11} - 1$). Also

$$\left(\alpha+\frac{1}{\alpha}\right)\cdot\left(\alpha^2+\frac{1}{\alpha^2}\right) = \left(\alpha^3+\frac{1}{\alpha^3}\right)+\left(\alpha+\frac{1}{\alpha}\right)$$

which is same as saying

$$2\cos\frac{2\pi}{11}\cdot 2\cos\frac{4\pi}{11} = 2\cos\frac{6\pi}{11} + 2\cos\frac{2\pi}{11}$$

or

$$a\cdot b = -a-c$$

which is one of the 15 relations (among $a, b, c, d,$ and e) that Vandermonde indicates for the process of reduction (to be explained soon). The following is an easily verifiable list of relations among $a, b, c, d,$ and e given by Vandermonde

$a^2 = -b+2$	$a\cdot b = -a-c$	$b\cdot c = -a-e$	$c\cdot d = -a-d$
$b^2 = -d+2$	$a\cdot c = -b-d$	$b\cdot d = -b-e$	$c\cdot e = -b-c$
$c^2 = -e+2$	$a\cdot d = -c-e$	$b\cdot e = -c-d$	$d\cdot e = -a-b$
$d^2 = -c+2$	$a\cdot e = -d-e$		
$e^2 = -a+2.$			

Since the permutation (a, b, d, c, e) induces the permutations

$$a \cdot b \to b \cdot d \to d \cdot c \to c \cdot e \to e \cdot a \to a \cdot b$$

$$a \cdot c \to b \cdot e \to d \cdot a \to c \cdot b \to e \cdot d \to a \cdot c$$

all the 15 relations among $a, b, c, d,$ and e given by Vandermonde can be derived by applying the permutation (a, b, d, c, e) to the one of the three relations, namely

$$a^2 = -b + 2, \qquad a \cdot b = -a - c, \qquad a \cdot c = -b - d$$

We observe that any polynomial in $a, b, c, d,$ and e with rational coefficients can be reduced to a linear polynomial in the same by using the 15 relations among them given by Vandermonde. We refer to this procedure as the *process of reduction.*

Let $\omega = e^{\frac{2\pi i}{5}} = \cos \frac{2\pi}{5} + i \sin \frac{2\pi}{5}$ be a primitive fifth root of unity. It is easy to see that

$$\cos \frac{2\pi}{5} = \frac{-1+\sqrt{5}}{4}, \qquad \sin \frac{2\pi}{5} = \frac{\sqrt{10+2\sqrt{5}}}{4}$$

For a later use we shall write down explicitly the values of ω^j for $1 \leq j \leq 4$.

$$\omega = e^{\frac{2\pi i}{5}} = \frac{-1+\sqrt{5}}{4} + \frac{\sqrt{10+2\sqrt{5}}}{4} i$$

$$\omega^2 = e^{\frac{4\pi i}{5}} = \frac{-1-\sqrt{5}}{4} + \frac{\sqrt{10-2\sqrt{5}}}{4} i$$

$$\omega^3 = e^{\frac{6\pi i}{5}} = \frac{-1-\sqrt{5}}{4} - \frac{\sqrt{10-2\sqrt{5}}}{4} i$$

$$\omega^4 = e^{\frac{8\pi i}{5}} = \frac{-1+\sqrt{5}}{4} - \frac{\sqrt{10+2\sqrt{5}}}{4} i.$$

Since $a, b, c, d,$ and e are roots of (4) we also have

$$a + b + c + d + e = 1$$

Let V_1, V_2, V_3, and V_4 be defined as follows:

$$\begin{aligned}
V_1 &= a + b\omega + d\omega^2 + c\omega^3 + e\omega^4 \\
V_2 &= a + b\omega^2 + d\omega^4 + c\omega + e\omega^3 \\
V_3 &= a + b\omega^3 + d\omega + c\omega^4 + e\omega^2 \\
V_4 &= a + b\omega^4 + d\omega^3 + c\omega^2 + e\omega
\end{aligned}$$

We note that V_i, for $1 \le i \le 4$ lie in the field $\mathbb{K} = \mathbb{Q}(\cos\frac{2\pi}{11}, \omega)$ which is of degree 20 over $\mathbb{Q}$. It is easily seen that

$$\begin{aligned}
a &= \tfrac{1}{5}(1 + V_1 + V_2 + V_3 + V_4) \\
b &= \tfrac{1}{5}(1 + \omega^4 V_1 + \omega^3 V_2 + \omega^2 V_3 + \omega V_4) \\
c &= \tfrac{1}{5}(1 + \omega^3 V_1 + \omega V_2 + \omega^4 V_3 + \omega^2 V_4) \\
d &= \tfrac{1}{5}(1 + \omega^2 V_1 + \omega^4 V_2 + \omega V_3 + \omega^3 V_4) \\
e &= \tfrac{1}{5}(1 + \omega V_1 + \omega^2 V_2 + \omega^3 V_3 + \omega^4 V_4).
\end{aligned}$$

Since we already have radical expression for ω, it is enough to give a radical expression for V_i's for $1 \le i \le 4$. This can be done by showing that the fifth powers of V_i's lie in the field $\mathbb{Q}(\omega)$ and we shall have the radical expressions for the roots of (4) as soon as we have an explicit expression for the fifth powers of V_i's as members of $\mathbb{Q}(\omega)$. Consider the following diagram of the subfields of $\mathbb{K} = \mathbb{Q}(\cos\frac{2\pi}{11}, \omega)$.

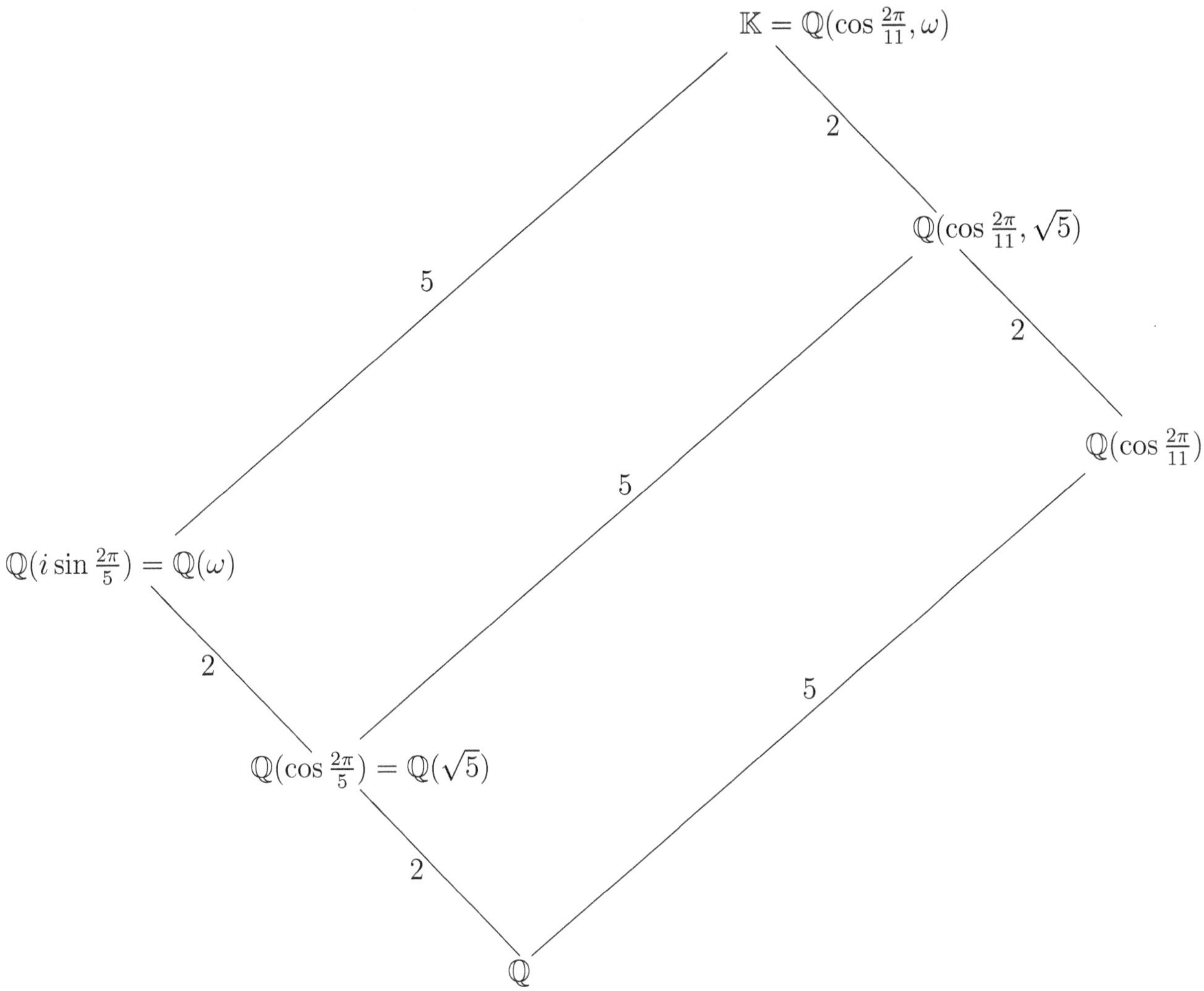

The Galois group of $G = Aut(\mathbb{K}/\mathbb{Q})$ is cyclic of order 20 as it is isomorphic to direct product of a group of order five and a cyclic group of order four. Let σ denotes the automorphism of $\mathbb{K}$ which fixes ω but permutes the five roots of (4) cyclically as:

$$a \to b \to d \to c \to e \to a.$$

In other words σ is the cyclic permutation (a, b, d, c, e). Clearly the fixed field of the subgroup $\langle\sigma\rangle$ of $G = Aut(\mathbb{K}/\mathbb{Q})$ is $\mathbb{Q}(\omega)$. In fact if we let η be the automorphism of $\mathbb{K}$ which maps ω to ω^2 and fixes a (and therefore the subfield $\mathbb{Q}(\cos\frac{2\pi}{11})$), then $G = Aut(\mathbb{K}/\mathbb{Q}) = \langle\sigma\eta\rangle$. Table I below gives the effect of σ and η on a, b, c, d, e, and ω and Table II gives the Galois correspondence between the subfields of $\mathbb{K}$ and the subgroups of $G = Aut(\mathbb{K}/\mathbb{Q})$.

Table I: Effects of σ and η on a, b, c, d, e, and ω.

x	$\sigma(x)$	$\eta(x)$	$\sigma\eta(x)$
ω	ω	ω^2	ω^2
a	b	a	b
b	d	b	d
d	c	d	c
c	e	c	e
e	a	e	a

Table II: Galois correspondence between subfields of $\mathbb{K}$ and subgroups of $G = Aut(\mathbb{K}/\mathbb{Q})$.

Subgroup	Order	Fixed Subfield	Degree over $\mathbb{Q}$
$G = \langle\sigma\eta\rangle$	20	$\mathbb{Q}$	1
$\langle\sigma\eta^2\rangle$	10	$\mathbb{Q}(\cos\frac{2\pi}{5}) = \mathbb{Q}(\sqrt{5})$	2
$\langle\sigma\rangle$	5	$\mathbb{Q}(i\sin\frac{2\pi}{5}) = \mathbb{Q}(\omega)$	4
$\langle\eta\rangle$	4	$\mathbb{Q}(\cos\frac{2\pi}{11}) = \mathbb{Q}(a)$	5
$\langle\eta^2\rangle$	2	$(\cos\frac{2\pi}{11}, \cos\frac{2\pi}{5}) = \mathbb{Q}(a, \sqrt{5})$	10
$\{e\}$	1	$\mathbb{Q}(\cos\frac{2\pi}{11}, \omega)$	20

Now as $V_1 = a + b\omega + d\omega^2 + c\omega^3 + e\omega^4$, we have

$$\sigma(V_1) = b + d\omega + c\omega^2 + e\omega^3 + a\omega^4 = \omega^4(a + b\omega + d\omega^2 + c\omega^3 + e\omega^4) = \omega^4 V_1$$

Therefore $\sigma(V_1^5) = V_1^5$ and so the fixed subfield of $\langle\sigma\rangle$ is $\mathbb{Q}(\omega)$. In other words $V_1^5 \in \mathbb{Q}(\omega)$. Likewise $V_2^5, V_3^5, V_4^5 \in \mathbb{Q}(\omega)$.

Although we know that $V_1^5 \in \mathbb{Q}(\omega)$ but nevertheless it is quite cumbersome to find an actual expression for it as an element of $\mathbb{Q}(\omega)$. It is hard to imagine how Vandermonde may have carried out this computation. While doing the calculations by hand it would be unwise to write down the $5^5 = 3125$ terms of $V_1^5 = (a + b\omega + d\omega^2 + c\omega^3 + e\omega^4)^5$ and then attempt to simplify these using the 15 relations from Vandermonde's paper which were indicated earlier (and the fact that $\omega^5 = 1$ and $a + b + c + d + e = 1$). The right approach would be to find the values of V_1^2, V_1^3, V_1^4, and V_1^5 in succession and simplifying each (i.e., expressing it as a linear polynomial in a, b, c, d, and e with coefficients in $\mathbb{Q}(\omega)$) before attempting to find the next higher power. Undoubtedly, Vandermonde followed this approach although he neither so claims and nor does he give the expressions for V_1^2, V_1^3, V_1^4, and V_1^5 in his paper. Despite the minor correction needed in the final result we are amazed that how very patiently Vandermonde carried out the enormous computation by hand. Since the automorphism η of $\mathbb{K}$ permutes $V_i's$ cyclically as $V_1 \to V_2 \to V_4 \to V_3 \to V_1$, we only need to compute V_1^5. Using the information that we have it is not difficult to carry out the computation of V_1^5 by hand. Here we shall simply provide the explicit expressions for V_1^2, V_1^3, V_1^4, and V_1^5. We have checked their correctness with the help of **MATHEMATICA**. For the purposes of calculations we shall find it more convenient to write each power of V_i as a polynomial in ω with coefficients which are linear in a, b, c, d, and e.

$$
\begin{aligned}
V_1 &= a + b\omega + d\omega^2 + c\omega^3 + e\omega^4 \\
V_1^2 &= (b - 2d + 2e) + (-2a + 2d + e)\omega + (2a - 2c + d)\omega^2 + (a - 2b + 2c)\omega^3 \\
&\quad + (2b + c - 2e)\omega^4 \\
V_1^3 &= (2 - 8a - 2b + 2c - 5d + 4e) + (2 + 2a + 4b - 2c - 8d - 5e)\omega \\
&\quad +(2 - 2a - 5b + 4c + 2d - 8e)\omega^2 + (2 + 4a - 8b - 5c - 2d + 2e)\omega^3 \\
&\quad +(2 - 5a + 2b - 8c + 4d - 2e)\omega^4 \\
V_1^4 &= (-16 + 34a + 28b + 18c - 7d + 8e) + (-16 + 28a - 7b + 8c + 18d + 34e)\omega \\
&\quad +(-16 - 7a + 18b + 34c + 8d + 28e)\omega^2 + (-16 + 18a + 8b + 28c + 34d - 7e))\omega^3 \\
&\quad (-16 + 8a + 34b - 7c + 28d + 18e)\omega^4 \\
V_1^5 &= 196 + 130\omega - 255\omega^2 + 20\omega^3 - 90\omega^4.
\end{aligned}
$$

In a like manner (or preferably by simply applying the automorphism η) we obtain similar expressions for V_2^5, V_3^5, and V_4^5. These expressions are:

$$V_2^5 = 196 + 20\omega + 130\omega^2 - 90\omega^3 - 255\omega^4$$

$$V_3^5 = 196 - 255\omega - 90\omega^2 + 130\omega^3 + 20\omega^4$$

$$V_4^5 = 196 - 90\omega + 20\omega^2 - 255\omega^3 + 130\omega^4.$$

By substituting the values of ω^i for $1 \le i \le 4$ in radical form we get the following radical expressions for V_j^5 for $1 \le j \le 4$.

$$V_1^5 = \frac{11}{4}\left(89 + 25\sqrt{5}\right) + \frac{11}{4}\left(20\sqrt{10+2\sqrt{5}} - 25\sqrt{10-2\sqrt{5}}\right) i$$

$$V_2^5 = \frac{11}{4}\left(89 - 25\sqrt{5}\right) + \frac{11}{4}\left(25\sqrt{10+2\sqrt{5}} + 20\sqrt{10-2\sqrt{5}}\right) i$$

$$V_3^5 = \frac{11}{4}\left(89 - 25\sqrt{5}\right) - \frac{11}{4}\left(25\sqrt{10+2\sqrt{5}} + 20\sqrt{10-2\sqrt{5}}\right) i$$

$$V_4^5 = \frac{11}{4}\left(89 + 25\sqrt{5}\right) - \frac{11}{4}\left(20\sqrt{10+2\sqrt{5}} - 25\sqrt{10-2\sqrt{5}}\right) i.$$

So we conclude that the following correction needs to be applied to the values of $\Delta', \Delta'', \Delta'''$, Δ^{iv} given by Vandermonde.

Vandermonde's Δ'^5 should be replace by our V_1^5.

Vandermonde's Δ''^5 should be replace by our V_4^5.

Vandermonde's Δ'''^5 should be replace by our V_3^5.

Vandermonde's Δ^{iv5} should be replace by our V_2^5.

We observe that $V_1 \cdot V_4 = V_2 \cdot V_3 = 11$. With appropriate choice of fifth roots of V_1^5, V_2^5, V_3^5, and V_4^5 we can find the radical expressions for the roots of the Equation (4). However, to check the correctness of our expressions for V_1^5, V_2^5, V_3^5, and V_4^5 we proceed as follows.

Let θ_1 and θ_2 be defined as follows:

$$\theta_1 = arctan\left(\frac{20\sqrt{10+2\sqrt{5}} - 25\sqrt{10-2\sqrt{5}}}{89 + 25\sqrt{5}}\right) \approx 6.8107268^o$$

$$\theta_2 = arctan\left(\frac{25\sqrt{10+2\sqrt{5}} + 20\sqrt{10-2\sqrt{5}}}{89 - 25\sqrt{5}}\right) \approx 76.890834^o$$

then

$$a = -2\cos\frac{2\pi}{11} = \frac{1}{5}\left[1 + 2\sqrt{11}\left\{\cos\frac{\theta_1+6\pi}{5} + \cos\frac{\theta_2+6\pi}{5}\right\}\right]$$
$$b = -2\cos\frac{4\pi}{11} = \frac{1}{5}\left[1 + 2\sqrt{11}\left\{\cos\frac{\theta_1+4\pi}{5} + \cos\frac{\theta_2+2\pi}{5}\right\}\right]$$
$$c = -2\cos\frac{6\pi}{11} = \frac{1}{5}\left[1 + 2\sqrt{11}\left\{\cos\frac{\theta_1}{5} + \cos\frac{\theta_2+4\pi}{5}\right\}\right]$$
$$d = -2\cos\frac{8\pi}{11} = \frac{1}{5}\left[1 + 2\sqrt{11}\left\{\cos\frac{\theta_1+2\pi}{5} + \cos\frac{\theta_2+8\pi}{5}\right\}\right]$$
$$e = -2\cos\frac{10\pi}{11} = \frac{1}{5}\left[1 + 2\sqrt{11}\left\{\cos\frac{\theta_1+8\pi}{5} + \cos\frac{\theta_2}{5}\right\}\right].$$

The correctness of these expression can be easily checked with the help of a calculator. Finally for the curious we have the following explicit radical expression for $\cos\frac{2\pi}{11}$.

$$\begin{aligned}\cos\frac{2\pi}{11} &= -\frac{1}{2}a = -\frac{1}{10}(1 + V_1 + V_2 + V_3 + V_4)\\
&= -\frac{1}{10}\left[1 + \left\{\frac{11}{4}\left(89+25\sqrt{5}\right) + \frac{11}{4}\left(20\sqrt{10+2\sqrt{5}} - 25\sqrt{10-2\sqrt{5}}\right)i\right\}^{\frac{1}{5}}\right.\\
&\quad + \left\{\frac{11}{4}\left(89-25\sqrt{5}\right) + \frac{11}{4}\left(25\sqrt{10+2\sqrt{5}} + 20\sqrt{10-2\sqrt{5}}\right)i\right\}^{\frac{1}{5}}\\
&\quad + \left\{\frac{11}{4}\left(89-25\sqrt{5}\right) - \frac{11}{4}\left(25\sqrt{10+2\sqrt{5}} + 20\sqrt{10-2\sqrt{5}}\right)i\right\}^{\frac{1}{5}}\\
&\quad \left. + \left\{\frac{11}{4}\left(89+25\sqrt{5}\right) - \frac{11}{4}\left(20\sqrt{10+2\sqrt{5}} - 25\sqrt{10-2\sqrt{5}}\right)i\right\}^{\frac{1}{5}}\right].\end{aligned}$$

This indeed gives correct value of $\cos\frac{2\pi}{11}$ with a proper choice of the fifth roots of unity in the above expression.

CHAPTER 10

A Natural Basis of $\mathbb{Q}(x)$ over $\mathbb{Q}(x^p)$

If p is a prime number then the field of rational functions $\mathbb{Q}(x)$ is a vector space over $\mathbb{Q}(x^p)$ of degree p. Although it can be easily argued that $1, x, x^2, \cdots, x^{p-1}$ is a basis of $\mathbb{Q}(x)$ over $\mathbb{Q}(x^p)$; however, it is not so clear how to express an element of $\mathbb{Q}(x)$ as a linear combination of $1, x, x^2, \cdots, x^{p-1}$ over $\mathbb{Q}(x^p)$. Here we shall do just that, namely give an (algorithmic) procedure to express an element of $\mathbb{Q}(x)$ as a linear combination of $1, x, x^2, \cdots, x^{p-1}$ with coefficients in $\mathbb{Q}(x^p)$. This algorithm is a simple consequence of the following:

Theorem 0.3. *Let p be a prime number and $f(x) \in \mathbb{Q}[x]$. Then there exists $g(x) \in \mathbb{Q}[x]$ such that $f(x) \cdot g(x] \in \mathbb{Q}[x^p]$.*

Proof. Let $f = f(x) \in \mathbb{Q}[x]$. Since

$$\mathbb{Q}[x] = \sum_{i=0}^{i=p-1} x^i \mathbb{Q}[x^p] = \mathbb{Q}[x^p] + x\mathbb{Q}[x^p] + x^2\mathbb{Q}[x^p] + \cdots\cdots + x^{p-1}\mathbb{Q}[x^p].$$

we can write

$$f = f_0 + f_1 + \cdots + f_{p-1}, \qquad \text{where} \qquad f_i \in x^i\mathbb{Q}[x^p].$$

Let $\varepsilon = e^{\frac{2\pi i}{p}}$, be a primitive p-th root of unity, and

$$\begin{aligned} g_0 = f &= f_0 + f_1 + f_2 + \cdots\cdots + f_{p-1} \\ g_1 &= f_0 + \varepsilon f_1 + \varepsilon^2 f_2 + \cdots\cdots + \varepsilon^{p-1} f_{p-1} \\ g_2 &= f_0 + \varepsilon^2 f_1 + \varepsilon^{2 \cdot 2} f_2 + \cdots\cdots + \varepsilon^{2 \cdot (p-1)} f_{p-1} \end{aligned}$$

$$\begin{aligned} g_3 &= f_0 + \varepsilon^3 f_1 + \varepsilon^{3\cdot 2} f_2 + \cdots\cdots + \varepsilon^{3\cdot(p-1)} f_{p-1} \\ \vdots & \;\vdots\; \vdots \\ g_i &= f_0 + \varepsilon^i f_1 + \varepsilon^{i\cdot 2} f_2 + \cdots\cdots + \varepsilon^{i\cdot(p-1)} f_{p-1} \\ \vdots & \;\vdots\; \vdots \\ g_{p-1} &= f_0 + \varepsilon^{p-1} f_1 + \varepsilon^{(p-1)\cdot 2} f_2 + \cdots\cdots + \varepsilon^{(p-1)^2} f_{p-1} \end{aligned}$$

or simply

$$g_i = \sum_{j=0}^{j=p-1} \varepsilon^{i\cdot j} f_j \qquad\qquad 0 \le i \le p-1.$$

We claim that

$$g(x) = g = g_1 \cdot g_2 \cdots g_{p-1} = \prod_{i=1}^{i=p-1} \left(\sum_{j=0}^{j=p-1} \varepsilon^{i\cdot j} f_j \right)$$

is the required polynomial, that is the polynomial $g = g(x)$ is such that $f(x) \cdot g(x) \in \mathbb{Q}[x^p]$. We shall show that

(i) $f \cdot g \in \mathbb{Q}(x^p)$ and therefore it lies in $\mathbb{Q}[x^p]$.
(ii) $g \in \mathbb{Q}(x)$ and therefore it lies in $\mathbb{Q}[x]$.

Clearly it is enough to prove (i) as (ii) would be an immediate consequence. As according to our notation $f = g_0$, we write

$$f \cdot g = g_0 \cdot g_1 \cdot g_2 \cdots g_{p-1} = \prod_{i=0}^{i=p-1} \left(\sum_{j=0}^{j=p-1} \varepsilon^{i\cdot j} f_j \right).$$

Let

$$\mathbb{K} = \mathbb{Q}(x, \varepsilon), \qquad \mathbb{K}_0 = \mathbb{Q}(x), \qquad \mathbb{L} = \mathbb{Q}(x^p, \varepsilon), \qquad \mathbb{F} = \mathbb{Q}(x^p).$$

Consider the following (partial) lattice of subfields of the field $\mathbb{K} = \mathbb{Q}(x, \varepsilon)$.

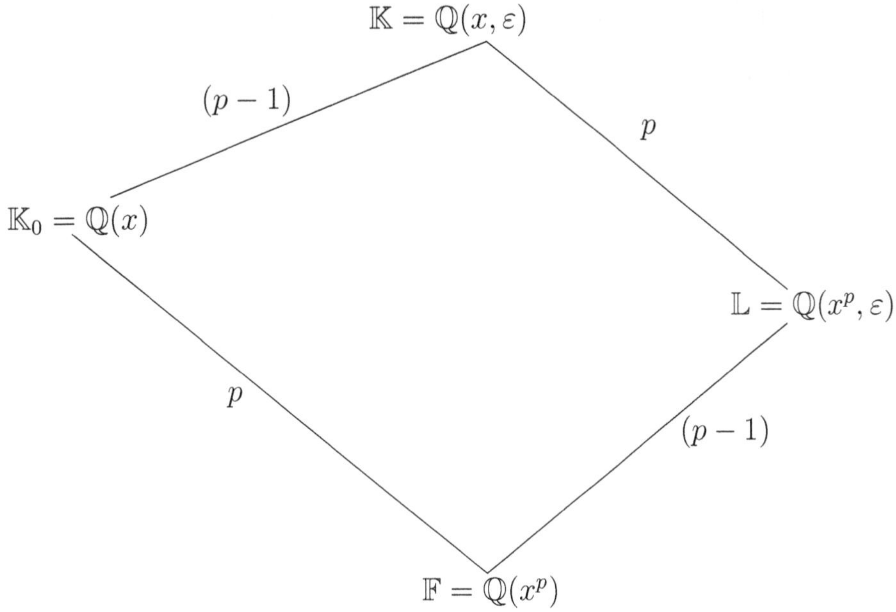

We point out that $y^p - x^p \in \mathbb{F}[y]$ is irreducible over $\mathbb{F} = \mathbb{Q}(x^p)$. This can be formally proved as follows. In $\mathbb{K} = \mathbb{Q}(x, \varepsilon)$, $y^p - x^p$ breaks up into linear factors as:

$$y^p - x^p = (y - x)(y - \varepsilon x)(y - \varepsilon^2 x) \cdots (y - \varepsilon^{p-1} x)$$

The "constant" term of the product of any of the j linear factors of $y^p - x^p$ (where $1 \leq j < p$) will be of the form $(-1)^j \varepsilon^j x^j$, or $(-1)^j \varepsilon^{j-1} x^j$, which is certainly not a member of $\mathbb{Q}(x^p)$. The Galois group of $\mathbb{K}$ over $\mathbb{F}$, namely $Aut(\mathbb{K}/\mathbb{F})$ has order $p(p-1)$. (Note: This group is so-called a Frobenius group and is isomorphic to the Sylow-p normalizer in the symmetric group S_p on p letters.) The group $G = Aut(\mathbb{K}/\mathbb{F})$ is generated by the automorphisms σ and η where,

$$\sigma : \mathbb{Q}(x, \varepsilon) \longrightarrow \mathbb{Q}(x, \varepsilon)$$
$$\sigma(x) = \varepsilon \cdot x, \qquad \text{and} \qquad \sigma(\varepsilon) = \varepsilon$$

and

$$\eta : \mathbb{Q}(x, \varepsilon) \longrightarrow \mathbb{Q}(x, \varepsilon).$$
$$\sigma(x) = x, \qquad \text{and} \qquad \eta(\varepsilon) = \varepsilon^r,$$

where r is a primitive root modulo p (i.e., a generator of the cyclic group $\mathbb{Z}_p^*$).

So $G = Gal(\mathbb{K}/\mathbb{F}) = Aut(\mathbb{K}/\mathbb{F}) = \langle \sigma, \eta \rangle$. It is easy to see the following:

(A) The automorphism σ permutes the p "polynomials" $g_0, g_1, \cdots, g_{p-1}$ among themselves. This is so because a permutation on $g_0, g_1, \cdots, g_{p-1}$ the effect of σ can be written as

$$\sigma \equiv (g_0, g_1, \cdots, g_{p-1}).$$

This can be seen as follows. As $f_j \in x^j\mathbb{Q}[x^p]$, $\sigma(x^j) = \varepsilon^j x^j$, and σ acts trivially on $\mathbb{Q}[x^p]$, so $\sigma(f_j) = \varepsilon^j f_j$. Now as

$$g_i = \sum_{j=0}^{j=p-1} \varepsilon^{i \cdot j} f_j \qquad\qquad 0 \leq i \leq p-1$$

we have

$$\sigma(g_i) = \sum_{j=0}^{j=p-1} \varepsilon^{i \cdot j}\sigma(f_j) = \sum_{j=0}^{j=p-1} \varepsilon^{i \cdot j}\varepsilon^j f_j = \sum_{j=0}^{j=p-1} \varepsilon^{j\cdot(i+1)} f_j \qquad\qquad 0 \leq i \leq p-1 = g_{i+1(modulo \;\; p)}$$

Therefore $\sigma(g_i) = g_{i+1(modulo \;\; p)}$ and so

$$\sigma \equiv (g_0, g_1, \cdots, g_{p-1}).$$

(B) The automorphism η fixes $f = g_0$ and permutes the $p-1$ "polynomials" $g_1, g_2, \cdots, g_{p-1}$ among themselves.

This is an easy consequence of the fact that r is a primitive root modulo p.

It follows (from A and B above) that the Galois group $G = \langle \sigma, \eta \rangle$ fixes the product

$$f \cdot g = f \cdot g_1 \cdot g_2 \cdots g_{p-1} = g_0 \cdot g_1 \cdot g_2 \cdots g_{p-1}$$

By the Fundamental Theorem of Galois Theory, $f \cdot g$ lies in the fixed field of the Galois group $G = \langle \sigma, \eta \rangle$ of $\mathbb{K}$ over $\mathbb{F}$. Hence $f \cdot g \in \mathbb{F} = \mathbb{Q}(x^p)$. This completes the proof of our assertion. □

Two Examples.

(1) Let $p = 3$ and $f(x) = 2 + 3x + 4x^2$. Here $f_0 = 2, f_1 = 3x, f_2 = 4x^2$. Let $\omega = e^{\frac{2\pi i}{3}}$, and

$$\begin{aligned} g_0 &= f_0 + f_1 + f_2 = f \\ g_1 &= f_0 + \omega f_1 + \omega^2 f_2 \\ g_2 &= f_0 + \omega^2 f_1 + \omega f_2. \end{aligned}$$

Therefore

$$\begin{aligned} g(x) = g &= (f_0 + \omega f_1 + \omega^2 f_2) \cdot (f_0 + \omega^2 \cdot f_1 + \omega \cdot f_2) \\ &= (2 + 3\omega x + 4\omega^2 x^2) \cdot (2 + 3\omega^2 x + 4\omega x^2) \\ &= 4 - 6x + x^2 - 12x^3 + 16x^4. \end{aligned}$$

So we have

$$\begin{aligned} f(x) \cdot g(x) &= (2 + 3x + 4x^2) \cdot (4 - 6x + x^2 - 12x^3 + 16x^4) \\ &= 8 + 3x^3 + 64x^6 \in \mathbb{Q}[x^3]. \end{aligned}$$

Now suppose we wish to express the rational function $\dfrac{1+x}{2+3x+4x^2} \in \mathbb{Q}(x)$ as a linear combination of $1, x, x^2$ over $\mathbb{Q}(x^p)$. We do it as follows.

$$\begin{aligned} \frac{1+x}{2+3x+4x^2} &= \frac{1+x}{2+3x+4x^2} \cdot \frac{(4 - 6x + x^2 - 12x^3 + 16x^4)}{(4 - 6x + x^2 - 12x^3 + 16x^4)} \\ &= \frac{4 - 2x - 5x^2 - 11x^3 + 4x^4 + 16x^5}{8 + 3x^3 + 64x^6} \\ &= \frac{(4 - 11x^3) + x(-2 + 4x^3) + x^2(-5 + 16x^3)}{8 + 3x^3 + 64x^6} \\ &= \mathbf{1} \cdot \frac{4 - 11x^3}{8 + 3x^3 + 64x^6} + \mathbf{x} \cdot \frac{-2 + 4x^3}{8 + 3x^3 + 64x^6} + \mathbf{x^2} \cdot \frac{-5 + 16x^3}{8 + 3x^3 + 64x^6} \end{aligned}$$

(2) In this example we take $p = 5$ and we shall express $\dfrac{1+x}{2+3x+4x^2} \in \mathbb{Q}(x)$ as a linear combination of $1, x, x^2, x^3, x^4$ over $\mathbb{Q}(x^5)$. We shall find rational functions $A_i(x) \in \mathbb{Q}(x^5)$ for $0 \le i \le 4$ such that

$$\frac{1+x}{2+3x+4x^2} = A_0(x) + A_1(x) \cdot x + A_2(x) \cdot x^2 + A_3(x) \cdot x^3 + A_4(x) \cdot x^4.$$

To do this we proceed as follows. Let

$$f(x) = 2 + 3x + 4x^2 = f_0 + f_1 + f_2 + f_3 + f_4$$

where $f_0 = 2$, $f_1 = 3x$, $f_2 = 4x^3$, $f_3 = f_4 = 0$. Let $\alpha = e^{\frac{2\pi i}{5}} = \cos\frac{2\pi}{5} + i\sin\frac{2\pi}{5}$, be a primitive fifth roots of unity. We let

$$\begin{aligned} g_0 &= 2 + 3x + 4x^2 \\ g_1 &= 2 + 3x\alpha + 4x^2\alpha^2 \\ g_2 &= 2 + 3x\alpha^2 + 4x^2\alpha^4 \\ g_3 &= 2 + 3x\alpha^3 + 4x^2\alpha \\ g_4 &= 2 + 3x\alpha^4 + 4x^2\alpha^3. \end{aligned}$$

Using the fact $\alpha^5 = 1$ and $\alpha + \alpha^2 + \alpha^3 + \alpha^4 = -1$, we deduce that

$$g_1 \cdot g_2 \cdot g_3 \cdot g_4 = 16 - 24x + 4x^2 + 42x^3 - 74x^4 + 84x^5 + 16x^6 - 192x^7 + 256x^8$$

and so

$$\begin{aligned} g_0 \cdot g_1 \cdot g_2 \cdot g_3 \cdot g_4 &= (2 + 3x + 4x^2) \cdot g_1 \cdot g_2 \cdot g_3 \cdot g_4 \\ &= 32 + 123x^5 + 1024x^{10} \in \mathbb{Q}[x^5]. \end{aligned}$$

We note that the product

$$\begin{aligned} (1 + x) \cdot g_1 \cdot g_2 \cdot g_3 \cdot g_4 = (1 + x) \cdot (16 &- 24x + 4x^2 + 42x^3 - 74x^4 \\ &+ 84x^5 + 16x^6 - 192x^7 + 256x^8) \end{aligned}$$

is equal to

$$16 - 8x - 20x^2 + 46x^3 - 29x^4 + 13x^5 + 100x^6 - 176x^7 + 64x^8 + 256x^9$$

which we write as

$$16 + (100x^5 - 8)x - (176 + 20)x^2 + (64x^5 + 46)x^3 + (256x^5 - 29)x^4.$$

We let

$$\begin{aligned} A_0(x) &= \frac{16}{32+135x^5+1024x^{10}} \\ A_1(x) &= \frac{100x^5-8}{32+135x^5+1024x^{10}} \\ A_2(x) &= -\frac{176x^5+20}{32+135x^5+1024x^{10}} \\ A_3(x) &= \frac{64x^5+46}{32+135x^5+1024x^{10}} \\ A_4(x) &= \frac{256x^5-29}{32+135x^5+1024x^{10}} \end{aligned}$$

Then $A_i(x) \in \mathbb{Q}(x^5)$, for $0 \le i \le 4$, and we have

$$\frac{1+x}{2+3x+4x^2} = \sum_{i=0}^{i=4} A_i(x) \cdot x^i$$

Note. Using the method used in proving the theorem in this chapter it can be shown that for any integer $N > 1$ the set $\{1, x, x^2, x^3, \cdots, x^{N-1}\}$ is a basis of $\mathbb{Q}(x)$ over $\mathbb{Q}(x^N)$. This can be proved on the induction on the number of prime factors of N (counting with multiplicity).

Let N be a composite natural number. Let p be one of its prime factors and $n = \frac{N}{p}$. Since n has fewer prime factors than N by induction hypothesis the set $\{1, x, x^2, \cdots, x^{n-1}\}$ is a basis of $\mathbb{Q}(x)$ over $\mathbb{Q}(x^n) = \mathbb{Q}(x^{\frac{N}{p}})$. Also $\mathbb{Q}(x^n) = \mathbb{Q}(x^{\frac{N}{p}})$ is a vector space over $\mathbb{Q}(x^N)$ of dimension p with basis $\{1, x^n, x^{2n}, x^{3n}, \cdots, x^{(p-1)n}\}$. Therefore $\mathbb{Q}(x)$ is a vector space over $\mathbb{Q}(x^N)$ of dimension $n \times p = N$ with a basis consisting of $\{x^i, 0 \le i \le N-1\}$.

CHAPTER 11

Some Interesting Irreducible Polynomials over $\mathbb{Z}$

It is well know that there are polynomials in $\mathbb{Z}[x]$ which are irreducible over $\mathbb{Z}$ but reducible modulo p for every prime p. The purpose of this chapter is to explain this phenomena and also give a method to produce such polynomials and provide several examples. It is easy to see that the Galois group of $x^4 + 1$ (over $\mathbb{Q}$) is a subgroup of A_4, the alternating group on four letters, and is isomorphic to the Klein-Four group. The polynomial $x^4 + 1$ factors (that is it is *reducible*) modulo p for every prime p. We shall show that this is a consequence of the fact that the Galois group of $x^4 + 1$ is a subgroup of A_4.

We again recall the following important result from *Modern Algebra* by Van Der Waerden [1953].

> Let $f(x) \in \mathbb{Z}[x]$ be a polynomial of degree n and p be a prime such that p does not divide the discriminant of $f(x)$. Let $G \subset S_n$ be the Galois group of $f(x)$ over the field $\mathbb{Q}$ of rational numbers. Suppose that $\bar{f} = f$ (modulo p) $\in \mathbb{Z}_p[x]$ factors as:
>
> $$\bar{f} = \prod_{i=1}^{k} \bar{f}_i$$
>
> where $\bar{f}_i$ are distinct monic irreducible polynomials in $\mathbb{Z}_p[x]$ with degree $(f_i) = d_i$ for $1 \leq i \leq k$. Then there exists an automorphism $\sigma \in G$ which when considered as a permutation on the zeros of $f(x)$ (in some splitting field of $f(x)$ over $\mathbb{Q}$) is a product of disjoint cycles of lengths $d_1, d_2, \cdots, d_k$.

We now prove the following theorem:

Theorem 11.1. *Let $f \in \mathbb{Z}[x]$ be a monic, irreducible polynomial of degree $n = 2k$, $k > 1$. Suppose that $Disc(f)$ is a perfect square in $\mathbb{Z}$. Then if p is a prime which does not divide $Disc(f)$ then the polynomial $f(x)$ is reducible modulo p.*

Proof. If the prime p divides $Disc(f(x))$ then the polynomial $f(x)$ has repeated roots modulo p. Such a polynomial cannot be irreducible modulo prime p.

Now suppose that $Disc(f(x))$ is not divisible by prime p and let $G = Gal(f)$ be the Galois group of $f(x)$ over $\mathbb{Q}$. Since $Disc(f)$ is a square in $\mathbb{Z}$, G is a subgroup of the alternating group A_n. If $f(x)$ is irreducible modulo p then its Galois group G will contain a cycle of length $n = 2k$, which is an odd permutation. But this is not possible as the Galois group G of $f(x)$ over $\mathbb{Q}$ is a subgroup of A_n which consists of even permutations. So $f(x)$ cannot be irreducible modulo prime p. □

We shall now give five examples of polynomials in $\mathbb{Z}[x]$ which are irreducible over $\mathbb{Z}$ but are reducible modulo p for every prime p—providing factorization modulo p for primes $p \leq 19$, and p not dividing the discriminant of the polynomial.

Example 1.

$$f(x) = x^4 + 1. \qquad Disc(f) = 2^8$$

Galois group $\approx C_2 \times C_2$(Klein-Four group)

Prime p	Factors of $f(x)$ Modulo p
3	$(x^2 + x + 2)(x^2 + 2x + 2)$
5	$(x^2 + 2)(x^2 + 3)$
7	$(x^2 + 3x + 1)(x^2 + 4x + 1)$
11	$(x^2 + 3x + 10)(x^2 + 8x + 10)$
13	$(x^2 + 5)(x^2 + 8)$
17	$(x + 2)(x + 8)(x + 9)(x + 15)$

Example 2.

$$f(x) = x^4 + 4x^3 + 12x^2 + 24x + 24. \qquad Disc(f) = 2^{12} \cdot 3^4$$

Galois group $\approx A_4$

Prime p	Factors of $f(x)$ Modulo p
5	$(x+4)(x^3+2x+1)$
7	$(x+5)(x^3+6x^2+3x+2)$
11	$(x+3)(x^3+x^2+9x+8)$
13	$(x+12)(x^3+5x^2+4x+2)$
17	$(x^2+x+16)(x^2+3x+10)$
19	$(x^2+11x+14)(x^2+12x+18)$

Example 3.

$$f(x) = x^6 + 10x^5 + 55x^4 + 140x^3 + 175x^2 + 170x + 25. \qquad Disc(f) = 2^{36} \cdot 5^8$$

Galois group $\approx A_5$

Prime p	Factors of $f(x)$ Modulo p
3	$(x+2)(x^5+2x^4+2x^2+2)$
7	$(x^3+4x^2+3)(x^3+6x^2+3x+6)$
11	$(x^3+x^2+6x+4)(x^3+9x^2+7x+9)$
13	$(x+8)(x^5+2x^4+10x^2+4x+8)$
17	$(x^3+3x^2+11)(x^3+7x^2+10)$
19	$(x+4)(x^5+6x^4+12x^3+16x^2+16x+11)$

Example 4.

$$f(x) = x^6 + 24x - 20. \qquad Disc(f) = 2^{16} \cdot 3^6 \cdot 5^6$$

Galois group $\approx A_6$

Prime p	Factors of $f(x)$ Modulo p
7	$(x+3)(x^5+4x^4+2x^3+x^2+4x+5)$
11	$(x+8)(x^5+3x^4+9x^3+5x^2+4x+3)$
13	$(x+3)(x^5+10x^4+9x^3+12x^2+3x+2)$
17	$(x+12)(x^5+5x^4+8x^3+6x^2+13x+4)$
19	$(x+5)(x^5+14x^4+6x^3+8x^2+17x+15)$

Example 5.

$$f(x) = x^8 + 8x^7 + 56x^6 + 336x^5 + 1680x^4 + 6720x^3 + 20160x^2 + 40320x + 40320.$$

$$Disc(f) = 2^{56} \cdot 3^{16} \cdot 5^8 \cdot 7^8.$$

Galois group $\approx A_8$

Prime p	Factors of $f(x)$ Modulo p
11	$(x^3+8x^2+10x+1)(x^5+2x^3+10x+5)$
13	$(x+1)(x+8)(x+11)(x^5+x^4+7x^3+x^2+4x+2)$
17	$(x+8)(x+14)(x^2+9x+2)(x^4+11x^3+15x^2+11x+10)$
19	$(x^4+2x^3+13x^2+9x+8)(x^4+6x^3+12x^2+16x+5)$

The rest of this chapter is devoted to the study of irreducible polynomials of odd degree in $\mathbb{Z}[x]$ which are reducible modulo p for every prime p. It is an immediate consequence of the Cebotarev Density Theorem (Stevenhagen and Lenstra 1996) that an irreducible polynomial of prime degree $p > 2$ in $\mathbb{Z}[x]$ will be irreducible modulo infinitely many primes. We shall prove that for any composite N there exist irreducible monic polynomials of degree N in $\mathbb{Z}[x]$ which are reducible modulo every prime. Such polynomials have been discussed in Brandl (1986) and Guralnick, Schacher, and Sonn (2005).

We shall first enumerate some known results which we shall use in what follows. Several of these results are true more generally but we shall state them as needed in this chapter.

1.0 For a given positive integer N every group of order N is cyclic if and only if $gcd(N, \phi(N)) = 1$, (Dixon 1967), here $\phi(N)$ denotes Euler's phi-function.

1.1 Let $f(x) \in \mathbb{Q}[x]$ be a nonconstant polynomial. Then the Galois group of $f(x)$ over $\mathbb{Q}$ acts transitively on its roots if and only if $f(x)$ is a power of an irreducible polynomial over $\mathbb{Q}$.

1.2 Let $\mathbb{K}_1/\mathbb{Q}$ and $\mathbb{K}_2/\mathbb{Q}$ be finite normal extensions, that is, splitting fields of some polynomial. Let $\mathbb{K} = \mathbb{K}_1\mathbb{K}_2$ denote the compositum of the fields $\mathbb{K}_1$ and $\mathbb{K}_2$, that is the smallest subfield of $\mathbb{C}$ containing $\mathbb{K}_1 \cup \mathbb{K}_2$. Then $\mathbb{K}$ is a normal extension of $\mathbb{Q}$ and if $[\mathbb{K}_1 : \mathbb{Q}]$ and $[\mathbb{K}_2 : \mathbb{Q}]$ are co-prime then

$$Aut(\mathbb{K}/\mathbb{Q}) \approx Aut(\mathbb{K}_1/\mathbb{Q}) \times Aut(\mathbb{K}_2/\mathbb{Q})$$

see Lang (1993).

1.3 Every finite solvable group can be realized as a Galois group of some polynomials over $\mathbb{Q}$. The same is true of the symmetric groups S_n and alternating groups A_n (Malle and Matzat 1999; Safarevic 1954; Schur 1930).

1.4 Let $f(x) \in \mathbb{Q}[x]$ be an irreducible polynomials of degree n. Let $\alpha_1, \alpha_2, \cdots, \alpha_n$ be its roots in $\mathbb{C}$. Let r be an integer, $1 < r < n-1$ and $\binom{n}{r} = m$. Let $f_r(x)$ denote the polynomial whose roots are all sums of r different $\alpha_i's$. Then $f_r(x) \in \mathbb{Q}[x]$ and $f(x)$ and $f_r(x)$ have the same splitting field. (We have discussed this in an earlier chapter.)

1.5 Let $f(x) \in \mathbb{Z}[x]$ be a monic irreducible polynomial of degree n and p a prime which does not divide the discriminant of $f(x)$. Let $G = Gal(f)$ be the Galois group of $f(x)$ over $\mathbb{Q}$. Suppose that modulo p the polynomial $f(x)$ factors into irreducible polynomials of degrees $n_1, n_2, \cdots, n_t$ so $n_1 + n_2 + \cdots + n_t = n$. Then there exists $\sigma \in G$ such that as a permutation on the n roots of $f(x)$, $\sigma = \sigma_1\sigma_2\cdots\sigma_t$, where σ_i is a cyclic permutation of length n_i for $1 \leq i \leq n$ (see Lang 1993).

1.6 Let n be any composite natural number and $f(x) \in \mathbb{Z}[x]$ be a monic irreducible polynomial of degree n whose Galois group over $\mathbb{Q}$ does not have any element of order n. Then $f(x)$ is reducible modulo every prime. (This is an immediate consequence of **1.5** above.)

1. Theorem and Proof

Theorem 11.2. For every composite natural number N there exists a monic irreducible polynomial $f(x) \in \mathbb{Z}[x]$ of degree N which is reducible modulo every prime

Proof.

Case I N is *not* square-free.

We write $N = p^t m$ where $t > 1$ and p is a prime which does not divide m. Let G_1 be any noncyclic group of order p^t and G_2 a cyclic group of order m. Let $f_1(x) \in \mathbb{Z}[x]$ be an irreducible polynomial of degree p^t with Galois group isomorphic to G_1 and $f_2(x) \in \mathbb{Z}[x]$ be an irreducible polynomial of degree m with Galois group isomorphic to G_2. Let $\mathbb{K}_1$ and $\mathbb{K}_2$ be splitting fields of $f_1(x)$ and $f_2(x)$ respectively. Let $\mathbb{K} = \mathbb{K}_1\mathbb{K}_2$ be the compositum of the fields $\mathbb{K}_1$ and $\mathbb{K}_2$. Then $\mathbb{K}$ is of degree N over $\mathbb{Q}$ and $G = Aut(\mathbb{K}/\mathbb{Q})$ is isomorphic to $G_1 \times G_2$ and so it does not have any element of order N. Let α be any algebraic integer such that $\mathbb{K} = \mathbb{Q}(\alpha)$. Let $f(x)$ be the minimum polynomial of α. Then $f(x) \in \mathbb{Z}[x]$ is a monic irreducible polynomial of degree N and its Galois group does not have any element of order N and therefore $f(x)$ has the desired property.

We will discuss an example to illustrate this construction. Let $N = 12 = 2^2 \cdot 3$. Let $\mathbb{K}_1 = \mathbb{Q}(\sqrt{2}, \sqrt{3})$ and $\mathbb{K}_2 = \mathbb{Q}(2\cos\frac{2\pi}{7})$. Then $Aut(\mathbb{K}_1/\mathbb{Q})$ is isomorphic to $C_2 \times C_2$, the Klein-Four group and $Aut(\mathbb{K}_2/\mathbb{Q})$ is isomorphic to C_3 the cyclic group of order three. Let

$$\mathbb{K} = \mathbb{K}_1\mathbb{K}_2 = \mathbb{Q}\left(\sqrt{2}, \sqrt{3}, 2\cos\frac{2\pi}{7}\right) = \mathbb{Q}\left(\sqrt{2} + \sqrt{3} + 2\cos\frac{2\pi}{7}\right).$$

Using Mathematica we get the following polynomial $f(x) \in \mathbb{Z}[x]$ as the minimal polynomial of the algebraic integer $\alpha = \sqrt{2} + \sqrt{3} + 2\cos\frac{2\pi}{7}$

$$x^{12}+4x^{11}-32x^{10}-124x^9+282x^8+1100x^7-850x^6-3648x^5+941x^4+4812x^3-174x^2-1976x-239.$$

As proved above the polynomial $f(x)$ is irreducible in $\mathbb{Z}[x]$ but is reducible modulo every prime. The following table gives the factors of $f(x)$ modulo primes $p < 50$.

$$f(x) = x^{12} + 4x^{11} - 32x^{10} - 124x^9 + 282x^8 + 1100x^7 - 850x^6 - 3648x^5 + 941x^4 + 4812x^3 - 174x^2 - 1976x - 239.$$

Prime p	Factors of $f(x)$ Modulo p
2	$(1+x+x^3)^4$
3	$(1+2x+x^2+x^3+2x^5+x^6)^2$
5	$(1+3x+x^3+x^4+2x^5+x^6)(1+x+3x^2+2x^3+3x^4+2x^5+x^6)$
7	$(5+2x+x^2)^3(1+4x+x^2)^3$
11	$(2+10x+3x^2+3x^3+8x^4+5x^5+x^6)(7+7x^2+x^3+9x^4+10x^5+x^6)$
13	$(8+x+x^2)(12+2x+x^2)(2+4x+x^2)(1+5x+x^2)(7+7x+x^2)(12+11x+x^2)$
17	$(11+2x+6x^2+2x^3+9x^4+x^6)(3+13x+8x^2+8x^3+10x^4+4x^5+x^6)$
19	$(17+6x+x^2+10x^3+9x^4+2x^5+x^6)(15+7x+11x^2+14x^3+12x^4+2x^5+x^6)$
23	$(7+14x+7x^2+x^3)(3+15x+11x^2+x^3)(7+17x+14x^2+x^3)(22+6x+18x^2+x^3)$
29	$(2+15x+x^2)(28+15x+x^2)(13+22x+x^2)(16+22x+x^2)(17+23x+x^2)(20+23x+x^2)$
31	$(10+x+20x^2+24x^3+16x^5+x^6)(4+19x+13x^2+20x^3+5x^4+19x^5+x^6)$
37	$(33+17x+31x^2+36x^3+x^4+18x^5+x^6)(32+x+9x^2+19x^3+34x^4+23x^5+x^6)$
41	$(28+x+x^2)(6+6x+x^2)(2+15x+x^2)^2(15+20x+x^2)(33+29x+x^2)$
43	$(26+5x+x^2)(41+5x+x^2)(19+13x+x^2)(34+13x+x^2)(2+27x+x^2)(30+27x+x^2)$
47	$(37+38x+11x^2+x^3)(45+36x+16x^2+x^3)(3+16x+33x^2+x^3)(25+9x+38x^2+x^3)$

Case II N is square-free and $gcd(N, \phi(N)) > 1$.

In this case we can write $N = pqm$ where p, q are primes, p divides $q-1$ and $gcd(pq, m) = 1$. Let G_1 be a nonAbelian group of order pq and G_2 a cyclic group of order m. Just as in the previous case we get a monic irreducible polynomial in $\mathbb{Z}[x]$ of degree N whose Galois group does not contain an element of order N.

Case III, N is square-free and $gcd(N, \phi(N)) = 1$.

In this case N is necessarily odd. First we assume that N is a product of just two primes. So let $N = pq$, where p and q are primes and neither p divides $q-1$ nor q divides $p-1$. Let t be the order of p modulo q. So $t > 1$ is the smallest integer such that $p^t \equiv 1$ (modulo q). Let G_1 be an elementary Abelian p-group of order p^t and G_2 be a group of order q. We note that $Aut(G_1)$ is isomorphic to $GL(t, p)$ and so its order is divisible by q. Let $G = G_1 \rtimes G_2$ be the semi-direct product of G_1 by G_2. A nonidentity element of G_2 induces a fixed-point-free automorphism of the group G_1. We also observe that G_2 as a subgroup of G is disjoint from its conjugates and is its own normalizer in G. Therefore G is a Frobenius group of order $p^t q$. Let $\mathbb{K}$ be a normal extension of $\mathbb{Q}$ with Galois group isomorphic to G. Then $[\mathbb{K} : \mathbb{Q}] = p^t q$. Let H be a subgroup of G of order p^{t-1} and let $\mathbb{F} \subseteq \mathbb{K}$ be its fixed subfield. Then by FTGT (Fundamental Theorem of Galois Theory) the field $\mathbb{F}$ is of degree $\frac{p^t q}{p^{t-1}} = pq$ over $\mathbb{Q}$. We also note that as H is not a normal subgroup of G, $\mathbb{F}$ is not a normal extension of $\mathbb{Q}$. Let α be an algebraic integer such that $\mathbb{F} = \mathbb{Q}(\alpha)$ and let $f(x)$ be its minimal polynomial over $\mathbb{Q}$. Then $f(x) \in \mathbb{Z}[x]$ is irreducible of degree pq. We claim that $\mathbb{K}$ is the splitting field of $f(x)$ (i.e., it is the normal closure of the field $\mathbb{F}$) and G is its Galois group over $\mathbb{Q}$. If the normal closure of $\mathbb{F}$ were a proper subfield of $\mathbb{K}$ then it would imply that G has a proper normal subgroup of order p^r where $r < t$, but this is not possible, as G is a Frobenius group. So $f(x) \in \mathbb{Z}[x]$ is a monic irreducible polynomial of degree $N = pq$ and its Galois group over $\mathbb{Q}$ does not have any element of order $N = pq$.

Finally assume that N and $\phi(N)$ are co-prime and N is a product of more than two primes. We write $N = pqm$, where p, q are primes and $gcd(pq, m) = 1$. Let $t =$ order of p modulo q. As discussed in the previous case let $f_1(x) \in \mathbb{Z}[x]$ be a monic irreducible polynomial of degree pq whose Galois group is the semidirect product of an elementary abelian group of order p^t by a cyclic group of order q and is a Frobenius group. Let G_1 denote this Frobenius group of order $p^t q$ and $\mathbb{K}_1$ denote the splitting field of $f_1(x)$. Let G_2 be a cyclic group of order m and $f_2(x) \in \mathbb{Z}[x]$ be a monic irreducible polynomial of degree m whose splitting field is $\mathbb{K}_2$ and Galois group over $\mathbb{Q}$ is G_2. Let $\mathbb{K} = \mathbb{K}_1\mathbb{K}_2$ be the compositum of the fields $\mathbb{K}_1$ and $\mathbb{K}_2$. Let $pq = n$ (so $N = n \cdot m$) and

$$f_1(x) = \prod_{i=1}^{n}(x - \alpha_i), \qquad f_2(x) = \prod_{j=1}^{m}(x - \beta_j), \qquad f(x) = \prod_{j=1}^{m}\prod_{i=1}^{n}(x - \alpha_i\beta_j).$$

We note the following:

(i) $[\mathbb{K}_1 : \mathbb{Q}] = p^t q, \qquad [\mathbb{K}_2 : \mathbb{Q}] = m, \quad$ and $\quad [\mathbb{K} : \mathbb{Q}] = p^t qm$

(ii) $\mathbb{K}_1 = \mathbb{Q}(\alpha_1, \alpha_2, \cdots, \alpha_n)$

(iii) $\mathbb{K}_2 = \mathbb{Q}(\beta_1, \beta_2, \cdots, \beta_m)$

(iv) $\mathbb{K} = \mathbb{Q}(\alpha_1, \alpha_2, \cdots, \alpha_n, \beta_1, \beta_2, \cdots, \beta_m)$

(v) $G = Aut(\mathbb{K}/\mathbb{Q})$ is a group of order $p^t qm$ isomorphic to the direct product of a Frobenius group of order $p^t q$ and a cyclic group of order m. Therefore it *does not* have an element of order $N = pqm$. Note that this Frobenius group does not have any subgroup of order pq.

(vi) The group G transitively permutes the nm algebraic numbers $\alpha_i\beta_j$, $1 \le i \le n$, $1 \le j \le m$. So $f(x) \in \mathbb{Z}[x]$ is an irreducible polynomial of degree $N = pqm$, whose Galois group does not have any element of order N.

This completes the proof of our theorem.

2. Alternate Method

As we noticed the construction of irreducible polynomials in $\mathbb{Z}[x]$ of odd composite degree N where $gcd(N, \phi(N)) = 1$, and whose Galois group does not contain an element of order N is not so straight forward. In some case such as $N = 15$ or $N = 35$ there is another interesting method of construction of such polynomials. In fact it works for most $N's$ (with very few exceptions) which are such that $\binom{n}{r} = N$ for some n and r such that $1 < r < n-1$. The method we are about to describe fails in cases where $\binom{n}{r} = N$ but the symmetric group S_n does have an element of order N, as it happens when $n = 15$, $r = 2$, and $N = 105$. As the symmetric group on 15 letters does have elements of order $105 = \binom{15}{2}$, namely any permutation which is a product of $3, 5$, and 7-cycle.

Let $f(x) \in \mathbb{Z}[x]$ be a monic irreducible polynomial of degree $n > 4$ whose Galois group is isomorphic to either A_n or S_n. Let r be such that $1 < r < n-1$ and $\binom{n}{r} = N$. Further assume that S_n does not have any element order N. We know that S_n is n-transitive and A_n is $(n-2)$-transitive on n letters. Let $f_r(x)$ denote a polynomial of degree $N = \binom{n}{r}$ whose roots are sum of all r different roots of $f(x)$. Let the roots of $f_r(x)$ be β_i, where $1 \le i \le N$.

The polynomials $f(x)$ and $f_r(x)$ have the same splitting field as shown in an earlier chapter. Since both S_n and A_n transitively permute the N roots of $f_r(x)$ this polynomial is irreducible. So the polynomial $f_r(x)$ is the required polynomial of degree N whose Galois group does not have any element of order N.

3. Examples

Example 1. The first interesting case is $N = 15$. Let $f(x) \in \mathbb{Z}[x]$ be an irreducible monic polynomial of degree six whose Galois group over $\mathbb{Q}$ is isomorphic to symmetric or alternating group on five or six letters. Then $f_2(x) \in \mathbb{Z}[x]$ is a irreducible monic polynomial whose Galois group is same as that of $f(x)$ and so does not have any element of order 15, and therefore $f_2(x)$ is reducible modulo every prime.

For instance let $f(x) = x^6 + 24x - 20$ whose discriminant is $2^{16} \cdot 3^6 \cdot 5^6$. We note that

$$f(x) \equiv (x+3)(x^5 + 4x^4 + 2x^3 + x^2 + 4x + 5) \text{ (modulo 7)}$$

$$f(x) \equiv (x+7)(x+12)(x+21)(x^3 + 6x^2 + 13x + 16) \text{ (modulo 23)}$$

$$f(x) \equiv (x^2 + 26x + 10)(x^4 + 3x^3 + 28x^2 + 25x + 27) \text{ (modulo 29).}$$

It follows that $f(x)$ is irreducible over $\mathbb{Q}$ and its Galois group G over $\mathbb{Q}$ is 2−transitive on its roots and has a 3−cycle. Therefore G is isomorphic to A_6 the alternating group on six letters (Wielandt 1964). We know that A_6 is 4−transitive on six letters. Let $f_2(x)$ represent the polynomial of degree 15 whose $15 = \binom{6}{2}$ roots are the sums of the roots of $f(x)$ taken two at a time. This polynomial turns out to be

$$x^{15} - 240x^{10} + 520x^9 - 6912x^5 - 8640x^4 - 10800x^3 - 13824.$$

The Galois group of this polynomial is same as that of $f(x)$ and so is isomorphic to A_6. As A_6 has no element of order 15 this polynomial is reducible modulo every prime.

Example 2. The second example is $N = 35$. As $\binom{7}{3} = 35$, we start with a some monic polynomial $f(x)$ of degree 7 with Galois group isomorphic to S_7 or A_7. The polynomial of degree 35 whose roots are the sums of three different roots of $f(x)$ is the required polynomial whose Galois group (being isomorphic to S_n or A_n) does not have any element of order 35. To illustrate this we begin with the polynomial $f(x) = x^7 - 2x^6 + 2x + 2$ of degree 7. We observe that the discriminant of the polynomial is $50808364 = 2^6 \cdot 3^8 \cdot 11^2$ and $f(x)$ is irreducible modulo 5. Also

$$f(x) \equiv (x^2 + 7x + 1)(x^2 + 11x + 8)(x^3 + 6x^2 + x + 10). \text{ (modulo 13)}.$$

So $f(x)$ is an irreducible polynomial of degree 7 whose discriminant is a square and Galois group G has a 3−cycle. So G is isomorphic to A_7 (Wielandt 1964). Suppose that the roots of $f(x)$ are α_i, $1 \leq i \leq 7$. Incidentally the polynomial $f_2(x)$ of degree $\binom{7}{2} = 21$ whose roots are $\alpha_i + \alpha_j$ $(1 \leq i < j \leq 7)$ is

$$x^{21} - 12x^{20} + 60x^{19} - 160x^{18} + 240x^{17} - 192x^{16} + 14x^{15} + 282x^{14} - 384x^{13}$$
$$-896x^{12} + 3456x^{11} - 4032x^{10} + 1452x^9 + 936x^8 - 3348x^7 + 8208x^6 - 10800x^5 + 6912x^4$$
$$- 1944x^3 + 648x^2 - 648x + 216$$

This polynomial is irreducible over $\mathbb{Q}$. As its Galois group is isomorphic to A_7 which does not have any element of order 21 this polynomial is reducible modulo every prime.

The polynomial $f_3(x)$ of degree $\binom{7}{3} = 35$ whose roots are $\alpha_i + \alpha_j + \alpha_k$ $(1 \leq i < j < k \leq 7)$ is

$$x^{35} - 30x^{34} + 420x^{33} - 3640x^{32} + 21840x^{31} - 96096x^{30} + 320400x^{29} - 824892x^{28}$$
$$+1651824x^{27} - 2520656x^{26} + 2467968x^{25} + 1014144x^{24} - 13570744x^{23} + 43939464x^{22}$$
$$- 97466448x^{21} + 165719040x^{20} - 229091136x^{19} + 279559296x^{18} - 328973632x^{17}$$
$$+ 369175728x^{16} - 339989856x^{15} + 197554480x^{14} - 25543680x^{13} + 5507328x^{12}$$
$$- 229676208x^{11} + 582038592x^{10} - 837493056x^9 + 855433568x^8 - 666645072x^7$$
$$+ 405962976x^6 - 192746432x^5 + 69432960x^4 - 17666304x^3 + 2572800x^2 - 58368x$$
$$- 18432$$

This polynomial is irreducible over $\mathbb{Q}$ and its Galois group is isomorphic to A_7. As A_7 does not have any element of order 35 this polynomial is reducible modulo every prime. Incidentally its discriminant is the following 311−digit number

$$2^{296} \cdot 3^{154} \cdot 11^{20} \cdot 2260889^2 \cdot 73504388212873^2 \cdot 307711591051853^6.$$

Note. The composite natural numbers N below 100 which are such that $gcd(N, \phi(N)) = 1$ are

$$15, 33, 35, 51, 65, 69, 77, 85, 91, 95.$$

Among these numbers the method described above works for $N = 15, 35$, and 91. This is so because $\binom{6}{2} = 15$, $\binom{7}{3} = 35$, and $\binom{14}{2} = 91$. As starting with a polynomial of degree 7 with Galois group isomorphic to A_7 or S_7, we constructed an irreducible polynomial of degree 35 which is reducible modulo every prime, likewise starting with a polynomial of degree 14 with Galois group isomorphic to A_{14} or S_{14} we can construct an irreducible polynomial of degree 91 which is reducible modulo every prime.

Example 3. The method discussed in the previous examples above does not work for $N = 33$. For this we proceed as in the proof of our theorem. As the order of 11 modulo 3 is 2, we construct a Frobenius group G of order $11^2 \cdot 3$ which is a semidirect product of $C_{11} \times C_{11}$ by a group of order 3.

More specifically we extend the group $C_{11} \times C_{11} = \langle x \rangle \times \langle y \rangle$ by the group $\langle \phi \rangle$ of order three where ϕ is the automorphism of $\langle x \rangle \times \langle y \rangle$ given by $\phi(x) = x^6 y$, $\phi(y) = xy^4$. It is easily seen that this automorphism has order three and is fixed-point-free. The resulting group, the semidirect product of $C_{11} \times C_{11}$ by $\langle \phi \rangle$ is a Frobenius group of order 363 having the subgroup $C_{11} \times C_{11}$ as its kernel and the group $\langle \phi \rangle$ as its complement. This Frobenius group of order 363 does not have any subgroup of order 33. Let $\mathbb{K}/\mathbb{Q}$ be a normal extension whose Galois group is isomorphic to G. Let H be a subgroup of G of order 11 and $\mathbb{F}$ be its fixed subfield. By FTGT (Fundamental Theorem of Galois Theory) the field $\mathbb{F}$ has degree 33 over $\mathbb{Q}$. Let α be an algebraic integer such that $\mathbb{F} = \mathbb{Q}(\alpha)$ and let $f(x) \in \mathbb{Z}[x]$ be its minimum polynomial. As proved in the theorem this polynomial has degree 33 and its Galois group is a Frobenius group of order $11^2 \times 3 = 363$ which does not have any subgroup of order 33 and therefore the irreducible polynomials $f(x)$ is reducible modulo every prime.

Appendix 1

Linear Independence of $\sqrt{p_1}, \sqrt{p_2}, \cdots, \sqrt{p_n}$ over $\mathbb{Q}$

Let $p_1, p_2, \cdots, p_n$ be distinct prime numbers. We shall prove that $\sqrt{p_1}, \sqrt{p_2}, \cdots, \sqrt{p_n}$ are linearly independent over $\mathbb{Q}$, the field of rational numbers. In fact as we shall see that it is easier to prove a little more. It is true that the following set of 2^n numbers

$$\begin{aligned}
&1, \sqrt{p_1}, \sqrt{p_2}, \sqrt{p_3}, \cdots, \sqrt{p_n}\\
&\sqrt{p_i p_j}, \quad 1 \le i < j \le n\\
&\sqrt{p_i p_j p_k}, \quad 1 \le i < j < k \le n\\
&\cdots\\
&\cdots\\
&\sqrt{p_1 p_2 p_3 \cdots p_n}
\end{aligned}$$

is linearly independent over $\mathbb{Q}$. To show this we prove the following slightly more general result.

Theorem 0.5. *Let $a_1, a_2, a_3, \cdots, a_n$ be integers such that none of which is a square and $gcd(a_i, a_j) = 1$ for $1 \le i < j \le n$. Let*

$$\mathbb{K} = \mathbb{Q}(\sqrt{a_1}, \sqrt{a_2}, \sqrt{a_3}, \cdots, \sqrt{a_n}).$$

Then $\mathbb{K}$ is a vector space over $\mathbb{Q}$ of dimension 2^n and has a basis consisting of the following elements:

$$\begin{aligned}
&1, \sqrt{a_1}, \sqrt{a_2}, \sqrt{a_3}, \cdots, \sqrt{a_n}\\
&\sqrt{a_i a_j}, \quad 1 \le i < j \le n\\
&\sqrt{a_i a_j a_k}, \quad 1 \le i < j < k \le n\\
&\cdots\\
&\cdots\\
&\sqrt{a_1 a_2 a_3 \cdots a_n}.
\end{aligned}$$

Proof. We shall prove this theorem by induction on n. The theorem is evidently true for $n = 1$. We now suppose that it is true for $n \leq k$. We shall show that it holds for $n = k + 1$. We observe that it is sufficient to prove that

$$\sqrt{a_{k+1}} \notin \mathbb{Q}(\sqrt{a_1}, \sqrt{a_2}, \sqrt{a_3}, \cdots, \sqrt{a_k}).$$

Suppose to the contrary

$$\sqrt{a_{k+1}} \in \mathbb{Q}(\sqrt{a_1}, \sqrt{a_2}, \sqrt{a_3}, \cdots, \sqrt{a_k}).$$

Since

$$\mathbb{Q}(\sqrt{a_1}, \sqrt{a_2}, \sqrt{a_3}, \cdots, \sqrt{a_k}) = \mathbb{Q}(\sqrt{a_1}, \sqrt{a_2}, \sqrt{a_3}, \cdots, \sqrt{a_{k-1}})(\sqrt{a_k})$$

we can write

$$\sqrt{a_{k+1}} = \alpha + \beta\sqrt{a_k}$$

where

$$\alpha, \beta \in \mathbb{Q}(\sqrt{a_1}, \sqrt{a_2}, \sqrt{a_3}, \cdots, \sqrt{a_{k-1}}).$$

Squaring $\sqrt{a_{k+1}} = \alpha + \beta\sqrt{a_k}$ we get

$$a_{k+1} = \alpha^2 + 2\alpha\beta\sqrt{a_k} + \beta^2 a_k.$$

If $\alpha\beta \neq 0$ then this would imply that

$$\sqrt{a_k} = \frac{a_{k+1} - \alpha^2 - \beta^2 a_k}{2\alpha\beta} \in \mathbb{Q}(\sqrt{a_1}, \sqrt{a_2}, \sqrt{a_3}, \cdots, \sqrt{a_{k-1}})$$

which will contradict the induction hypothesis.
So we must have $\alpha\beta = 0$. But $\beta = 0$ would imply that

$$\sqrt{a_{k+1}} = \alpha \in \mathbb{Q}(\sqrt{a_1}, \sqrt{a_2}, \sqrt{a_3}, \cdots, \sqrt{a_{k-1}})$$

which again contradicts the induction hypothesis. So we must have $\alpha = 0$ and so $\sqrt{a_{k+1}} = \beta\sqrt{a_k}$. Now we have

$$\beta = \frac{\sqrt{a_{k+1}}}{\sqrt{a_k}} = \frac{\sqrt{a_k a_{k+1}}}{a_k}.$$

Therefore

$$\sqrt{a_k \cdot a_{k+1}} = \beta \cdot a_k \in \mathbb{Q}(\sqrt{a_1}, \sqrt{a_2}, \sqrt{a_3}, \cdots, \sqrt{a_{k-1}})$$

which again contradicts the induction hypothesis for the k numbers $a_1, a_2, \cdots, a_{k-1}$, and $a_k a_{k+1}$.

This completes the proof of the assertion that

$$\sqrt{a_{k+1}} \notin \mathbb{Q}(\sqrt{a_1}, \sqrt{a_2}, \sqrt{a_3}, \cdots, \sqrt{a_k})$$

and so $\mathbb{K}$ is a vector space over $\mathbb{Q}$ of dimension 2^n.

Now about the basis. By induction hypothesis, $\mathbb{Q}(\sqrt{a_1}, \sqrt{a_2}, \sqrt{a_3}, \cdots, \sqrt{a_k})$—as a vector space over $\mathbb{Q}$—has a basis consisting of the following 2^k elements:

$$\begin{array}{l} 1, \sqrt{a_1}, \sqrt{a_2}, \sqrt{a_3}, \cdots, \sqrt{a_k} \\ \sqrt{a_i a_j}, \quad 1 \le i < j \le k \\ \sqrt{a_i a_j a_k}, \quad 1 \le i < j \le k \\ \cdots \\ \cdots \\ \sqrt{a_1 a_2 a_3 \cdots a_k}. \end{array}$$

These together with the 2^k elements obtained by multiplying each of the above by $\sqrt{a_{k+1}}$ will constitute a basis of $\mathbb{Q}(\sqrt{a_1}, \sqrt{a_2}, \sqrt{a_3}, \cdots, \sqrt{a_k}, \sqrt{a_{k+1}})$ over $\mathbb{Q}$. Evidently this set has $2^k + 2^k = 2^{k+1}$ elements. This completes the proof of the theorem. □

Appendix 2

Ruler and Compass Construction

In this appendix we shall discuss what are known as constructible numbers and give some description of the real numbers $c \in \mathbb{R}$ which are such that starting with a line segment of one unit length, we can construct (find, isolate, determine) a line segment of length $|c|$ by using a sequence of operations involving straight edge ruler and compass alone.

Definition. Constructible Number

A real number c is said to be a constructible number if we can construct a line segment of length $|c|$, starting from some predefined unit length in a finite number of steps using straight edge ruler and a compass alone.

We shall argue that this is same thing as saying that the number c is given by an expression which involves members of $\mathbb{Z}$, four basic arithmetical operations of addition, subtraction, multiplication, division, and (multiple, nested) extraction of square root of positive numbers therein. For example, as seen in an earlier chapter we have the following expression for $\cos\dfrac{2\pi}{17}$.

$$\frac{1}{16}\left\{-1+\sqrt{17}+\sqrt{(34-2\sqrt{17})}\right\}+\frac{1}{16}\sqrt{\left\{68+12\sqrt{17}-16\sqrt{(34+2\sqrt{17})}+2(\sqrt{17}-1)\sqrt{(34-2\sqrt{17})}\right\}}.$$

So accordingly $\cos\frac{2\pi}{17}$ will be considered a constructible number which in particular implies that it is possible to draw a regular 17-gon as was first shown by Gauss.

We shall assume the familiarity with following elementary geometric constructions which can be performed by using straight edge ruler and compass:

(1) Bisecting a given line segment.

(2) Bisecting a given angle.

(3) Drawing a 90 degree angle and a 60 degree angle.

(4) Given a straight line L and a point P outside L, we can draw a straight line parallel to L and passing through the point P.

(5) Given a line L and a point P on it. We can draw a line passing through P and perpendicular to the line L.

It is rather obvious that if a and b are constructible numbers then so are the numbers $a+b$ and $a-b$. We shall show given the line segments of length a and b how to construct line segments corresponding to the numbers ab, a/b $(b \neq 0)$, and $\sqrt{a}$, for $a > 0$.

- $a \cdot b$ is constructible.

  Consider the following diagram:

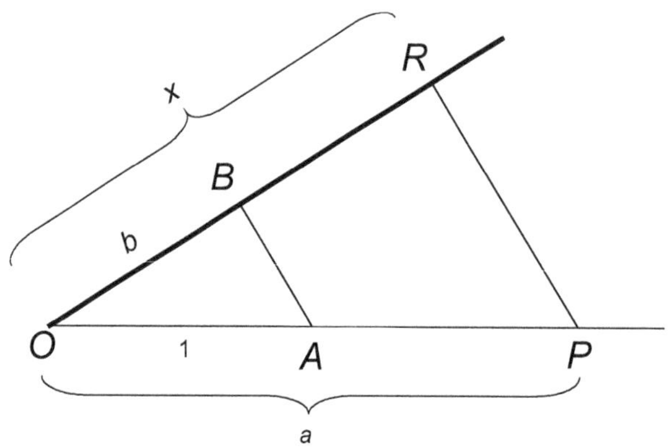

In this diagram, $OA = 1$, $OP = a$, $OB = b$. The line PR is drawn parallel to the line AB which can be done using ruler and compass alone and we let $x = OR$. Since the triangles OPR and OAB are similar, we have

$$\frac{OA}{OP} = \frac{OB}{OR}, \qquad \text{or} \qquad \frac{1}{a} = \frac{b}{x}, \qquad \text{or } x = ab.$$

In the above diagram we have assumed that $a > 1$. The case of $a < 1$ can be dealt with likewise.

- $\dfrac{1}{b}$ is constructible.

  Consider the following diagram:

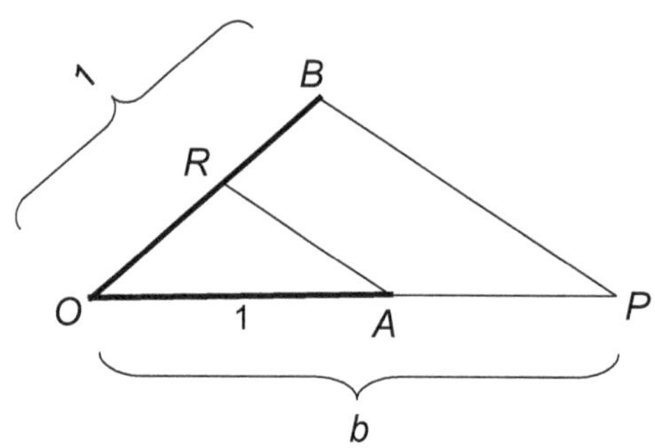

In this diagram $OA = 1$, $OP = b$, $OB = 1$, and the line PB is drawn parallel to the line AR. Let $OR = x$. Since the triangles OAR and OPB are similar we have

$$\frac{OA}{OP} = \frac{OR}{OB}, \qquad \text{or} \qquad \frac{1}{b} = \frac{x}{1}, \qquad \text{or } x = 1/b.$$

Since $a/b = a \times \frac{1}{b}$ and as both a and b are constructible numbers so is the number a/b.

- If $a > 0$ is constructible then so is $\sqrt{a}$.

  To show this is possible using ruler and compass alone we take a line segment OP with length a units and extend it by one unit to the point B. We draw a (semi) circle with OB as diameter. We draw a line through P perpendicular to the line OB. Let Q be the point of intersection of this line with the semicircle as shown in the following diagram. As the $\angle OQB$ is a right angle, the triangles OPQ and PBQ are similar triangles.

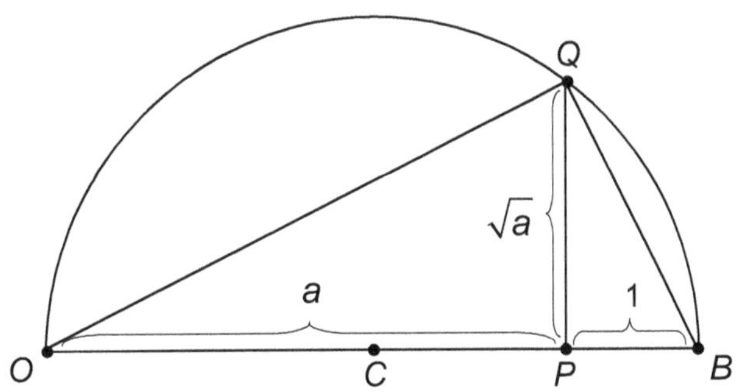

So we have

$$\frac{a}{PQ} = \frac{PQ}{1} \text{ that is } PQ^2 = a, \text{ or } PQ = \sqrt{a}.$$

It is easy to see from the following diagram that angle θ can be drawn by ruler and compass if and only if the number $\cos\theta$ is a constructible number.

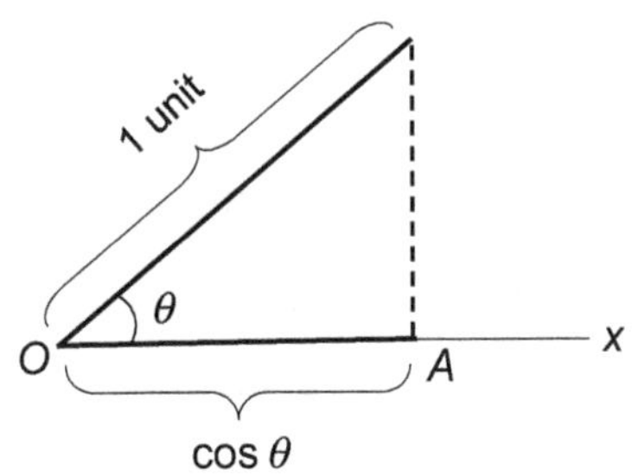

Starting with a unit length we can construct a line segment of any length n where n is any natural number. This in turn implies that we can construct a line segment whose length is any positive rational number. So every rational number is constructible. We shall now show that how the field $\mathbb{Q}$ whose every element is a constructible number can be extended by one usage of compass and straight edge ruler to a larger field of constructible numbers.

- Let L_1 and L_2 be two straight lines such that

$$L_1: \quad a_1x + b_1y + c_1 = 0, \qquad L_2: \quad a_2x + b_2y + c_2 = 0$$

  where $a_i, b_i, c_i \in \mathbb{Q}$ for $1 \leq i \leq 2$. If these two lines are not parallel then their point of intersection will be such that its both coordinates are members of $\mathbb{Q}$.
- Consider the line L_1 given above and a circle with center (a, b) and radius r where a, b, and r are all members of $\mathbb{Q}$. So the equation of this circle is $(x-a)^2+(y-b)^2 = r^2$. If the line L_1 intersects this circle in real points then the coordinates of these points will be roots of a quadratic equation and therefore will involve (at most) one square root of some positive number in $\mathbb{Q}$ in their representation.
- Suppose that C_1 and C_2 are two circles. Assume that the center of C_i is (a_i, b_i) and radius is r_i, and, $a_i, b_i, r_i \in \mathbb{Q}$ for $1 \leq i \leq 2$. Assume that these two circles intersect in real points—in particular these circles have different centers. Let

$$C_1: \quad (x-a_1)^2 + (y-b_1)^2 = r_1^2, \qquad C_2: \quad (x-a_2)^2 + (y-b_2)^2 = r_2^2.$$

  Subtracting the above two equations and simplifying we get

$$2x(a_2 - a_1) + 2y(b_2 - b_1) + (a_1^2 + b_1^2 - r_1^2) - (a_2^2 + b_2^2 - r_2^2) = 0,$$

  which represents a straight line passing through the points of intersection of the two circles. The points of intersection of the two circles are same as the point of intersection of the above straight line with any of the two circles. As argued before, the coordinates of these points of intersection are solution of some quadratic equation with rational coefficients. As we are assuming that the two circles intersect in real points, the coordinates of these points involve at most one square root of some positive (constructible) number along with basic arithmetical operation.

  When the coordinates of these points are added to the field $\mathbb{Q}$ and all the numbers which can be rationally expressed in terms of these new points and the members of

$\mathbb{Q}$, a field $\mathbb{K}$ possibly larger than the field $\mathbb{Q}$ is formed. The resulting field $\mathbb{K}$ has degree at most two over $\mathbb{Q}$.

So starting with $\mathbb{Q}$, if with one usage of straight edge rule and compass we do get a new constructible number then when this number is added to $\mathbb{Q}$, a quadratic field is generated which can be thought of as being generated by the square root of some positive number in $\mathbb{Q}$. If this positive number is denoted by $w \in \mathbb{Q}$, then every element of the field $\mathbb{Q}(\sqrt{w}) = \{a + b\sqrt{w}\}$ is a constructible number and this field has dimension two over $\mathbb{Q}$.

What we have said above can be just well applied to any field $\mathbb{K}$ of constructible numbers instead of $\mathbb{Q}$. Let $\mathcal{C} \subset \mathbb{R}$ represent the set of all (real) constructible numbers. Then $\mathcal{C}$ is a field which is closed with respect to taking the square roots of its positive elements—this means that $\alpha \in \mathcal{C} \Longrightarrow \sqrt{|\alpha|} \in \mathcal{C}$.

Let $\alpha \in \mathcal{C}$. Therefore $\mathbb{Q} \subset \mathbb{Q}(\alpha) \subset \mathcal{C}$. Let us try to look at how do we arrive at α beginning with elements of $\mathbb{Q}$ by successively applying the various steps of the usage of straight edge ruler and compass. For example, as can be easily checked

$$\cos\frac{\pi}{10} = \cos 18^\circ = \frac{1}{2}\sqrt{2 + \sqrt{\frac{3+\sqrt{5}}{2}}} = \frac{1}{2}\sqrt{2 + \frac{1+\sqrt{5}}{2}} = \frac{1}{2}\sqrt{\frac{5+\sqrt{5}}{2}}.$$

We have the following chain of fields:

$$\mathbb{Q} \subset \mathbb{Q}(\sqrt{5}) \subset \mathbb{Q}\left(\sqrt{\frac{5+\sqrt{5}}{2}}\right) = \mathbb{Q}\left(\cos\frac{\pi}{10}\right).$$

Following chain of field shows the three main steps in the constructibility of $\cos\dfrac{2\pi}{17}$.

$$\mathbb{Q} \subset \mathbb{Q}(\sqrt{17}) \subset \mathbb{Q}\left(\sqrt{\frac{17-\sqrt{17}}{2}}\right) \subset \mathbb{Q}\left(\cos\frac{2\pi}{17}\right).$$

Likewise for a constructible number $\alpha \in \mathcal{C}$, we will have a chain of fields:

$$\mathbb{Q} = \mathbb{K}_0 \subset \mathbb{K}_1 \subset \mathbb{K}_2 \subset \cdots\cdots \subset \mathbb{K}_{n-1} \subset \mathbb{K}_n = \mathbb{K}_{n-1}(\alpha).$$

From this we conclude that a positive real number α is constructible if it is an algebraic number of degree 2^n and there is a chain of fields $\mathbb{K}_i$ $0 \le i \le n$ as above where each field K_i, $1 \le i \le n$ is of dimension 2 over the preceding field and is generated by the square root of some (positive) element in the previous field. As we observed above in the expression for

$\cos\dfrac{\pi}{10}$, there are two fields in the chain leading to the field which contains $\cos\dfrac{\pi}{10}$ and so it is a (constructible) algebraic number of degree $2^2 = 4$.

Also in the expression for $\cos\dfrac{2\pi}{17}$ there are three intermediate fields and so it is a constructible number of degree $2^3 = 8$.

It is clear from the above discussion that for a number $\alpha \in \mathbb{R}$ to be constructible it is necessary (but not sufficient) that it is algebraic of degree 2^n for some integer $n \geq 0$. So if the possibility of some construction using ruler and compass implies encountering some line segment in the process whose length is either a nonalgebraic numbers or is an algebraic number of degree which is not a power of two then such ruler and compass constructions are deemed impossible.

(1) **The problem of doubling a cube**

This problem is about constructing, using a ruler and compass alone, a cube whose volume is twice that of a given cube.

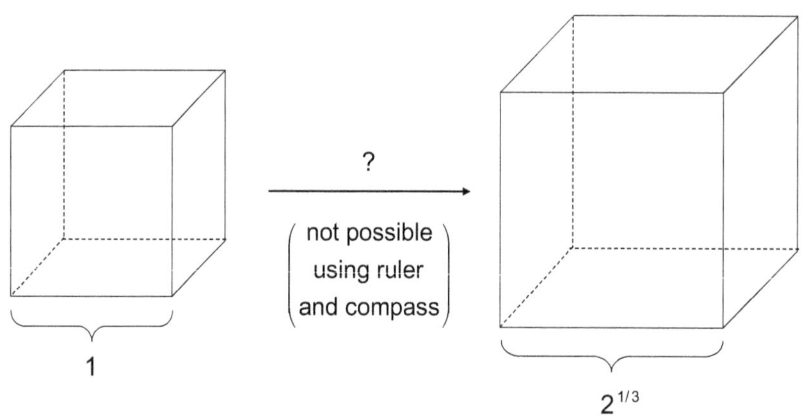

Without loss of generality we may assume that we are give a unit cube—that is each of its side is one unit length and so it has one unit volume. The cube whose volume is twice of this cube will have volume two units and therefore each of its side will have length $2^{1/3}$. If it were possible to find the side of cube whose volume is twice that of a unit cube by using straight edge and compass alone, then it would imply that starting with a segment of one unit length it is possible to find a line segment whose length is $\alpha = 2^{1/3}$. But we know that $\alpha = 2^{1/3}$ is algebraic of degree three as the polynomial $x^3 - 2 \in \mathbb{Q}[x]$ is irreducible over $\mathbb{Q}$. This proves that it is not possible to double a cube.

(2) The problem of squaring a circle

This problem is about constructing, using a ruler and compass alone, a square whose area is equal to the area of a given circle.

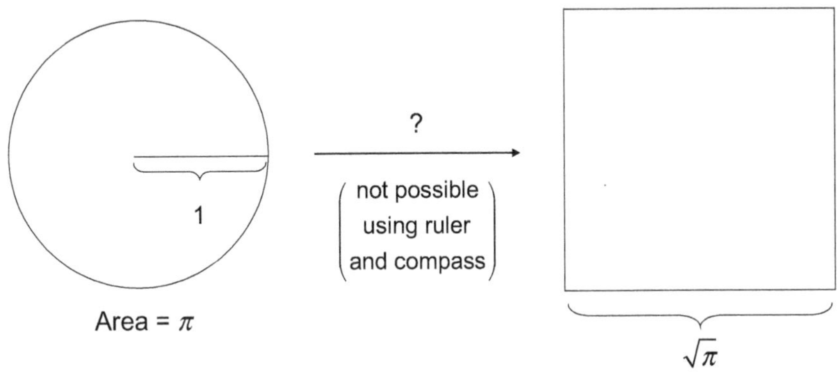

Consider a circle with one unit radius. Its area will be π units. A square whose area is equal to π would have its side equal to $\sqrt{\pi}$. So the possibility of being able to square a circle by ruler and compass would imply that the number $\sqrt{\pi}$ is algebraic which in turn implies that π is algebraic. But as we have shown in an earlier chapter that π is a transcendental number, it is not possible to square a circle.

(3) The problem of trisection of an angle

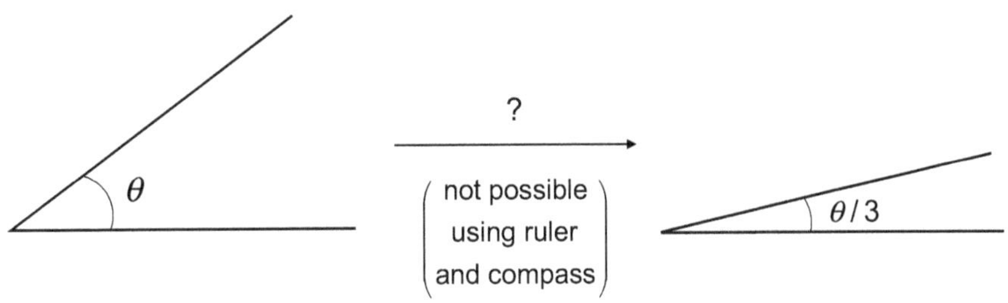

If it were possible to trisect an arbitrary angle then it should be possible to trisect a 60-degree angle which we can easily draw. This would imply that it is possible to draw an angle of 20 degrees. This in turn would imply that $\cos 20°$ is a constructible number. We know that $\cos 3\theta = 4\cos^3\theta - 3\cos\theta$. Substituting $\theta = \frac{\pi}{18} = 20°$ we get

$$\frac{1}{2} = 4\cos^3 20° - 3\cos 20°.$$

Letting $\cos 20° = x$ and simplifying we get $8x^3 - 6x - 1 = 0$. The polynomial $f(x) = 8x^3 - 6x - 1$ is easily seen to be irreducible by Rational Root Theorem as it does not have any rational roots or by seeing that it does not factor modulo five. So $\alpha = \cos 20^\circ$ is a root of an irreducible cubic equation in $\mathbb{Z}[x]$ and so it is an algebraic number of degree three. Therefore $\cos 20^\circ$ is not a constructible number. Since the possibility of trisection of an angle by ruler and compass would imply an incorrect conclusion namely that $\cos 20^\circ$ is an algebraic number of degree of power of two we conclude that it is not possible to trisect an angle by using ruler and compass.

We recall that in an earlier chapter we showed that

$$\cos 20^\circ = \frac{1}{2}\left(\frac{1}{2} + \frac{\sqrt{-3}}{2}\right)^{\frac{1}{3}} + \frac{1}{2}\left(\frac{1}{2} - \frac{\sqrt{-3}}{2}\right)^{\frac{1}{3}}$$

which shows that not only $\cos 20^\circ$ is *not* a constructible number but despite the fact that it is a real number it requires taking cube roots of (nonreal) complex numbers for its expression in radical form.

(4) **The problem of drawing regular n-gon.**

As we know we can draw a regular n-gon when $n = 3, 4, 5, 6$ and $n = 17$. Gauss showed that when p is a prime of the form $2^n + 1$, then it is possible to draw a regular p-gon by ruler and compass. Primes of the form $2^n + 1$ are called Fermat primes. The following three statements are equivalent regarding the ruler and compass construction of regular n-gons.

- It is possible to draw a regular n-gon in a circle.
- It is possible to draw an angle of $\dfrac{360}{n}$ degrees.
- $\cos\dfrac{2\pi}{n} = \cos\left(\dfrac{360}{n}\right)^\circ$ is a constructible number.

We know that it is possible to draw an equilateral triangle. Since

$$\cos\frac{2\pi}{5} = \cos 72^\circ = \frac{-1+\sqrt{5}}{4},$$

we know that it is possible to draw a regular pentagon. Since

$$2\cdot\frac{2\pi}{3} - 3\cdot\frac{2\pi}{5} = \frac{2\pi}{15} \qquad \text{or} \qquad 2\times 120^\circ - 3\times 72^\circ = 24^\circ,$$

it follows that we can draw a regular 15-gon as well. In fact we have

$$\cos\frac{2\pi}{15} = \cos 24^\circ = \frac{1+\sqrt{5}}{8} + \frac{\sqrt{3}}{4}\cdot\sqrt{\frac{5-\sqrt{5}}{2}}$$

which shows that $\cos 24^\circ$ is a constructible number and so it is possible to draw an angle of 24 degrees by ruler and compass and thereby construct a regular 15-gon.

Let m and n be two natural numbers such that $gcd(m,n) = 1$. Then we can find integers r and s such that $mr + ns = 1$. Therefore

$$r \cdot \frac{2\pi}{n} + s \cdot \frac{2\pi}{m} = \frac{2\pi}{mn}$$

which implies that if the angles $\frac{2\pi}{n}$ and $\frac{2\pi}{m}$ are constructible then the angle $\frac{2\pi}{mn}$ is also constructible. In other words, if $gcd(m,n) = 1$ and if a regular m-gon and n-gon are constructible, then a regular mn-gon is also constructible.

In general from what Gauss proved it follows that it is possible to draw a regular N-gon by using ruler and compass alone if and only if $N = 2^k p_1 p_2 p_3 \cdots p_r$, where $p_i's$ are all distinct Fermat primes. In other words, $\cos \frac{2\pi}{N}$ is a constructible number if and only if $N = 2^k p_1 p_2 p_3 \cdots p_r$, where $p_i's$ are all *distinct* Fermat primes.

Appendix 3

Tate's Proof of a Theorem of Dedekind

In this appendix we shall give a complete proof of the Theorem of Dedekind.

Let $f = f(x) \in \mathbb{Z}[x]$ be square-free monic polynomial of degree n and p be a prime such that p does not divide the discriminant of $f(x)$. Let $G \subset S_n$ be the Galois group of $f(x)$ over the field $\mathbb{Q}$ of rational numbers. Suppose that $f_p = \overline{f} = f$ (modulo p) $\in \mathbb{Z}_p[x]$ factors as:

$$f_p = \overline{f} = \prod_{i=1}^{r} \overline{f_i}$$

where $\overline{f_i}$ are distinct monic irreducible polynomials in $\mathbb{Z}_p[x]$, degree $(\overline{f_i}) = d_i$, $1 \leq i \leq r$, and $d_1 + d_2 + \cdots d_r = n$. Then there exists an automorphism $\sigma \in G$ which when considered as a permutation on the zeros of $f(x)$ is a product of disjoint cycles of lengths $d_1, d_2, \cdots, d_r$.

Note. Our proof is a somewhat detailed version of the one given in *Basic Algebra* by Jacobson (2014) and we shall keep the notations used in that proof. We shall assume that the reader is familiar with the following well-known results:

(1) Let $\mathbb{F}$ be a field and $f(x) \in \mathbb{F}[x]$ be a polynomial of degree $n \geq 2$. Then any two splitting fields of $f(x)$ are isomorphic.

(2) A finitely generated Abelian group is direct sum of (finitely many) cyclic groups. (This is the fundamental theorem of finitely generated Abelian groups.)

(3) A system of n homogeneous equation in $m > n$ variables has a nontrivial solutions.

(4) Let $\mathbb{E}/\mathbb{F}$ be an algebraic extension. Then any subring of $\mathbb{E}$ containing $\mathbb{F}$ is a subfield of $\mathbb{E}$.

    Proof. Let K be a ring such that $\mathbb{F} \subset K \subset \mathbb{E}$. Let $\alpha \in K - \mathbb{F}$. As α is algebraic over $\mathbb{F}$, $\mathbb{F}(\alpha) = \mathbb{F}[\alpha] \subset K$. So $\alpha^{-1} \in \mathbb{F}(\alpha) \subset K$. So K is a field.

(5) **Dedekind's Independence Theorem.** Distinct characters of a monoid (a set with associative binary operation with an identity element) into a field are linearly

independent. That is, if $\chi_1, \chi_2, \cdots, \chi_n$ are distinct characters of a monoid H into a field $\mathbb{F}$, then the only elements $a_i \in \mathbb{F}$, $1 \leq i \leq n$ such that

$$a_1\chi_1(h) + a_2\chi_2(h) + \cdots\cdots + a_n\chi_n(h) = 0$$

for all $h \in H$ are $a_i = 0$, $1 \leq i \leq n$.

(6) Let p be a prime and $GF(p^m)$ be a finite field with p^m elements. Then the group $\text{Aut}(GF(p^m)) = \langle\sigma\rangle$ is cyclic of order m and the generating automorphism σ maps $\alpha \in GF(p^m)$ to α^p.

(7) If R is a commutative ring with identity and M is a maximal ideal of R then R/M is a field.

(8) Let $\sigma, \eta \in S_n$. Then σ and $\eta^{-1}\sigma\eta$ have same cyclic structure.

Let $f(x) \in \mathbb{Z}[x]$ be a polynomial of degree $n \geq 1$, and p a prime number. Then $f_p(x) \in \mathbb{Z}_p[x]$ will denote the polynomial obtained by reducing the coefficients of $f(x)$ modulo p.

Theorem. Let $f(x) \in \mathbb{Z}[x]$ be a monic polynomial of degree $n \geq 1$ and p be a prime number which does not divide the discriminant of $f(x)$. Let $\mathbb{E}$ be a splitting field of $f(x)$ over $\mathbb{Q}$. Let $\mathbb{E}_p$ be a splitting field of $f_p(x)$ over $\mathbb{Z}_p = \mathbb{Z}/(p)$. Let

$$f(x) = (x - r_1)(x - r_2)\cdots\cdots(x - r_n), \qquad r_i \in \mathbb{E} \subset \mathbb{C}, 1 \leq i \leq n$$

$$R = \{r_1, r_2, \cdots, r_n\}, \qquad R_p = \{\overline{r_1}, \overline{r_2}, \cdots, \overline{r_n}\} \subset \mathbb{E}_p$$

where $\overline{r_i}$, $1 \leq i \leq n$ are the roots of $f_p(x) \in \mathbb{Z}_p[x]$ and

$$\mathbb{E} = \mathbb{Q}(r_1, r_2, \cdots, r_n), \qquad \mathbb{E}_p = \mathbb{Z}_p(\overline{r_1}, \overline{r_2}, \cdots, \overline{r_n}).$$

Let $D = \mathbb{Z}[r_1, r_2, \cdots, r_n]$ be the subring generated by the roots $r_1, r_2, \cdots, r_n$ of $f(x)$ in $\mathbb{C}$. Then

(a) There exists a homomorphism ψ of D onto $\mathbb{E}_p$.

(b) Any such homomorphism ψ gives a bijection of the set R of the roots of $f(x)$ in $\mathbb{E}$ onto the set R_p of the roots of the $f_p(x)$ in $\mathbb{E}_p$.

(c) If ψ and ψ' are two such homomorphisms, then there exist $\sigma \in Aut(\mathbb{E}/\mathbb{Q}) = Gal(f(x))$, such that $\psi' = \psi \cdot \sigma$. (Note that the restriction of σ to D is an automorphism of D.)

Proof.

(a)

$$\mathbb{E} = \mathbb{Q}(r_1, r_2, \cdots, r_n) = \mathbb{Q}[r_1, r_2, \cdots r_n].$$

We claim that $D = \mathbb{Z}[r_1, r_2, \cdots, r_n]$ is a finitely generated (additive) Abelian group. Since each r_i is a root of the monic polynomial $f(x) \in \mathbb{Z}[x]$ of degree n any positive power of r_i, $1 \leq i \leq n$ can be expressed as an integral linear combination of $1, r_i, r_i^2, \cdots r_i^{n-1}$. It follows that

$$D = \sum_{0 \leq e_i \leq n-1} \mathbb{Z} r_1^{e_1} r_2^{e_2} \cdots r_n^{e_n}.$$

Therefore D is a finitely generated (additive) Abelian group generated by at most n^n elements. By the *Fundamental Theorem for Finitely Generated Abelian Groups*, D is a direct sum of finitely many cyclic groups. Since $D \subset \mathbb{C}$, none of these cyclic groups is finite. So D is a direct sum of finitely many infinite cyclic groups. Let $\{u_1, u_2, \cdots, u_N\}$ be a set consisting of an independent generating system of D. We have

$$D = \mathbb{Z}u_1 \oplus \mathbb{Z}u_2 \oplus \cdots \oplus \mathbb{Z}u_N, \qquad N \leq n^n.$$

We claim that $\{u_1, u_2, \cdots, u_N\}$ is a basis of $\mathbb{E}/\mathbb{Q}$. Obviously $\{u_1, u_2, \cdots, u_N\}$ is linearly independent over $\mathbb{Q}$. Let $\mathbb{Q}D = \sum_{1 \leq i \leq N} \mathbb{Q}u_i$. Then $\mathbb{Q}D$ is a ring and $\mathbb{Q} \subset \mathbb{Q}D \subset \mathbb{E}$ therefore by (4) $\mathbb{Q}D$ is a field. Since $r_i \in D$ for $1 \leq i \leq n$, $\mathbb{Q}D = \mathbb{E}$ and $\{u_1, u_2, \cdots, u_N\}$ is a basis of $E/\mathbb{Q}$. As $D = \mathbb{Z}u_1 \oplus \mathbb{Z}u_2 \oplus \cdots \oplus \mathbb{Z}u_N$,

$$pD = \mathbb{Z}(pu_1) \oplus \mathbb{Z}(pu_2) \oplus \cdots \oplus \mathbb{Z}(pu_N)$$

is an ideal of D and

$$D/pD = \{\overline{a_1u_1 + a_2u_2 + \cdots + a_Nu_N} : 0 \leq a_i \leq p-1\}.$$

Therefore the D/pD is finite of order p^N. Let M be a maximal ideal of D containing pD. That is, $pD \subset M \subset D$ and D/M is a finite field of characteristic p and so it has a subfield isomorphic to $\mathbb{Z}_p = \mathbb{Z}/p\mathbb{Z}$ which we will identify as $\mathbb{Z}_p$ in what follows. As

$$D/M \approx \frac{D/pD}{M/pD}$$

the order of D/M is p^m, $1 \leq m \leq N$. Consider the canonical epimorphism

$$\nu : D \longrightarrow D/M$$

whose kernel is M and $p\mathbb{Z} \subset M$. Therefore, $\nu(\mathbb{Z}) = \mathbb{Z}_p$. We note that as $D = \mathbb{Z}[r_1, r_2, \cdots, r_n]$, we have for $1 \leq i \leq n$

$$\nu(r_i) = r_i + M = \overline{r_i}, \qquad \nu(D) = \mathbb{Z}_p[\overline{r_1}, \overline{r_2}, \cdots, \overline{r_n}].$$

As ν is an epimorphism, we have

$$\nu(D) = D/M = \mathbb{Z}_p[\overline{r_1}, \overline{r_2}, \cdots, \overline{r_n}]$$

is a splitting field of $f_p(x)$ over $\mathbb{Z}_p$. As both D/M and $\mathbb{E}_p$ are splitting fields of $f_p(x)$ over $\mathbb{Z}_p$, they are isomorphic. Let

$$\phi : D/M \longrightarrow \mathbb{E}_p$$

be such an isomorphism. Then $\psi = \phi \cdot \nu$ is a homomorphism of D onto $\mathbb{E}_p$.

(b) Let $\psi : D \longrightarrow \mathbb{E}_p$ be a homomorphism. So $\psi(1) = 1$. As $\mathbb{Z} \subset D$ and $\mathbb{E}_p$ has characteristic p, $\psi(p) = 0$, so $\psi(\mathbb{Z}) = \mathbb{Z}_p \subset \mathbb{E}_p$. ψ can be extended to homomorphism of the polynomial rings $D[x] \longrightarrow \mathbb{E}_p[x]$. Under this mapping $f(x) \longrightarrow f_p(x)$. As

$$f(x) = (x - r_1)(x - r_2) \cdot \cdots \cdot (x - r_n)$$

$$\psi(f(x)) = f_p(x) = (x - \psi(r_1))(x - \psi(r_2)) \cdot \cdots \cdot (x - \psi(r_n)),$$

$\psi(r_i)$, $1 \leq i \leq n$ are the roots of the $f_p(x)$ in $\mathbb{E}_p$ and therefore the restriction of ψ to R

$$\psi_{|R} : \{r_1, r_2, \cdots, r_n\} \longrightarrow \{\overline{r_1}, \overline{r_2}, \cdots, \overline{r_n}\}$$

is a bijection of the set R of roots of $f(x)$ in $\mathbb{E}$ to the set R_p of the roots of $f_p(x)$ in $\mathbb{E}_p$.

(c) We have seen that given a homomorphism $\psi : D \longrightarrow \mathbb{E}_p$ and $\sigma \in Gal(f) = Aut(\mathbb{E}/\mathbb{Q})$ $\psi' = \psi \cdot \sigma$ is also a homomorphism from D to $\mathbb{E}_p$ (note that the restriction of $\sigma \in Aut(\mathbb{E}/\mathbb{Q})$ to $D = \mathbb{Z}[r_1, r_2, \cdots, r_n]$ is also an automorphism of the ring D). Since $[\mathbb{E} : \mathbb{Q}] = N$, the group $Aut(\mathbb{E}/\mathbb{Q})$ has order N. Let

$$Aut(\mathbb{E}/\mathbb{Q}) = \{\sigma_1, \sigma_2, \cdots, \sigma_N\}.$$

So given a nontrivial homomorphism $\psi : D \longrightarrow \mathbb{E}_p$, we get N distinct homomorphisms $\psi_j = \psi \cdot \sigma_j$, $1 \leq j \leq N$, from D to $\mathbb{E}_p$. We claim that these are all the homomorphisms from D to $\mathbb{E}_p$. Suppose that there is a homomorphism from D to $\mathbb{E}_p$ which is different from ψ_j, $1 \leq j \leq N$. Let us denote it by ψ_{N+1}. By *Dedekind Independence Theorem* the set

$\{\psi_1, \psi_2, \cdots, \psi_N, \psi_{N+1}\}$ of $N+1$ homomorphisms from D to $\mathbb{E}_p$ is linearly independent over the field $\mathbb{E}_p$.

Consider the following system of N homogeneous equations in $N+1$ variables $\{x_1, x_2, \cdots, x_N, x_{N+1}\}$,

$$\sum_{i=1}^{N+1} x_i \psi_i(u_j) = 0, \qquad 1 \leq j \leq N.$$

Since there are more variables than the equations, this system of equations has a nontrivial solution. Let this nontrivial solution be $x_i = a_i \in \mathbb{E}_p$, $1 \leq i \leq N+1$. So we have

$$\sum_{i=1}^{i=N+1} a_i \psi_i(u_j) = 0, \qquad 1 \leq j \leq N.$$

Let $y \in D = \mathbb{Z}u_1 \oplus \mathbb{Z}u_2 \oplus \cdots \oplus \mathbb{Z}u_N$. So $y = n_1u_1 + n_2u_2 + \cdots n_Nu_N$, $n_k \in \mathbb{Z}$, $1 \leq k \leq N$. Then for $1 \leq i \leq N+1$ we have

$$\psi_i(y) = \overline{n_1}\psi_i(u_1) + \overline{n_2}\psi_i(u_2) + \cdots \overline{n_N}\psi_i(u_N) = \sum_{j=1}^{j=N} \overline{n_j}\psi_i(u_j)$$

where $\overline{n_j} = n_j + (p)$. We shall show that

$$\sum_{i=1}^{i=N+1} a_i \psi_i(y) = 0,$$

which will contradict the linear independence of $\{\psi_1, \psi_2, \cdots, \psi_N, \psi_{N+1}\}$ over $\mathbb{E}_p$.

$$\begin{aligned}
\sum_{i=1}^{i=N+1} a_i \psi_i(y) &= \sum_{i=1}^{i=N+1} a_i \sum_{j=1}^{j=N} \overline{n_j}\psi_i(u_j) \\
&= \sum_{i=1}^{i=N+1} a_i \left(\overline{n_1}\psi_i(u_1) + \overline{n_2}\psi_i(u_2) + \cdots + n_N\psi_i(u_N)\right) \\
&= \overline{n_1} \sum_{i=1}^{i=N+1} a_i\psi_i(u_1) + \overline{n_2} \sum_{i=1}^{i=N+1} a_i\psi_i(u_2) + \cdots + \overline{n_N} \sum_{i=1}^{i=N+1} a_i\psi_i(u_N) \\
&= 0.
\end{aligned}$$

Proof of the main theorem

Since the field $\mathbb{E}_p$ has order p^m, the group $Aut(\mathbb{E}_p)$ has order m and $\pi : \mathbb{E}_p \longrightarrow \mathbb{E}_p$ where $\pi(a) = a^p$ for all $a \in \mathbb{E}_p$, is the generating automorphism of $Aut(\mathbb{E}_p)$. So if $\psi : D \longrightarrow \mathbb{E}_p$ is any homomorphism then so is $\pi \cdot \psi$. Since ψ and $\pi \cdot \psi$ are two homomorphisms from D to

$\mathbb{E}_p$, there exist $\sigma \in Aut(\mathbb{E}/\mathbb{Q})$ such that $\pi \cdot \psi = \psi \cdot \sigma$ or $\psi^{-1} \cdot \pi \cdot \psi = \sigma$. This proves that the action on σ on $\{r_1, r_2, \cdots, r_n\}$ is similar to the action of π on $\{\overline{r_1}, \overline{r_2}, \cdots, \overline{r_n}\}$.

Note. In the following diagram, the mapping

$$D \xrightarrow{\sigma} D$$

is the restriction of $\sigma \in Aut(\mathbb{E}/\mathbb{Q})$ to D and we are only concerned with the effect of the mappings σ, ψ, and π on $\{r_1, r_2, \cdots, r_n\}$, and $\{\overline{r_1}, \overline{r_2}, \cdots, \overline{r_n}\}$. Clearly

$$\{r_1, r_2, \cdots, r_n\} \xrightarrow{\sigma} \{r_1, r_2, \cdots, r_n\}$$
$$\{\overline{r_1}, \overline{r_2}, \cdots, \overline{r_n}\} \xrightarrow{\pi} \{\overline{r_1}, \overline{r_2}, \cdots, \overline{r_n}\}$$
$$\{r_1, r_2, \cdots, r_n\} \xrightarrow{\psi} \{\overline{r_1}, \overline{r_2}, \cdots, \overline{r_n}\}$$

$$\begin{array}{ccc} D & \xrightarrow{\sigma} & D \\ \psi \downarrow & & \downarrow \psi \\ \mathbb{E}_p & \xrightarrow{\pi} & \mathbb{E}_p \end{array}$$

As $\psi^{-1} \cdot \pi \cdot \psi = \sigma$ and $\psi \cdot \sigma \cdot \psi^{-1} = \pi$, the effect of σ on $\{r_1, r_2, \cdots, r_n\}$ is similar to the effect of π on $\{\overline{r_1}, \overline{r_2}, \cdots, \overline{r_n}\}$. This is further illustrated by the following:

- $\sigma(r_i) = r_j \Longrightarrow \pi(\overline{r_i}) = \overline{r_j}$

$$\overline{r_i} \xrightarrow{\psi^{-1}} r_i \xrightarrow{\sigma} r_j \xrightarrow{\psi} \overline{r_j}.$$

- $\pi(\overline{r_i}) = \overline{r_j} \Longrightarrow \sigma(r_i) = r_j$

$$r_i \xrightarrow{\psi} \overline{r_i} \xrightarrow{\pi} \overline{r_j} \xrightarrow{\psi^{-1}} r_j.$$

Appendix 4

About Solvable Quintics

In this appendix we shall show that to every irreducible quintic (an irreducible polynomial of degree five) $f(x) \in \mathbb{Q}[x]$ we can associate a polynomial $F(x) \in \mathbb{Q}[x]$ of degree six (called the sextic resolvent for the polynomial $f(x)$) which is such that the equation $f(x) = 0$ is solvable by radicals (in other words the Galois group of $f(x)$ over $\mathbb{Q}$ is solvable) if and only if its sextic resolvent polynomial $F(x)$ has a rational root. Before stating the main theorem we will discuss some facts about some subgroups of S_5, the symmetric group on five letters.

- **About the subgroups of S_5**

  Let us write

$$S_3 = \{\varepsilon, (1,2,3), (1,3,2), (1,2), (2,3), (1,3)\}.$$

  Let us number these six permutations in the group S_3 as follows:

$$h_1 = \varepsilon, \quad h_2 = (1,2,3), \quad h_3 = (1,3,2), \quad h_4 = (1,2), \quad h_5 = (2,3), \quad h_6 = (1,3).$$

  The six Sylow-5 subgroups of S_5 are as follows (*we multiply permutations from left to right*):

$$P_1 = \langle(1,2,3,4,5)\rangle, \quad P_2 = h_2^{-1}P_1h_2 = \langle(1,4,5,2,3)\rangle, \quad P_3 = h_3^{-1}P_1h_3 = \langle(1,2,4,5,3)\rangle$$

$$P_4 = h_4^{-1}P_1h_4 = \langle(1,3,4,5,2)\rangle, \quad P_5 = h_5^{-1}P_1h_5 = \langle(1,3,2,4,5)\rangle, \quad P_6 = h_6^{-1}P_1h_6 = \langle(1,4,5,3,2)\rangle.$$

- The normalizer of a Sylow-5 subgroup of S_5 has order 20 and it is a Frobenius group. This Frobenius group in turn has a subgroup of order 10 which is a dihedral group. For $1 \le i \le 6$ let $F_i =$ be the normalizer of the Sylow-5 subgroup, H_i denote the dihedral subgroup of order 10 contained in F_i. So the following inclusions hold:

$$P_1 \subset \langle(1,2,3,4,5), (2,5)(3,4)\rangle = H_1 \subset \langle(1,2,3,4,5), (2,3,5,4)\rangle = F_1$$
$$P_2 \subset \langle(1,4,5,2,3), (1,4)(3,5)\rangle = H_2 \subset \langle(1,2,3,4,5), (1,3,4,5)\rangle = F_2$$

$$P_3 \subset \langle (1,2,4,5,3),(1,5)(2,4)\rangle = H_3 \subset \langle (1,2,4,5,3),(1,2,5,4)\rangle = F_3$$

$$P_4 \subset \langle (1,3,4,5,2),(1,5)(3,4)\rangle = H_4 \subset \langle (1,3,4,5,2),(1,3,5,4)\rangle = F_4$$

$$P_5 \subset \langle (1,3,2,4,5),(3,5)(2,4)\rangle = H_5 \subset \langle (1,3,2,4,5),(3,2,5,4)\rangle = F_5$$

$$P_6 \subset \langle (1,4,5,3,2),(1,4)(2,5)\rangle = H_6 \subset \langle (1,4,5,3,2),(1,5,4,2)\rangle = F_6.$$

- Let $\theta_1 = \theta_1(x_1,x_2,x_3,x_4,x_5)$, for $1 \le i \le 6$ be the following polynomials in $x_i's$

$$\theta_1(x_1,x_2,x_3,x_4,x_5) = (x_1x_2+x_2x_3+x_3x_4+x_4x_5+x_5x_1-x_1x_3-x_3x_5-x_5x_2-x_2x_4-x_4x_1)^2.$$

We let $\theta_1 = \theta$, and we define (by abuse of notation) θ_i, $2 \le i \le 6$ as follows:

$$\theta_2 = \theta(1,2,3), \quad \theta_3 = \theta(1,3,2), \quad \theta_4 = \theta(1,2), \quad \theta_5 = \theta(2,3), \quad \theta_6 = \theta(1,3).$$

Here $\theta_2 = \theta(1,2,3)$ means that θ_2 is obtained by applying the permutation $(1,2,3)$ to the subscripts of $x_i's$ appearing in θ. The polynomials θ_3, θ_4, θ_5, θ_6 are defined likewise.

Then we have

$$\theta_2 = (x_2x_3+x_3x_1+x_1x_4+x_4x_5+x_5x_2-x_2x_1-x_1x_5-x_5x_3-x_3x_4-x_4x_2)^2$$

$$\theta_3 = (x_3x_1+x_1x_2+x_2x_4+x_4x_5+x_5x_3-x_3x_2-x_2x_5-x_5x_1-x_1x_4-x_4x_3)^2$$

$$\theta_4 = (x_2x_1+x_1x_3+x_3x_4+x_4x_5+x_5x_1-x_2x_3-x_3x_5-x_5x_1-x_1x_4-x_4x_2)^2$$

$$\theta_5 = (x_1x_3+x_3x_2+x_2x_4+x_4x_5+x_5x_1-x_1x_2-x_2x_5-x_5x_3-x_3x_4-x_4x_1)^2$$

$$\theta_6 = (x_3x_2+x_2x_1+x_1x_4+x_4x_5+x_5x_3-x_3x_1-x_1x_5-x_5x_2-x_2x_4-x_4x_3)^2.$$

We define resolvent sextic $F(x)$ as follows:

$$F(x) = \prod_{i=1}^{i=6}(x-\theta_i) = \prod_{i=1}^{i=6}\left(x-\theta_i(x_1,x_2,x_3,x_4,x_5)\right).$$

We note that the coefficients of $F(x)$ are elementary symmetric functions of $\theta_i's$ and since these are permuted by the symmetric group S_5, these coefficients are symmetric in $x_i's$. We emphasize that θ_i are polynomials in x_i and the symmetric group S_5 permutes $\theta_i's$.

- Let $f(x) \in \mathbb{Q}[x]$ be an irreducible polynomial of degree five and α_i, $1 \le i \le 5$ be its zeros in $\mathbb{C}$. Let $\mathbb{K} = \mathbb{Q}(\alpha_1, \alpha_2, \alpha_3, \alpha_4, \alpha_5)$ be its splitting field. Let $G = Gal(f) = Aut(\mathbb{K}/\mathbb{Q})$ be the Galois group of $f(x)$ over $\mathbb{Q}$. We define the polynomial $F_\alpha(x)$ as the sextic polynomial obtained by replacing $x_i's$ by the corresponding $\alpha_i's$. In other words

$$F_\alpha(x) = \prod_{i=1}^{i=6}(x - \theta_i(\alpha_1, \alpha_2, \alpha_3, \alpha_4, \alpha_5)) = \prod_{i=1}^{i=6}(x - \Theta_i) \in \mathbb{Q}[x]$$

where $\Theta_i = \theta_i(\alpha_1, \alpha_2, \alpha_3, \alpha_4, \alpha_5)$ for $1 \le i \le 6$.

Theorem. The Galois group (over $\mathbb{Q}$) of the irreducible quintic polynomial $f(x) \in \mathbb{Q}[x]$ namely $G = Gal(f) = Aut(\mathbb{K}/\mathbb{Q})$ is a solvable group if and only if one of the roots of the polynomial

$$F_\alpha(x) = \prod_{i=1}^{i=6}(x - \theta_i(\alpha_1, \alpha_2, \alpha_3, \alpha_4, \alpha_5, \alpha_6)) = \prod_{i=1}^{i=6}(x - \Theta_i) \in \mathbb{Q}[x]$$

is a rational number.

Proof. We first note that the symmetric group S_5 permutes $\theta_1, \theta_2, \theta_3, \theta_4, \theta_5$, and θ_6 transitively. Let us denote the stabilizer of θ_i in S_5 by $Fix(\theta_i)$, for $1 \le i \le 6$. We have

$$Fix(\theta_1) = \langle (1, 2, 3, 4, 5), (2, 3, 4, 5) \rangle = F_1,$$

$$Fix(\theta_2) = h_2^{-1} Fix(\theta_1) h_2 = F_2,$$

$$Fix(\theta_3) = h_3^{-1} Fix(\theta_1) h_3 = F_3,$$

$$Fix(\theta_4) = h_4^{-1} Fix(\theta_1) h_4 = F_4,$$

$$Fix(\theta_5) = h_4^{-1} Fix(\theta_1) h_5 = F_5,$$

$$Fix(\theta_6) = h_6^{-1} Fix(\theta_1) h_6 = F_6.$$

First suppose that G is solvable. Every solvable transitive subgroup of S_5 is contained in one of the $F_i's$. So $G \subset F_i$ for some i and therefore it G fixes $\Theta_i = \theta_i(\alpha_1, \alpha_2, \alpha_3, \alpha_4, \alpha_5)$. Hence Θ_i is a rational number.

Conversely, suppose that one of the $\Theta_i = \theta_i(\alpha_1, \alpha_2, \alpha_3, \alpha_4, \alpha_5)$ is a rational number. Without loss of generality we may assume that it is Θ_1. We know that F_1 is the stabilizer of θ_1 in S_5 and so it is the largest subgroup of S_5 which fixes θ_1. Therefore $G \subset F_1$. Since F_1 is solvable so is G.

Note. Dummit (1991) has used this method to explicitly compute the resolvent sextic $F(x)$ for the quintic polynomial $f(x) = x^5 + px^3 + qx^2 + rx + s$. However, he has used the following polynomial in x_1, x_2, x_3, x_4, x_5

$$x_1^2x_2x_5 + x_1^2x_3x_4 + x_2^2x_1x_3 + x_2^2x_1x_5 + x_3^2x_1x_5 + x_3^2x_2x_4 + x_4^2x_1x_2 + x_4^2x_3x_5 + x_5^2x_1x_4 + x_5^2x_2x_3$$

for this purpose to compute the sextic resolvent instead of the polynomial

$$\theta_1(x_1, x_2, x_3, x_4, x_5) = (x_1x_2 + x_2x_3 + x_3x_4 + x_4x_5 + x_5x_1 - x_1x_3 - x_3x_5 - x_5x_2 - x_2x_4 - x_4x_1)^2.$$

It can be shown that

$$\begin{aligned}&(x_1x_2 + x_2x_3 + x_3x_4 + x_4x_5 + x_5x_1 - x_1x_3 - x_3x_5 - x_5x_2 - x_2x_4 - x_4x_1)^2\\ &= 3(x_1^2x_2x_5 + x_1^2x_3x_4 + x_2^2x_1x_3 + x_2^2x_1x_5 + x_3^2x_1x_5 + x_3^2x_2x_4 + x_4^2x_1x_2 + x_4^2x_3x_5 + x_5^2x_1x_4 + x_5^2x_2x_3)\\ &+ \text{a symmetric polynomial in } x_1, x_2, x_3, x_4, x_5.\end{aligned}$$

Problems

(1) Prove that

$$\mathbb{Q}(\sqrt{2}, \sqrt{3}) = \mathbb{Q}(\sqrt{2} + \sqrt{3}).$$

Express $\sqrt{2}$ and $\sqrt{3}$ as a polynomial in $\sqrt{2} + \sqrt{3}$ with rational coefficients.

(2) Let $p \neq q$ be prime numbers. Prove that

$$\mathbb{Q}(\sqrt{p}, \sqrt{q}) = \mathbb{Q}(\sqrt{p} + \sqrt{q}).$$

Express $\sqrt{p}$ and $\sqrt{q}$ as a polynomial in $\sqrt{p} + \sqrt{q}$ with rational coefficients.

(3) Show that

$$\mathbb{Q}(\sqrt{2}, \sqrt{3}, \sqrt{5}) = \mathbb{Q}(\sqrt{2} + \sqrt{3} + \sqrt{5}).$$

Find the minimal polynomial of $\sqrt{2} + \sqrt{3} + \sqrt{5}$ over $\mathbb{Q}$.

(4) Show that $x^3 + 3x + 2$ is irreducible in $\mathbb{Q}[x]$. Suppose that $\alpha \in \mathbb{C}$ is a root of $x^3 + 3x + 2$. Express $\dfrac{1}{1+\alpha}$ as a polynomial in α of degree at most two.

(5) Let G be a finite group and $x \in G$ is an element of order n. Prove that

$$Order(x^d) = \frac{n}{gcd(n, d)}.$$

(6) Show that if α is a primitive n-th root of unity then so is α^d if and only if $gcd(n, d) = 1$.

(7) Show that the polynomial $x^6 - 17x^4 + 80x^2 - 100$ has a root modulo p for every prime p but has no root in $\mathbb{Z}$.

(8) Let $\mu(n)$ be the Mobius-μ function. Show that for any natural number N, $\mu(N)$ is equal to the sum of primitive n-the roots of unity.

(9) Show that minimal polynomial $f(x)$ of $\alpha = 2^{\frac{1}{4}} + i$ (over $\mathbb{Q}$) is

$$f(x) = x^8 + 4x^6 + 2x^4 + 28x^2 + 1$$

and

$$i = -\frac{1}{24}(5\alpha^7 + 19\alpha^5 + 5\alpha^3 + 127\alpha)$$

$$2^{\frac{1}{4}} = \frac{1}{24}(5\alpha^7 + 19\alpha^5 + 5\alpha^3 + 151\alpha).$$

(10) Let $\alpha = 2^{\frac{1}{3}} + \omega$, where $\omega = e^{\frac{2\pi i}{3}} = \dfrac{-1 + i\sqrt{3}}{2}$, is primitive cube root of unity. Show that the minimal polynomial of α over $\mathbb{Q}$ is: $f(x) = x^6 + 3x^5 + 6x^4 + 3x^3 + 9x + 9$. Express $2^{\frac{1}{3}}$ and ω as a polynomial in α.

(11) Let

$$\alpha = \sqrt[3]{2 + \frac{10}{9}\sqrt{3}} \qquad \text{and} \qquad \beta = \sqrt[3]{2 - \frac{10}{9}\sqrt{3}}.$$

Find the minimal polynomials (over $\mathbb{Q}$) of the following algebraic numbers:

(a) α

(b) β

(c) $\alpha + \beta$

(d) $\omega\alpha + \omega^2\beta$

(e) $\omega^2\alpha + \omega\beta$

(f) $\omega\alpha - \omega^2\beta$

(12) Show that

$$\mathbb{Q}(2^{\frac{1}{3}}, \sqrt{3}) = \mathbb{Q}(2^{\frac{1}{3}} + \sqrt{3}).$$

Find the minimal polynomial of $2^{\frac{1}{3}} + \sqrt{3}$.

(13) Let $n \geq 2$ be a natural number and

$$\mathbb{K} = \mathbb{Q}\left(\left\{\sqrt{i} : 2 \leq i \leq n\right\}\right).$$

Show that $[\mathbb{K} : \mathbb{Q}] = 2^k$, where $k = \#\text{primes} \leq n$.

(14) Find the minimal polynomials of $\cos\frac{2\pi}{7}$.

(15) Prove that the minimal polynomial of $\cos^2\frac{2\pi}{7}$ is $x^3 - \frac{5}{4}x^2 + \frac{3}{8}x - \frac{1}{64}$.

(16) Fill in the second column of the following table which gives the minimal polynomials of the algebraic numbers in the first column.

Algebraic Number	Minimal Polynomial
$\sin^2 \frac{2\pi}{7}$	
$\sin \frac{2\pi}{7}$	
$i \sin \frac{2\pi}{7}$	
$2 \sin \frac{2\pi}{7}$	

(17) As can be easily checked the two polynomials $f_1(x) = x^3 + x^2 + 1$ and $f_2(x) = x^3 + x + 1$ are irreducible over $\mathbb{F}_2$—the field of order two. Let $\mathbb{K}_1$ be the splitting field of the polynomial $f_1(x)$ and $\mathbb{K}_2$ be the splitting field of the polynomial $f_2(x)$. Each of these fields have order eight. It is known that any two finite fields of same order are isomorphic. Give an explicitly description of an isomorphism between these two splitting fields.

(18) Prove that

$$\tan \frac{3\pi}{11} + 4 \sin \frac{2\pi}{11} = \sqrt{11}.$$

(19) Let

$$\alpha = \sqrt{1985 + 31\sqrt{1985}}, \qquad \beta = \sqrt{3970 + 64\sqrt{1985}}.$$

Prove that $\mathbb{Q}(\alpha) = \mathbb{Q}(\beta)$.

(20) Let $f(x) \in \mathbb{Q}[x]$ be an irreducible polynomial of degree three whose Galois group is cyclic (of order three). Prove that all the roots of $f(x)$ are real. Is the converse of this true?

(21) Let $f(x) \in \mathbb{Q}[x]$ be a polynomial of degree $n \geq 2$. Let $\alpha \in \mathbb{C}$ be a root of $f(x)$. Let $g(x) \in \mathbb{Q}[x]$ be a nonzero polynomial of degree m where $0 < m < n$. Show how to express $\dfrac{1}{g(\alpha)}$ as an element of $\mathbb{Q}[\alpha]$.

(22) Let $\alpha \neq 1$ be a root of the equation $x^{17} - 1 = 0$, and

$$\begin{aligned} \beta_1 &= \alpha + \alpha^4 + \alpha^{13} + \alpha^{16} \\ \beta_2 &= \alpha^2 + \alpha^8 + \alpha^9 + \alpha^{15} \\ \beta_3 &= \alpha^3 + \alpha^5 + \alpha^{12} + \alpha^{14} \\ \beta_4 &= \alpha^6 + \alpha^7 + \alpha^{10} + \alpha^{11}. \end{aligned}$$

Find an equation of degree four whose roots are β_1, β_2, β_3, and β_4.

(23) Let $\alpha \neq 1$ be a root of the equation $x^{19} - 1 = 0$, and

$$\begin{aligned} \beta_1 &= \alpha + \alpha^8 + \alpha^7 + \alpha^{18} + \alpha^{11} + \alpha^{12} \\ \beta_2 &= \alpha^2 + \alpha^{16} + \alpha^{14} + \alpha^{17} + \alpha^3 + \alpha^5 \\ \beta_3 &= \alpha^4 + \alpha^{13} + \alpha^9 + \alpha^{15} + \alpha^6 + \alpha^{10}. \end{aligned}$$

Find an equation of degree three whose roots are β_1, β_2, and β_3.

(24) Given that $a + b + c = 3$, $a^2 + b^2 + c^2 = 5$, and $a^3 + b^3 + c^3 = 7$, find the values of $a^4 + b^4 + c^4$ and $a^5 + b^5 + c^5$.

(25) If α, β, and γ be three numbers (real or complex) such that

$$\alpha + \beta + \gamma = 1, \quad \alpha^2 + \beta^2 + \gamma^2 = 2, \quad \alpha^3 + \beta^3 + \gamma^3 = 3.$$

(a) Find an equation whose roots are α, β, and γ.

(b) Find the value of $\alpha^4 + \beta^4 + \gamma^4$ and $\alpha^5 + \beta^5 + \gamma^5$.

(26) As can be easily checked the polynomial $x^4 + x + 1 \in \mathbb{F}_2[x]$ is irreducible. Let α be such that $\alpha^4 + \alpha + 1 = 0$ and $\mathbb{K} = \mathbb{F}_2[\alpha]$. Find the minimal polynomial of α^3 in $\mathbb{F}_2[x]$.

(27) Prove that

$$\alpha = \frac{1}{3}\left(1 + 10^{\frac{1}{3}} + 10^{\frac{2}{3}}\right)$$

is an algebraic integer by showing that its minimal polynomial is $x^3 - x^2 - 3x - 3$.

(28) As can be easily seen the polynomial $f(x) = x^3 - 2x + 2 \in \mathbb{Q}[x]$ is irreducible. Let $\alpha \in \mathbb{C}$ be a root of $f(x)$. Find the minimal polynomial over $\mathbb{Q}$ of $\alpha^2 + \alpha$, $\alpha^2 - \alpha$, and $\alpha^2 + \alpha + 1$.

(29) Let $\alpha = 2^{\frac{1}{2}} + 2^{\frac{1}{3}}$. Show that the minimal polynomial of α over $\mathbb{Q}$ is $x^6 - 6x^4 - 4x^3 + 12x^2 - 24x - 4$.

(30) Let p and q be two different prime numbers. Prove that

$$\mathbb{Q}(p^{\frac{1}{2}}, q^{\frac{1}{3}}) = \mathbb{Q}(p^{\frac{1}{2}} + q^{\frac{1}{3}}) = \mathbb{Q}(p^{\frac{1}{2}} \cdot q^{\frac{1}{3}}).$$

Find the minimal polynomial of $p^{\frac{1}{2}} + q^{\frac{1}{3}}$ over $\mathbb{Q}$.

(31) Let $f(x) = x^6 - 2x^3 - 1 \in \mathbb{Q}[x]$. Prove that the splitting field of $f(x)$ is a radical extension of degree 12 by showing that

$$\mathbb{Q} \subset \mathbb{Q}(\sqrt{2}) \subset \mathbb{Q}(\sqrt{2}, (1+\sqrt{2})^{\frac{1}{3}}) \subset \mathbb{Q}(\sqrt{2}, (1+\sqrt{2})^{\frac{1}{3}}, \omega)$$

where $\omega \neq 1$ is a cube root of unity.

(32) Let $\alpha \in \mathbb{C}$ be an algebraic number of degree p, where p is a prime number. Let $f(x) \in \mathbb{Q}[x]$ be a polynomial of degree k where $1 \leq k < p$. Prove that α can be expressed as a polynomial in $f(\alpha)$.

(33) Let p be a prime number. Show that $x^p - x - 1 \in \mathbb{F}_p[x]$ is irreducible over $\mathbb{F}_p$. Let $\mathbb{K} = \mathbb{F}_p[\alpha]$ be a splitting field of $x^p - x - 1$. Let $n = 1 + p + p^2 + \cdots + p^{p-1} = \dfrac{p^p - 1}{p - 1}$. Prove that $\alpha^n = 1$.

(34) Prove that the Galois group of $x^4 + 8x + 12$ over $\mathbb{Q}$ is isomorphic to A_4, the alternating group on four letters.

(35) Find the Galois group of $x^4 + 3x^3 - 3x - 2$ over $\mathbb{Q}$.

(36) Prove that the Galois group of $x^4 - 3x^2 + 4$ over $\mathbb{Q}$ is isomorphic to $C_2 \times C_2$, the Klein-Four group.

(37) Prove that the Galois group of $x^3 + x + 1 \in \mathbb{Q}[x]$ over $\mathbb{Q}$ is isomorphic to S_3—the symmetric group on three letters.

(38) Find a polynomial $f(x) \in \mathbb{Z}[x]$ of degree four whose Galois group over $\mathbb{Q}$ is isomorphic to S_4. Let $\alpha \in \mathbb{C}$ be a root of this polynomial. Argue that although α has degree $4 = 2^2$ but it is not a constructible number—namely α cannot be expressed as radical number involving square root radicals only.

(39) Show that for any prime p the polynomial $f(x) = x^4 + 1$ is *not* irreducible over the finite field $\mathbb{F}_p$ for any prime p.

(40) Let $m, n > 1$ be two natural numbers such that $gcd(m, n) = 1$ and p be any prime number. Assume that n is odd. Show that $f(x) = x^m + p^n$ is irreducible over $\mathbb{Q}$.

(41) Let $k \geq 2$ be any integer and p is a prime number. Prove that the Galois group of the polynomial $x^5 - kpx + p \in \mathbb{Z}[x]$ over $\mathbb{Q}$ is isomorphic to S_5, the symmetric group on five letters.

(42) Let p and q be distinct primes. Prove that

$$\mathbb{Q}\left(\sqrt{p} + q^{\frac{1}{p}}\right) = \mathbb{Q}\left(\sqrt{p}, q^{\frac{1}{p}}\right).$$

If $p > 2$ then

$$\mathbb{Q}\left(\sqrt{p} \cdot q^{\frac{1}{p}}\right) = \mathbb{Q}\left(\sqrt{p}, q^{\frac{1}{p}}\right).$$

(43) Let

$$\alpha = \sqrt[3]{\sqrt[3]{2} - 1}, \qquad \beta = (\sqrt[3]{3} - \sqrt[3]{6} + \sqrt[3]{12}).$$

Show that $\mathbb{Q}(\alpha) = \mathbb{Q}(\beta)$. Find the minimal polynomial of α over $\mathbb{Q}$.

(44) Prove without using calculator that

$$\sqrt[3]{10 + 6\sqrt{3}} + \sqrt[3]{10 - 6\sqrt{3}} = 2.$$

(45) Let $p > 2$ be a prime number and $f(x) \in \mathbb{Q}[x]$ be an irreducible polynomial of degree p. Suppose that the Galois group of $f(x)$ is solvable. Prove that
  (a) $f(x)$ has either just one real root or p real roots.
  (b) If $p \equiv 3$ (modulo 4) then $f(x)$ has one real root if and only if $Disc(f) < 0$.

(46) Let $f(x) = ax^3 + bx^2 + cx + d \in \mathbb{C}[x]$ be a polynomial of degree three. Suppose that for four different rational numbers α, β, γ, and δ such that $f(\alpha)$, $f(\beta)$, $f(\gamma)$, and $f(\delta)$ are all rational. Prove that $a, b, c, d \in \mathbb{Q}$.

(47) Let $f(x) = \sum_{j=0}^{j=n} a_j x^{n-j} = a_0 x^n + a_1 x^{n-1} + \cdots + a_n \in \mathbb{C}[x]$ be polynomial of degree $n \geq 1$. Let α_i, $0 \leq i \leq n$ be $n+1$ algebraic numbers such that $(n+1)$ complex numbers $f(\alpha_i)$, $0 \leq i \leq n$ are algebraic. Prove that a_j, $0 \leq j \leq n$ are algebraic numbers.

(48) Let $\alpha \in \mathbb{C}$ be an algebraic number such that $[\mathbb{Q}[\alpha] : \mathbb{Q}] = p$, where p is a prime number. Let $\beta \in \mathbb{Q}[\alpha] - \mathbb{Q}$. Prove that $\mathbb{Q}[\alpha] = \mathbb{Q}[\beta]$.

(49) Prove that the discriminant of $x^4 + px^2 + r$ is $16r(p^2 - 4r)^2$.

(50) Prove that the discriminant of $x^4 + qx + r$ is $-27q^4 + 256r^3$.

(51) Let α, β, and γ be the roots of the equation $x^3 + px + q = 0$. Find the equation whose roots are $(\alpha - \beta)^2$, $(\beta - \gamma)^2$, $(\gamma - \alpha)^2$.

(52) Let $f(x) \in \mathbb{Q}[x]$ be an irreducible polynomial of degree four with Galois group isomorphic to S_4 or A_4. $\alpha \in \mathbb{C}$ be a root of $f(x)$. Prove that α is not constructible, that is there is no radical expression for α which only involves square roots radicals.

(53) Let $f(x) = x^3 - 4x + 2$. Prove the following:

(a) $f(x)$ is irreducible over $\mathbb{Q}$.

(b) $f(x)$ has three real roots.

(c) Discriminant of $f(x)$ is $148 = 2^2 \cdot 37$.

(d) The Galois group of $f(x)$ over $\mathbb{Q}$ is isomorphic to S_3—the symmetric group on three letters.

(e) Let

$$A = \left(-1 + \frac{1}{9}\sqrt{-111}\right)^{\frac{1}{3}}, \quad B = \left(-1 - \frac{1}{9}\sqrt{-111}\right)^{\frac{1}{3}}, \qquad A \cdot B = \frac{4}{3}.$$

Show that the roots α, β, and γ of $f(x)$ can be written as

$$\alpha = A + B, \qquad \beta = \omega A + \omega^2 B, \qquad \beta = \omega^2 A + \omega B.$$

Let $\mathbb{K} = \mathbb{Q}(\alpha, \beta, \gamma, \omega)$, where $\omega = \dfrac{-1 + \sqrt{-3}}{2}$ is primitive cube root of unity. So $\mathbb{K}$ is the field obtained by adjoining ω to the splitting field of $f(x)$. You can look at $\mathbb{K}$ as the splitting field of the polynomial $(x^3 - 4x + 2)(x^2 + x + 1)$. Prove that $[\mathbb{K} : \mathbb{Q}] = 12$. Let $G = Aut(\mathbb{K}/\mathbb{Q})$. Show that $G \approx S_3 \times C_2$—which in turn is isomorphic to dihedral group of order 12. Set up the Galois correspondence between the subgroups of G and the subfields of $\mathbb{K}$.

(54) Let $f(x) = x^3 - 4x + 2$ and $\alpha \in \mathbb{C}$ be one of its root. Find the minimal polynomial of $\alpha^2 + \alpha + 1$ over $\mathbb{Q}$.

(55) Let $f(x) = x^4 - 10x^2 + 20$ and $\alpha \in \mathbb{C}$ be one of its root. Find the minimal polynomial of $\alpha^3 + 7$ over $\mathbb{Q}$.

(56) Let $\alpha = \sqrt{(2 + \sqrt{2})(3 + \sqrt{3})}$. Show that the minimal polynomial of α is $f(x) = x^8 - 24x^6 + 144x^4 - 288x^2 + 144$. Express $\sqrt{2}$ and $\sqrt{3}$ as a polynomial in α.

(57) Let $f(x)$ and $g(x)$ be two nonconstant polynomials over a field of degree n and m respectively. Prove that

$$Disc(f(x)g(x)) = Disc(f(x)) \times Disc(g(x)) \times Res(f(x), g(x))^2.$$

(58) Let $\{f_i(x)\}$, $1 \leq i \leq n$, $n \geq 2$ be any polynomials over a field. Prove that

$$\begin{aligned} &Disc(f_1 f_2 \cdots f_n) \\ &= \prod_{i=1}^{i=n} Disc(f_i) \times \left(Res(f_1, f_2) \cdot Res(f_1 f_2, f_3) \cdot \cdots \cdot Res(f_1 f_2 \cdots f_{n-1}, f_n)\right)^2. \end{aligned}$$

(59) Let G be the Galois group of the polynomial $f(x) = x^{12} - x^{10} + x^8 - x^6 + x^4 - x^2 + 1$ over $\mathbb{Q}$. Show that G is isomorphic to $C_6 \times C_2$. Let $\mathbb{K}$ be the splitting field of $f(x)$. Exhibit the Galois correspondence between the subgroups of G and the subfields of $\mathbb{K}$.

(60) Let α be a root of the equation $x^4 + x^3 + x^2 + x + 1 = 0$. Prove that the minimal polynomial of $\alpha - \alpha^2$ is $x^4 - 5x + 5$ and the minimal polynomial of $\alpha + \alpha^2$ is $x^4 + 2x^3 + 4x^2 + 3x + 1$.

(61) Let p be a prime and $f(x) = x^p - a \in \mathbb{K}[x]$ where $\mathbb{K} \subseteq \mathbb{R}$ is a subfield of the field of real numbers. Prove that if $f(x)$ has no root in $\mathbb{K}$ then it is irreducible.

(62) Let $f(x) \in \mathbb{Q}[x]$ be an irreducible cubic and α, β, and $\gamma \in \mathbb{C}$ be three roots of $f(x)$. Let $\Delta = (\alpha - \beta)^2(\beta - \gamma)^2(\gamma - \alpha)^2 \in \mathbb{Q}$ be the discriminant of $f(x)$. Prove that

$$\mathbb{Q}(\alpha, \beta, \gamma) = \mathbb{Q}(\alpha, \sqrt{\Delta}) = \mathbb{Q}(\beta, \sqrt{\Delta}) = \mathbb{Q}(\gamma, \sqrt{\Delta}).$$

(63) Let $\mathbb{K} \subseteq \mathbb{L}$ be subfields of the field of complex numbers $\mathbb{C}$. Suppose that $\mathbb{L}$ is a finite extension of $\mathbb{K}$. Show that the following two statements are equivalent.

(**A**) If an irreducible polynomial $p(x) \in \mathbb{K}[x]$ has one root in $\mathbb{L}$ then it has all of its roots in $\mathbb{L}$.

(**B**) $\mathbb{L}$ is the splitting field of some polynomial in $\mathbb{K}[x]$.

(64) Let $f(x) \in \mathbb{Q}[x]$ be an irreducible cubic whose all the three roots are real. Show that its roots cannot be expressed using real radicals alone.

(65) Let $f(x) \in \mathbb{Z}[x]$ be the minimal polynomial of the algebraic number $\alpha = (4 + \sqrt{15})^{\frac{1}{3}} + (4 - \sqrt{15})^{\frac{1}{3}}$. Prove that $Gal(f)$ is isomorphic to S_3—the symmetric group on three letters.

(66) Find the minimal polynomial of the algebraic number

$$\alpha = 2\cos\frac{2\pi}{15} = 2\cos 24^o = \frac{1+\sqrt{5}}{4} + \frac{\sqrt{3}}{2}\sqrt{\frac{5-\sqrt{5}}{2}}.$$

(67) Let p and q be two different prime numbers and

$$\alpha = \sqrt{(p+\sqrt{p})(q+\sqrt{q})}.$$

Find the minimal polynomial of α over $\mathbb{Q}$.

(68) Prove that

$$\mathbb{Q}\left((2+\sqrt{2})(-1+i)\right) = \mathbb{Q}(\sqrt{2}, i).$$

(69) Let $\alpha = \sqrt{(2+\sqrt{2})(-1+\sqrt{-1})} = \sqrt{(2+\sqrt{2})(-1+i)}$. Prove the following:

(a) α is algebraic of degree eight over $\mathbb{Q}$.

(b) The minimal polynomial of α (over $\mathbb{Q}$) is $f(x) = x^8 + 8x^6 + 32x^4 + 32x^2 + 128$.

(c) Let $\mathbb{K}$ denote the splitting field of $f(x)$. Prove that $Gal(f) = Aut(\mathbb{K}/\mathbb{Q})$ is isomorphic to dihedral group of order eight.

(70) Let $f(x) = x^5 - x - 16$. Prove that $f(x)$ is irreducible over $\mathbb{Q}$ and its Galois group is isomorphic to S_5—the symmetric group on five letters.

(71) Prove that $\cos\left(\frac{1}{3}\cos^{-1}\frac{11}{6}\right)$ is a constructible number.

(72) Let $f(x) \in \mathbb{Q}[x]$ be an irreducible cubic with Galois group cyclic of order three. Prove that $f(x)$ has no nonreal root.

(73) Let $\alpha \in \mathbb{C}$ is an algebraic number of degree three. Prove that $\mathbb{Q}(\alpha) = \mathbb{Q}(\alpha^2)$.

(74) Let $\alpha \in \mathbb{C}$ is an algebraic number of odd degree. Prove that

(a) $\mathbb{Q}(\alpha) = \mathbb{Q}(\alpha^2)$.

(b) Show how α can be expressed as an element of $\mathbb{Q}(\alpha^2)$.

(75) Let $p > 2$ be a prime number $\mathbb{K}$ be the splitting field of the polynomial $x^p - 2 \in \mathbb{K}[x]$. Let $G = Aut\,(\mathbb{K}/\mathbb{Q})$ be the Galois group of $x^p - 2$. Prove the following:

(a) $\mathbb{K} = \mathbb{Q}(2^p, \varepsilon)$, where $\varepsilon = e^{\frac{2\pi i}{p}} = \cos\frac{2\pi}{p} + i\sin\frac{2\pi}{p}$.

(b) $G = \langle \sigma, \eta \rangle$, where

$$\sigma : \begin{cases} 2^{\frac{1}{p}} & \longrightarrow \quad \varepsilon 2^{\frac{1}{p}} \\ \\ \varepsilon & \longrightarrow \quad \varepsilon \end{cases} \quad \text{and} \quad \eta : \begin{cases} 2^{\frac{1}{p}} & \longrightarrow \quad 2^{\frac{1}{p}} \\ \\ \varepsilon & \longrightarrow \quad \varepsilon^r, \text{ where } r \neq 1, \;\; r^{p-1} \equiv 1 \text{ (modulo } p). \end{cases}$$

(c) The group G has order $p(p-1)$. It is nonabelian. $\langle \sigma \rangle$ is its subgroup of order p and $\langle \eta \rangle$ is its cyclic subgroup of order $p - 1$.

(76) Let $f(x) = x^n + a_1 x^{n-1} + \cdots + a_{n-1}x + a_n \in \mathbb{Q}[x]$ be an irreducible polynomial of degree $n > 1$. Let $\Omega = \{\alpha_1, \alpha_2, \cdots, \alpha_n\}$ be the set of roots of $f(x)$ in $\mathbb{C}$. Let $\alpha, \beta \in \Omega$, $\alpha \neq \beta$. Then prove that

(a) $\alpha - \beta \notin \mathbb{Q}$.

(b) If $n > 1$ is odd then $\alpha + \beta, \;\; \alpha\beta, \;\; \dfrac{\alpha}{\beta} \notin \mathbb{Q}$.

(77) Let $\mathbb{K} = \mathbb{Q}(\sqrt{2}, \sqrt{3})$. Let $0 \neq \alpha_0 = a + b\sqrt{2} + c\sqrt{3} + d\sqrt{6} \in \mathbb{K}$, where $a, b, c, d \in \mathbb{Q}$. Find $A, B, C, D \in \mathbb{Q}$ such that $\alpha_0^{-1} = A + B\sqrt{2} + C\sqrt{3} + D\sqrt{6}$. Using the result of this problem or otherwise find the inverse of $1 + \sqrt{2} + \sqrt{3} + \sqrt{6} \in \mathbb{K}$.

(78) Let p and q be distinct prime numbers. Let $\mathbb{K} = \mathbb{Q}(\sqrt{p}, \sqrt{q})$ and $0 \neq \alpha = a + b\sqrt{p} + c\sqrt{q} + d\sqrt{pq} \in \mathbb{K}$. Let $\alpha^{-1} = A + B\sqrt{p} + C\sqrt{q} + D\sqrt{pq}$, where $A, B, C, D \in \mathbb{Q}$. Express the rational numbers $A, B, C,$ and D in terms of a, b, c, d, p, q.

(79) Prove that the polynomial $p(x) = x^4 - 8x^3 + 2x^2 + 8x + 1 \in \mathbb{Q}[x]$ is the minimal polynomial of the algebraic number $\alpha = 2 + \sqrt{2} + \sqrt{3} + \sqrt{6}$. Convince yourself that the four roots of this polynomial are

$$\begin{aligned} \alpha &= 2 + \sqrt{2} + \sqrt{3} + \sqrt{6} \\ \beta &= 2 - \sqrt{2} + \sqrt{3} - \sqrt{6} \\ \alpha &= 2 + \sqrt{2} - \sqrt{3} - \sqrt{6} \\ \alpha &= 2 - \sqrt{2} - \sqrt{3} + \sqrt{6} \end{aligned}$$

and its splitting field is $\mathbb{K} = \mathbb{Q}(\sqrt{2}, \sqrt{6})$. Prove that the polynomial $p(x) \in \mathbb{Z}[x]$ is irreducible but it is reducible modulo p for every prime p.

(80) Let $\alpha \in \mathbb{C}$ is algebraic of degree four and assume that its minimal polynomial $f(x) \in \mathbb{Q}[x]$ has Galois group which is isomorphic to either the symmetric groups S_4, or the alternating group A_4. Prove that α is not a constructible number.

(81) Let $f(x) = x^3 + 2x + 1 \in \mathbb{Q}[x]$. Let $\alpha \in \mathbb{C}$ be a root of $f(x)$ and $\mathbb{K} = \mathbb{Q}(\alpha) = \{a + b\alpha + c\alpha^2, a, b, c \in \mathbb{Q}\}$. Find $A, B, C \in \mathbb{Q}$ such that

$$\frac{1}{\alpha^2 - \alpha + 1} = A\alpha^2 + B\alpha + C.$$

(82) Let $f(x) = x^5 + 20x + 16 \in \mathbb{Q}[x]$ and $\mathbb{K}$ be its splitting field. Prove that the Galois group of $f(x)$, $G = Aut(\mathbb{K}/\mathbb{Q})$ is isomorphic to A_5—the alternating group on five letters.

(83) Let $m \in \mathbb{N}$ be a natural number. Prove that an angle measuring m degrees can be constructed by ruler and compass if and only if m is divisible by 3.

(84) Let $f(x) = x^5 - 5x^3 + x^2 + 3x - 1 \in \mathbb{Q}[x]$. Let $\mathbb{K} \subset \mathbb{C}$ be its splitting field. Show that

(a) $f(x)$ in irreducible.

(b) It Galois group is isomorphic to S_5.

(c) $\mathbb{K} \subset \mathbb{R}$.

(85) Let $f(x) = x^5 - 6x^3 + x^2 + 4x + 1 \in \mathbb{Q}[x]$. Let $\mathbb{K} \subset \mathbb{C}$ be its splitting field. Show that

(a) $f(x)$ in irreducible.

(b) It Galois group is isomorphic to S_5.

(c) $\mathbb{K} \subset \mathbb{R}$.

(86) Let $f(x) = x^4 - 7x^2 - 3x + 1 \in \mathbb{Q}[x]$. Let $\mathbb{K} \subset \mathbb{C}$ be its splitting field. Show that

(a) $f(x)$ in irreducible.

(b) It Galois group is isomorphic to A_4.

(c) $\mathbb{K} \subset \mathbb{R}$.

(87) Let $f(x) = x^4 - 10x^2 - 4x + 6 \in \mathbb{Q}[x]$. Let $\mathbb{K} \subset \mathbb{C}$ be its splitting field. Show that

(a) $f(x)$ in irreducible.

(b) It Galois group is isomorphic to A_4.

(c) $\mathbb{K} \subset \mathbb{R}$.

(88) Let $f(x) = x^4 - 4x^2 - x + 1 \in \mathbb{Q}[x]$. Let $\mathbb{K} \subset \mathbb{C}$ be its splitting field. Show that

(a) $f(x)$ in irreducible.

(b) It Galois group is isomorphic to S_4.

(c) $\mathbb{K} \subset \mathbb{R}$.

(89) Let $f(x) = x^4 - 6x^2 - 2x + 1 \in \mathbb{Q}[x]$. Let $\mathbb{K} \subset \mathbb{C}$ be its splitting field. Show that

(a) $f(x)$ in irreducible.

(b) It Galois group is isomorphic to S_4.

(c) $\mathbb{K} \subset \mathbb{R}$.

(90) Let $\alpha = e^{\frac{2\pi i}{11}} = \cos\frac{2\pi}{11} + i\sin\frac{2\pi}{11}$. Prove that

$$\mathbb{Q}(\alpha) = \mathbb{Q}\left(i\sin\frac{2\pi}{11}\right).$$

Prove that the minimal polynomial of $2i\sin\frac{2\pi}{11}$ over $\mathbb{Q}$ is $x^{10} + 11x^8 + 44x^6 + 77x^4 + 55x^2 + 11$. Find the minimal polynomial of $2\sin\frac{2\pi}{11}$ over $\mathbb{Q}$. Show that $\sqrt{-11} \in \mathbb{Q}(\alpha)$ and $\mathbb{Q}(\alpha) = \mathbb{Q}\left(\cos\frac{2\pi}{11}, \sqrt{-11}\right)$.

(91) Let $\mathbb{K}$ = Splitting field of the 55-th cyclotomic polynomial $\Phi_{55}(x)$.

(a) Describe the field $\mathbb{K}$ and its subfields.

(b) Describe the group $G = Aut(\mathbb{K}/\mathbb{Q})$ and its subgroups.

(c) Exhibit the Galois correspondence.

(92) Let $\mathbb{K}$ = Splitting field of the 33-th cyclotomic polynomial $\Phi_{33}(x)$.

(a) Describe the field $\mathbb{K}$ and its subfields.

(b) Describe the group $G = Aut(\mathbb{K}/\mathbb{Q})$ and its subgroups.

(c) Exhibit the Galois correspondence.

(93) Recall that in the text in the discussion of Cardano's formula it was proved that the roots α_i, $1 \le i \le 3$ of the cubic $x^3 + px + q = 0$ are given by

$$\alpha_1 = A + B = \left(-\frac{q}{2} + \sqrt{\frac{q^2}{4} + \frac{p^3}{27}}\right)^{\frac{1}{3}} + \left(-\frac{q}{2} - \sqrt{\frac{q^2}{4} + \frac{p^3}{27}}\right)^{\frac{1}{3}}$$

$$\alpha_2 = \omega^2 A + \omega B = \omega^2\left(-\frac{q}{2} + \sqrt{\frac{q^2}{4} + \frac{p^3}{27}}\right)^{\frac{1}{3}} + \omega\left(-\frac{q}{2} - \sqrt{\frac{q^2}{4} + \frac{p^3}{27}}\right)^{\frac{1}{3}}$$

$$\alpha_3 = \omega A + \omega^2 B = \omega\left(-\frac{q}{2} + \sqrt{\frac{q^2}{4} + \frac{p^3}{27}}\right)^{\frac{1}{3}} + \omega^2\left(-\frac{q}{2} - \sqrt{\frac{q^2}{4} + \frac{p^3}{27}}\right)^{\frac{1}{3}}$$

where the cube roots in A and B in the following are so chosen that $A \cdot B = -\frac{p}{3}$.

$$A = \left(-\frac{q}{2} + \sqrt{\frac{q^2}{4} + \frac{p^3}{27}}\right)^{\frac{1}{3}}, \quad B = \left(-\frac{q}{2} - \sqrt{\frac{q^2}{4} + \frac{p^3}{27}}\right)^{\frac{1}{3}}.$$

Consider the cubic equation $x^3 - 3x = 0$, whose roots are evidently $0, \sqrt{3}$, and $-\sqrt{3}$. However if we decide to use the Cardano's formula we substitute $p = -3$ and $q = 0$ we get the following values of A and B

$$A = \sqrt[3]{i}, \qquad B = \sqrt[3]{-i}$$

and so the roots of the equation are given by $A + B == \sqrt[3]{i} + \sqrt[3]{-i}$. Explain how to explain this.

(94) Gauss proved that if $f(x) \in \mathbb{R}[x]$ is any polynomial with real coefficients then there exists $\alpha \in \mathbb{C}$ such that $f(\alpha) = 0$. Show that this implies the Fundamental Theorem of Algebra, namely that if $f(x) \in \mathbb{C}[x]$ is any polynomial with complex coefficients then there exists $\alpha \in \mathbb{C}$ such that $f(\alpha) = 0$.

(95) Let α be a root of the polynomial $x^4 + p$, where p is a prime number. Prove that the minimal polynomial of $\alpha^3 + \alpha^2 + \alpha + 1$ is $x^4 - 4x^3 + 6(p+1)x^2 - 4(p+1)^2x + (p+1)^3$.

(96) Let $m, n > 1$ be two natural numbers such that $gcd(m, n) = 1$. Let $\alpha = e^{\frac{2\pi i}{m}}$ and $\beta = e^{\frac{2\pi i}{n}}$ be primitive m-th and primitive n-th roots of unity. Prove the following:

(a) $\mathbb{Q}(\alpha\beta) = \mathbb{Q}(\alpha, \beta)$.

(b) $[\mathbb{Q}(\alpha\beta) : \mathbb{Q}] = \phi(m)\phi(n)$ and $\Omega = \{\alpha^i\beta^j : 0 \le i \le \phi(m),\ \ 0 \le j \le \phi(n)\}$ is a basis of $\mathbb{Q}(\alpha\beta)$ over $\mathbb{Q}$.

(c) $\mathbb{Q}(\alpha) \cap \mathbb{Q}(\beta) = \mathbb{Q}$.

(97) Let $f(x) = x^3 - 4x + 2 \in \mathbb{Q}[x]$. Convince yourself that

(a) $f(x)$ is irreducible over $\mathbb{Q}$.

(b) $f(x)$ has three real roots.

(c) Discriminant of $f(x)$ is $148 = 2^2 \cdot 37$.

(d) The Galois group of $f(x)$ is isomorphic to S_3, the symmetric group on three letters.

(e) Let

$$A = \left(-1 + \frac{1}{9}\sqrt{-111}\right)^{\frac{1}{3}}, \qquad B = \left(-1 - \frac{1}{9}\sqrt{-111}\right)^{\frac{1}{3}}$$

we have $AB = \dfrac{4}{3}$ and the three roots α, β, and γ are given by

$$\alpha = A + B, \qquad \beta = \omega A + \omega^2 B, \qquad \gamma = \omega^2 A + \omega B.$$

Let $g(x) = f(x)(x^2 + x + 1) = (x^3 - 4x + 2)(x^2 + x + 1)$. Show that the splitting field of $g(x)$ is a radical extension of $\mathbb{Q}$ whereas the splitting field of $f(x)$ is not.

(98) Let $f(x) = x^6 + x^3 + 1$ and $p \neq 3$ be a prime number. Then

(a) If p is a prime $p \equiv 1$ (modulo 9), then $f(x)$ splits into linear factors modulo p.

(b) If p is a prime $p \equiv 4, 7$ (modulo 9) then $f(x)$ factors as a product of two irreducible cubics modulo p.

(c) If $p \equiv 2, 5$ (modulo 9) then $f(x)$ is irreducible modulo p.

(d) If $p \equiv 8$ (modulo 9) then $f(x)$ splits as a product of three quadratic factors modulo p.

(99) The following table illustrates the factoring of the polynomial $\Phi_{27}(x) = x^{18} + x^9 + 1$ modulo prime $p \neq 3$.

p	Order of $\zeta \longrightarrow \zeta^p$	$\Phi_{27}(x)$ Factors as
$p \equiv 1$ (modulo 27)	1	Linear Factors
$p \equiv 26$ (modulo 27)	2	Nine quadratic Factors
$p \equiv 10, 19$ (modulo 27)	3	Six Cubic Factors
$p \equiv 8, 17$ (modulo 27)	6	Three Factors of degree six
$p \equiv 4, 16, 13, 25, 22, 7$ (modulo 27)	9	Two Factors of degree nine
$p \equiv 2, 5, 20, 23, 11, 14$ (modulo 27)	18	Irreducible

Justify each step/entry in the above table and provide an example or two of each.

(100) Prove that if $f(x) = x^5 - kx - 1 \in \mathbb{Z}[x]$ is irreducible then $k \notin \{-1, 0, 2\}$.

(101) Give an example of field $\mathbb{K}$, with subfields $\mathbb{E}$ and $\mathbb{F}$ such that $[\mathbb{K} : \mathbb{E}] = [\mathbb{K} : \mathbb{F}] = 2$ but $\mathbb{K}$ is not algebraic over $\mathbb{E} \cap \mathbb{F}$.

(102) Let $m, n > 1$ be natural numbers. Give an example of field $\mathbb{K}$, with subfields $\mathbb{L}$ and $\mathbb{M}$ such that $[\mathbb{K} : \mathbb{L}] = m$, $[\mathbb{K} : \mathbb{M}] = n$ but $\mathbb{K}$ is not algebraic over $\mathbb{L} \cap \mathbb{M}$.

Hints or Solutions to Problems

(1) Let $\alpha = \sqrt{2} + \sqrt{3}$. Evidently $\mathbb{Q}(\alpha) \subseteq \mathbb{Q}(\sqrt{2}, \sqrt{3})$. The minimal polynomial of α is $x^4 - 10x^2 + 1$. So $\alpha^{-1} = -\alpha^3 + 10\alpha$ and $\alpha^5 = 10\alpha^3 - \alpha$. Squaring $\alpha - \sqrt{2} = \sqrt{3}$ we get $\sqrt{2} = \frac{1}{2}(\alpha^2 - 1)\alpha^{-1}$. This shows that $\sqrt{2} \in \mathbb{Q}(\alpha)$ and $\sqrt{3} = \alpha - \sqrt{2} \in \mathbb{Q}(\alpha)$ Substituting $\alpha^{-1} = -\alpha^3 + 10\alpha$ and $\alpha^5 = 10\alpha^3 - \alpha$ we have

$$\begin{aligned}
\sqrt{2} &= -\frac{1}{2}(\alpha^2 - 1)(\alpha^3 - 10\alpha) \\
&= \frac{1}{2}(\alpha^5 - 11\alpha^3 + 10\alpha) \\
&= -\frac{1}{2}(10\alpha^3 - \alpha - 11\alpha^3 + 10\alpha) \\
&= \frac{1}{2}(\alpha^3 - 9\alpha)
\end{aligned}$$

and therefore $\sqrt{3} = \alpha - \sqrt{2} = \frac{1}{2}(11\alpha - \alpha^3)$

(2) Let $\alpha = \sqrt{p} + \sqrt{q}$. Obviously $\mathbb{Q}(\alpha) \subseteq \mathbb{Q}(\sqrt{p}, \sqrt{q})$. The minimal polynomial of α is easily seen to be $x^4 - 2(p+q)x^2 + (p-q)^2$ and

$$\sqrt{p} = \frac{1}{2\alpha}(\alpha^2 + p - q), \qquad \sqrt{q} = \alpha - \sqrt{p} = \frac{1}{2\alpha}(\alpha^2 + q - p)$$

and

$$\alpha^5 = 2(p+q)\alpha^3 - (p-q)^2\alpha, \qquad \frac{1}{\alpha} = -\frac{1}{(p-q)^2}\left[\alpha^3 - 2(p+q)\alpha\right]$$

As in the previous problem it can be shown that

$$\sqrt{p} = -\frac{1}{2(p-q)^2}\left[(p-q)\alpha^3 - (3p^2 - q^2 - 2pq)\alpha\right]$$

$$\sqrt{q} = \frac{1}{2(p-q)^2}\left[(p-q)\alpha^3 - (p^2 - 3q^2 + 2pq)\alpha\right]$$

(3) Let $\alpha = \sqrt{2} + \sqrt{3} + \sqrt{5}$ and $\mathbb{K} = \mathbb{Q}(\alpha)$. Evidently $\mathbb{K} \subseteq \mathbb{Q}(\sqrt{2}, \sqrt{3}, \sqrt{5})$. We shall show that $\mathbb{Q}(\sqrt{2}, \sqrt{3}, \sqrt{5}) \subseteq \mathbb{K}$.

$\alpha - \sqrt{2} = \sqrt{3} + \sqrt{5}$. Squaring we get $\alpha^2 - 6 - 2\sqrt{2}\alpha = 2\sqrt{15}$. Squaring again and solving for $\sqrt{2}$ we have $\sqrt{2} = \dfrac{\alpha^4 - 4\alpha^2 - 24}{4\alpha(\alpha^2 - 6)}$. Therefore $\sqrt{2} \in \mathbb{K}$. Similarly it can be shown that $\sqrt{3}, \sqrt{5} \in \mathbb{K}$.

The minimal polynomial of $\alpha = \sqrt{2}+\sqrt{3}+\sqrt{5}$ is $x^8 - 40x^6 + 352x^4 - 960x^2 + 576$.

(4) The cubic polynomial $x^3 + 3x + 2$ has no rational roots so it is irreducible. We have

$$x^3 + 3x + 2 = (x^2 - x + 4)(x + 1) - 2, \text{ therefore } \frac{1}{\alpha + 1} = \frac{1}{2}(\alpha^2 - \alpha + 4)$$

(5) Suppose that $Order(x^d) = k$. Since $gcd(n, d)$ divides d we have $(x^d)^{\frac{n}{gcd(n,d)}} = e \Longrightarrow k$ divides $\frac{n}{gcd(n,d)}$. Also $x^{dk} = e \Longrightarrow n$ divides dk. It follows that $\frac{n}{gcd(n,d)}$ divides $\frac{dk}{gcd(n,d)}$. Since $\frac{n}{gcd(n,d)}$ and $\frac{d}{gcd(n,d)}$ are relatively prime $\frac{n}{gcd(n,d)}$ divides k.

Now as k divides $\frac{n}{gcd(n,d)}$ and $\frac{n}{gcd(n,d)}$ divides k, therefore $\frac{n}{gcd(n,d)} = k$

(6) This follows from the previous problem.

(7) As

$$x^6 - 17x^4 + 80x^2 - 100 = (x^2 - 2)(x^2 - 5)(x^2 - 10)$$

the given polynomial has no roots in $\mathbb{Z}$. Evidently the given polynomial has a root modulo 2 and modulo 5. Let $p \neq 2, 5$ be a prime. By the Law of Quadratic Reciprocity if neither 2 nor 5 is a quadratic residue modulo p then 10 is a quadratic residue modulo p. So the given polynomial has a root modulo p for every p.

(8) Let m, n are two natural numbers such that $gcd(m, n) = 1$. Then if α is a primitive m-th root of unity and β is a primitive n-th root of unity then $\alpha\beta$ is a primitive mn-th root of unity. Let $f(N)$ =sum of primitive N-th roots of unity. In view of above remark f is a multiplicative arithmetical function. So it is enough to prove that for every prime power p^k, $f(p^k) = \mu(p^k)$. It is obviously true for $k = 1$ as $\mu(p) = f(p) = -1$. For $k > 1$ the p^k-th cyclotomic polynomial $\Phi_{p^k}(x)$ has degree $\phi(p^k) = p^k - p^{k-1}$ and is given by:

$$\Phi_{p^k}(x) = x^{(p-1)p^{k-1}} + x^{(p-2)p^{k-1}} + x^{(p-3)p^{k-1}} + \cdots\cdots + x^{p^{k-1}} + 1$$

Therefore for $k > 1$ the sum of the primitive p^k-th roots of unity is equal to zero and so is same as value of $\mu(p^k)$.

(9) **Hint.** Note that to prove the irreducibility of $f(x) = x^8 + 4x^6 + 2x^4 + 28x^2 + 1$ it is helpful to note that

$$f(x) \equiv (x^4 + 2x^3 + 4x^2 + 3x + 4)(x^4 + 3x^3 + 4x^2 + 2x + 4) \text{ (modulo 5)}$$

So it is enough to show that $f(x)$ does not factor as a product of two quartics (polynomials of degree four) over $\mathbb{Z}$

(10) $\alpha = 2^{\frac{1}{3}} + \omega$, so $\alpha - \omega = 2^{\frac{1}{3}}$. Cubing both sides and simplifying we get $2\alpha^3 + 3\alpha^2 - 3\alpha - 6 = 3\alpha(\alpha+1)i\sqrt{3}$. Squaring both sides we get $\alpha^6 + 3\alpha^5 + 6\alpha^4 + 3\alpha^3 + 9\alpha + 9 = 0$. So the minimal polynomial of $\alpha = 2^{\frac{1}{3}} + \omega$ is $f(x) = x^6 + 3x^5 + 6x^4 + 3x^3 + 9x + 9$ which is easily seen to be irreducible.

It is easily seen that

$$\omega = \frac{\alpha^3 - 3\alpha - 3}{3\alpha^2 + 3\alpha}, \qquad 2^{\frac{1}{3}} = \frac{2\alpha^3 + 3\alpha^2 + 3\alpha + 3}{3\alpha^2 + 3\alpha}$$

By Euclidean algorithm we find that

$$(x^2 + x)(8x^5 + 15x^4 + 30x^3 - 12x^2 + 9x + 63)$$
$$- (8x - 1)(x^6 + 3x^5 + 6x^4 + 3x^3 + 9x + 9) = 9$$

Substituting $x = \alpha$ in above we get

$$\frac{1}{\alpha^2 + \alpha} = \frac{1}{9}(8\alpha^5 + 15\alpha^4 + 30\alpha^3 - 12\alpha^2 + 9\alpha + 63).$$

Substituting $\frac{1}{\alpha^2 + \alpha} = \frac{1}{9}(8\alpha^5 + 15\alpha^4 + 30\alpha^3 - 12\alpha^2 + 9\alpha + 63)$ in

$$\omega = \frac{\alpha^3 - 3\alpha - 3}{3\alpha^2 + 3\alpha}, \qquad 2^{\frac{1}{3}} = \frac{2\alpha^3 + 3\alpha^2 + 3\alpha + 3}{3\alpha^2 + 3\alpha}$$

we get expressions for ω and $2^{\frac{1}{3}}$ as a polynomial in α.

Note. The irreducibility of $f(x)$ is an immediate consequence of the following:

$$f(x) \equiv (x^2 + x + 1)^3 \text{ (modulo 2)},$$
$$f(x) \equiv (x^3 + x^2 + 5x + 4)(x^3 + 2x^2 + 6x + 4) \text{ (modulo 7)}$$

(11) First note that $\alpha^3 = 2 + \frac{10}{9}\sqrt{3}$, we have

$$\alpha^3 = 2 + \frac{10}{9}\sqrt{3} = \frac{54 + 30\sqrt{3}}{27} = \frac{27 + 9\sqrt{3}(3 + \sqrt{3}) + 3\sqrt{3}}{27} = \left(\frac{3 + \sqrt{3}}{3}\right)^3$$

and so $\alpha = \frac{3+\sqrt{3}}{3}$. Likewise $\beta^3 = 2 - \frac{10}{9}\sqrt{3}$ and so $\beta = \frac{3-\sqrt{3}}{3}$. The following table gives the minimal polynomials of the various algebraic numbers.

Algebraic Number	Minimal Polynomial
$\alpha = \frac{3+\sqrt{3}}{3}$	$x^2 - 2x + \frac{2}{3}$
$\beta = \frac{3-\sqrt{3}}{3}$	$x^2 - 2x + \frac{2}{3}$
$\alpha + \beta = 2$	$x - 2$
$\alpha - \beta = \frac{2\sqrt{3}}{3}$	$x^2 - \frac{4}{3}$
$\omega\alpha + \omega^2\beta = -1 + i$	$x^2 + 2x + 2$
$\omega^2\alpha + \omega\beta = -1 - i$	$x^2 + 2x + 2$
$\omega\alpha - \omega^2\beta = \sqrt{3}i - \frac{1}{\sqrt{3}}$	$x^4 + \frac{16}{3}x^2 + \frac{100}{9}$
$\omega^2\alpha - \omega\beta = -\sqrt{3}i - \frac{1}{\sqrt{3}}$	$x^4 + \frac{16}{3}x^2 + \frac{100}{9}$

(12) Let $\alpha = 2^{\frac{1}{3}} + \sqrt{3}$. The first assertion is an immediate consequence of the fact that

$$\sqrt{3} = \frac{\alpha^3 + 9\alpha - 2}{3(\alpha^2 + 1)}, \qquad 2^{\frac{1}{3}} = \frac{2\alpha^3 - 6\alpha + 2}{3(\alpha^2 + 1)}$$

The minimal polynomial of $\alpha = 2^{\frac{1}{3}} + \sqrt{3}$ is $x^6 - 9x^4 - 4x^3 + 27x^2 - 36x - 23$.

(13) Let p_i, $1 \leq i \leq k$ be primes $\leq n$. Then

$$\mathbb{Q}\left(\left\{\sqrt{i} : 2 \leq i \leq n\right\}\right) = \mathbb{Q}(\sqrt{p_1}, \sqrt{p_2}, \cdots, \sqrt{p_k})$$

We have already shown in the text that the field $\mathbb{Q}(\sqrt{p_1}, \sqrt{p_2}, \cdots, \sqrt{p_k})$ has dimension 2^kover $\mathbb{Q}$.

(14) As we have seen in the text the minimal polynomial of $e^{\frac{2\pi i}{7}}$ is $x^6+x^5+x^4+x^3+x^2+x+1$. Therefore the minimal polynomial of $e^{\frac{2\pi i}{7}}+e^{-\frac{2\pi i}{7}} = 2\cos\frac{2\pi}{7}$ is x^3+x^2-2x-1. So the monic polynomial whose roots are one half of the roots of $x^3 + x^2 - 2x - 1$ is $x^3 + \frac{1}{2}x^2 - \frac{1}{2}x - \frac{1}{8}$ - and this is the minimal polynomial of $\cos\frac{2\pi}{7}$

(15) Wee need to find the monic polynomial whose roots are the squares of the roots of the minimal polynomials of $\cos\frac{2\pi}{7}$, namely $x^3 + \frac{1}{2}x^2 - \frac{1}{2}x - \frac{1}{8}$. Let the three roots

of $x^3+\frac{1}{2}x^2-\frac{1}{2}x-\frac{1}{8}$ be α,β and γ. Then we have

$$\begin{aligned}\alpha+\beta+\gamma &= -\frac{1}{2},\\ \alpha\beta+\beta\gamma+\gamma\alpha &= -\frac{1}{2},\\ \alpha\beta\gamma &= \frac{1}{8}\end{aligned}$$

Therefore

$$\begin{aligned}\alpha^2+\beta^2+\gamma^2 &= (\alpha+\beta+\gamma)^2-2(\alpha\beta+\beta\gamma+\gamma\alpha)=\frac{5}{4}\\ \alpha^2\beta^2+\beta^2\gamma^2+\gamma^2\alpha^2 &= (\alpha\beta+\beta\gamma+\gamma\alpha)^2-2\alpha\beta\gamma(\alpha+\beta+\gamma)=\frac{3}{8}\\ \alpha^2\beta^2\gamma^2 &= (\alpha\beta\gamma)^2=\frac{1}{64}.\end{aligned}$$

Therefore the minimal polynomial of $\cos^2\frac{2\pi}{7}$ is $x^3-\dfrac{5}{4}x^2+\dfrac{3}{8}-\dfrac{1}{64}$.

Another method of doing the same would be to write $x^3+\dfrac{1}{2}x = -\dfrac{1}{2}x^2+\dfrac{1}{8}$ which upon squaring gives $x^6-\dfrac{5}{4}x^4+\dfrac{3}{8}x^2-\dfrac{1}{64}=0$. Substituting $x^2=X$ we get $X^3-\dfrac{5}{4}X^2+\dfrac{3}{8}X-\dfrac{1}{64}$ as the minimal polynomial of $\cos^2\frac{2\pi}{7}$.

(16) The entries in this table are calculated by the methods discussed in the previous two problems and are as follows:

Algebraic Number	Minimal Polynomial
$\sin^2\frac{2\pi}{7}$	$x^3-\frac{7}{4}x^2+\frac{7}{8}x-\frac{7}{64}$
$\sin\frac{2\pi}{7}$	$x^6-\frac{7}{4}x^4+\frac{7}{8}x^2-\frac{7}{64}$
$i\sin\frac{2\pi}{7}$	$x^6+\frac{7}{4}x^4+\frac{7}{8}x^2+\frac{7}{64}$
$2\sin\frac{2\pi}{7}$	$x^6+7x^4+14x^2+7$

(17) Let α and β be such that $\alpha^3+\alpha^2+1=0$, $\beta^3+\beta+1=0$. Then elements of the two fields can be written as follows:

$$\mathbb{K}_1=\{0,1,\alpha,\alpha^2,1+\alpha,1+\alpha^2,\alpha+\alpha^2,1+\alpha+\alpha^2\}$$
$$\mathbb{K}_2=\{0,1,\beta,\beta^2,1+\beta,1+\beta^2,\beta+\beta^2,1+\beta+\beta^2\}$$

In the field $\mathbb{K}_1$ the roots of $f_1(x)$ are α,α^2 and $\alpha^4=\alpha^2+\alpha+1$ In the field $\mathbb{K}_1$ the roots of $f_2(x)$ are $1+\alpha,1+\alpha^2$ and $\alpha+\alpha^2$

In the field $\mathbb{K}_2$ the roots of $f_2(x)$ are β,β^2 and $\beta^4=\beta^2+\beta$. In the field $\mathbb{K}_2$ the roots of $f_1(x)$ are $1+\beta,1+\beta^2$ and $1+\beta+\beta^2$

Since an isomorphism between the fields $\mathbb{K}_1$ and $\mathbb{K}_2$ will extend to an isomorphism between the two polynomial rings $\mathbb{K}_1[x]$ and $\mathbb{K}_2[x]$, the element $\alpha \in \mathbb{K}_1$ being a root of $f_1(x)$ should be mapped to any one of the three roots of $f_1(x)$ in $\mathbb{K}_2$, namely $1+\beta, 1+\beta^2$ or $1+\beta+\beta^2$. So an isomorphism from $\mathbb{K}_1$ to $\mathbb{K}_2$ will be completely defined by mapping $\alpha \in \mathbb{K}_1$, to $1+\beta \in \mathbb{K}_2$. The following table gives the images of various elements of $\mathbb{K}_1$ in $\mathbb{K}_2$ - under the isomorphism which maps $\alpha \in \mathbb{K}_1$ to $1+\beta \in \mathbb{K}_2$.

Element	0	1	α	α^2	$1+\alpha$	$1+\alpha^2$	$\alpha+\alpha^2$	$1+\alpha+\alpha^2$
Image	0	1	$1+\beta$	$1+\beta^2$	β	β^2	$\beta+\beta^2$	$1+\beta+\beta^2$

(18) Let $\alpha = e^{\frac{2\pi i}{11}} = \cos\frac{2\pi}{11} + i\sin\frac{2\pi}{11}$. Then

$$\alpha+\alpha^2+\alpha^3+\alpha^4+\alpha^5+\alpha^6+\alpha^7+\alpha^8+\alpha^9+\alpha^{10} = -1$$

Let

$$A = \alpha+\alpha^4+\alpha^5+\alpha^9+\alpha^3 \ , \quad \text{and} \quad B = \alpha^2+\alpha^8+\alpha^{10}+\alpha^7+\alpha^6$$

Then $A+B=-1$ and $A\cdot B = 3$. So A and B are roots of the equation $x^2+x+3=0$ and so

$$A = \frac{-1+\sqrt{-11}}{2}, \qquad B = A = \frac{-1-\sqrt{-11}}{2}, \quad A-B = \sqrt{-11} = i\sqrt{11}$$

We have

$$\begin{aligned} i\tan\frac{3\pi}{11} &= -i\tan\frac{8\pi}{11} = -\frac{2i\sin\frac{8\pi}{11}}{2\cos\frac{8\pi}{11}} = -\frac{\alpha^4-\alpha^{-4}}{\alpha^4+\alpha^{-4}} = \frac{\alpha^{-4}-\alpha^4}{\alpha^{-4}+\alpha^4} \\ &= \frac{\alpha^{-8}-1}{\alpha^{-8}+1} = \frac{\alpha^3-1}{\alpha^3+1} = \frac{\alpha^3-\alpha^{33}}{\alpha^3+1} = -\frac{\alpha^3(\alpha^{30}-1)}{\alpha^3+1} \end{aligned}$$

Now

$$\begin{aligned} -\frac{\alpha^3(\alpha^{30}-1)}{\alpha^3+1} &= -\frac{\alpha^3(\alpha^{15}-1)(\alpha^{15}+1)}{\alpha^3+1} \\ &= \alpha^{10}+\alpha^9-\alpha^8-\alpha^7-\alpha^6+\alpha^5+\alpha^4+\alpha^3-\alpha^2-\alpha \\ &= (\alpha+\alpha^3+\alpha^4+\alpha^5+\alpha^9)-(\alpha^2+\alpha^6+\alpha^7+\alpha^8+\alpha^{10})-2(\alpha-\alpha^{10}) \\ &= A-B-4i\sin\frac{2\pi}{11} \\ &= i\sqrt{11}-4i\sin\frac{2\pi}{11} \end{aligned}$$

So

$$i\tan\frac{3\pi}{11} = i\sqrt{11}-4i\sin\frac{2\pi}{11}, \quad \text{hence} \quad \mathbf{\tan\frac{3\pi}{11}+4\sin\frac{2\pi}{11} = \sqrt{11}}$$

(19) Obviously $\mathbb{Q}(\sqrt{1985}) = \mathbb{Q}(\alpha^2) = \mathbb{Q}(\beta^2)$. It is enough to show that $\alpha\beta \in \mathbb{Q}(\sqrt{1985})$ as $\alpha\beta \in \mathbb{Q}(\alpha^2) \subseteq \mathbb{Q}(\alpha) \Longrightarrow \beta \in \mathbb{Q}(\alpha)$.

We have by direct calculation

$$\begin{aligned}\alpha\beta &= \sqrt{2 \cdot 1985^2 + 2 \cdot 31 \cdot 1985\sqrt{1985} + 1985 \cdot 64 \cdot \sqrt{1985} + 31 \cdot 64 \cdot 1985} \\ &= \sqrt{1985\left(2 \cdot 1985 + 1984 + 126\sqrt{1985}\right)} \\ &= \sqrt{1985\left(1985 + 3969 + 126 \cdot \sqrt{1985}\right)} \\ &= \sqrt{1985\left(63 + \sqrt{1985}\right)^2} \\ &= \sqrt{1985}(63 + \sqrt{1985}) \in \mathbb{Q}(\sqrt{1985})\end{aligned}$$

(20) Let $\mathbb{K}$ denotes the splitting field of the polynomial $f(x)$. The polynomial $f(x)$ has at least one real root. If it has only one real root (say α) then the complex conjugation in the field $\mathbb{K}$ will constitute an element of order two of the Galois group of $f(x)$. This is not possible as the Galois group of $f(x)$ is cyclic of order three. The polynomial $x^3 - 4x + 2 \in \mathbb{Q}[x]$ is irreducible and has discriminant $2^2 \cdot 37$. Therefore its Galois group is isomorphic to the symmetric group S_3. It can be easily seen that all the three roots of this polynomial are real. The following table gives some more examples of such polynomials.

Polynomial	Discriminant
$x^3 - 5x^2 + 5$	1825
$x^3 - 11x + 11$	2057
$x^3 - 17x + 17$	11849
$x^3 - 23x + 23$	34385
$x^3 - 29x + 29$	74849
$x^3 - 31x + 31$	93217
$x^3 - 6x^2 + 3$	2349

If p is a prime then the polynomial $x^3 - px + p \in \mathbb{Q}[x]$ is irreducible and has discriminant $p^2(4p-27)$. So if $4p-27$ is not a perfect square then such a polynomial has S_3 as a Galois group over $\mathbb{Q}$. If addition if $p > 7$ then the all the roots of the polynomial $x^3 - px + p$ are real.

(21) Since $f(x)$ is irreducible and $m < n$, $GCD(f(x), g(x)) = 1$. By Euclidean Algorithm as applied to the polynomial ring $\mathbb{Q}[x]$ there exist polynomials $a(x), b(x) \in \mathbb{Q}[x]$ such

that $a(x)f(x)+b(x)g(x) = 1$. Substituting $x = \alpha$ in this get $b(\alpha)g(\alpha) = 1$. Therefore $\dfrac{1}{g(\alpha)} = b(\alpha)$.

(22) **Answer**

The required equation is $x^4 + x^3 - 6x^2 - x + 1 = 0$.

The following observations will help in the calculation.

The mapping $\alpha \longrightarrow \alpha^3$ induces the mapping $\beta_1 \longrightarrow \beta_3 \longrightarrow \beta_2 \longrightarrow \beta_4 \longrightarrow \beta_1$

The following relations among the β's can be easily verified.

$$\beta_1\beta_2 = -1, \quad \beta_3\beta_4 = -1, \quad \beta_1\beta_3 = 2\beta_1 + \beta_2 + \beta_4,$$
$$\beta_3\beta_2 = 2\beta_3 + \beta_4 + \beta_1, \quad \beta_2\beta_4 = 2\beta_2 + \beta_1 + \beta_3,$$
$$\beta_4\beta_1 = 2\beta_4 + \beta_3 + \beta_2$$

Also

$$\beta_1\beta_2\beta_3 = -\beta_3, \quad \beta_3\beta_4\beta_2 = -\beta_2, \quad \beta_2\beta_1\beta_4 = -\beta_4, \quad \beta_4\beta_3\beta_1 = -\beta_1, \quad \beta_1\beta_2\beta_3\beta_4 = 1$$

(23) The required equation is easily seen to be $x^3 + x^2 - 6x - 7 = 0$.

(24) As $(a + b + c)^2 = a^2 + b^2 + c^2 + 2(ab + bc + ca)$, we get $ab + bc + ca = 2$ and from $(a + b + c)(a^2 + b^2 + c^2) = a^3 + b^3 + c^3 + a^2b + ab^2 + b^2c + bc^2 + c^2a + ca^2$, we get $a^2b + ab^2 + b^2c + bc^2 + c^2a + ca^2 = 8$. As $(a + b + c)(ab + bc + ca) = a^2b + ab^2 + b^2c + bc^2 + c^2a + ca^2 + 3abc$, we get $abc = -\frac{2}{3}$. So a, b, c are the roots of the equation $3x^3 - 9x^2 + 6x + 2 = 0$. Multiply this equation by x we get $3x^4 - 9x^3 + 6x^2 + 2x = 0$. Substituting $x = a, b, c$ in this equation and adding we get

$$a^4 + b^4 + c^4 = 3(a^3 + b^3 + c^3) - 2(a^2 + b^2 + c^2) - \frac{2}{3}(a + b + c) = 9.$$

Likewise

$$a^5 + b^5 + c^5 = 3(a^4 + b^4 + c^4) - 2(a^3 + b^3 + c^3) - \frac{2}{3}(a^2 + b^2 + c^2) = \frac{29}{3}.$$

(25) Proceeding as in the previous problem, we get $\alpha\beta+\beta\gamma+\gamma\alpha = -\frac{1}{2}$, $\alpha^2\beta+\alpha\beta^2+\beta^2\gamma+\beta\gamma^2 + \gamma^2\alpha + \gamma\alpha^2 = -1$, and $\alpha\beta\gamma = \frac{1}{6}$. So the equation whose roots are α, β, and γ is $6x^3 - 6x^2 - 3x - 1 = 0$. With the help of this equation we get $\alpha^4 + \beta^4 + \gamma^4 = \frac{25}{6}$ and $\alpha^5 + \beta^5 + \gamma^5 = 6$

(26) We note that $\alpha^4 = \alpha + 1$, $\alpha^5 = \alpha^2 + \alpha$ and so α has order 15 and α^3 has order 5. The minimal polynomial of α^3 over $\mathbb{F}_2$ is of degree 4. There are just three irreducible polynomial of degree four over $\mathbb{F}_2$, these are $x^4 + x + 1$, $x^4 + x^3 + 1$ and $x^4 + x^3 + x^2 + x + 1$. Since α^3 has order 5 the roots of its minimal polynomial are the four nonidentity elements of a group of order 5. Let $\alpha^3 = \beta$. Then $\beta^5 = 1$ and so $\beta^5 - 1 = (\beta - 1)(\beta^4 + \beta^3 + \beta^2 + \beta + 1)$. Since $\beta \neq 1$, $\beta^4 + \beta^3 + \beta^2 + \beta + 1 = 0$ and the minimal polynomial of $\beta = \alpha^3$ is $x^4 + x^3 + x^2 + x + 1$.

(27) Let $\alpha = \frac{1}{3}(1 + 10^{\frac{1}{3}} + 10^{\frac{2}{3}})$. So $3\alpha - 1 = 10^{\frac{1}{3}} + 10^{\frac{2}{3}}$. Cubing both sides and simplifying we get $\alpha^3 - \alpha^2 - 3\alpha - 3 = 0$. So the minimal polynomial of α is $x^3 - x^2 - 3x - 3$.

(28) Evidently $\alpha^2 + \alpha$ is an algebraic number of degree 3. Let $\alpha^2 + \alpha = y$. Eliminating α from $\alpha^3 - 2\alpha + 2 = 0$ and $\alpha^2 + \alpha = y$, we get $y^3 - 4y^2 + 8y - 6 = 0$. So the minimal polynomial of $\alpha^2 + \alpha$ is $y^3 - 4y^2 + 8y - 6$. Likewise it can be seen that the minimal polynomial of $\alpha^2 - \alpha$ is $y^3 - 4y^2 - 4y - 2$.

Finally as the minimal polynomial $\alpha^2 + \alpha$ is $y^3 - 4y^2 + 8y - 6$, we let $x = y + 1$. Substituting $y = x - 1$ in $y^3 - 4y^2 + 8y - 6 = 0$ we get $x^3 - 7x^2 + 19x - 19 = 0$. So the minimal polynomial of $\alpha^2 + \alpha + 1$ is $x^3 - 7x^2 + 19x - 19$.

(29) $\alpha - 2^{\frac{1}{2}} = 2^{\frac{1}{3}}$. Cubing both sides and simplifying this gives $\alpha^3 + 6\alpha - 2 = 2^{\frac{1}{2}}(3\alpha^2 + 2)$. Now squaring both sides we have $\alpha^6 - 6\alpha^4 - 4\alpha^3 + 12\alpha^2 - 24\alpha - 4 = 0$.

The polynomial $x^6 - 6x^4 - 4x^3 + 12x^2 - 24x - 4$ can be checked to be irreducible.

(30) Obviously $\mathbb{Q}(p^{\frac{1}{2}} \cdot q^{\frac{1}{3}}) \subseteq \mathbb{Q}(p^{\frac{1}{2}}, q^{\frac{1}{3}})$ and $\mathbb{Q}(p^{\frac{1}{2}} + q^{\frac{1}{3}}) \subseteq \mathbb{Q}(p^{\frac{1}{2}}, q^{\frac{1}{3}})$ We have

$$p^{\frac{1}{2}} \cdot q^{\frac{1}{3}} \in \mathbb{Q}(p^{\frac{1}{2}} \cdot q^{\frac{1}{3}}) \Longrightarrow p^{\frac{1}{2}}pq \in \mathbb{Q}(p^{\frac{1}{2}} \cdot q^{\frac{1}{3}}) \Longrightarrow p^{\frac{1}{2}} \in \mathbb{Q}(p^{\frac{1}{2}} \cdot q^{\frac{1}{3}}) \Longrightarrow q^{\frac{1}{3}} \in \mathbb{Q}(p^{\frac{1}{2}} \cdot q^{\frac{1}{3}})$$

so $\mathbb{Q}(p^{\frac{1}{2}}, q^{\frac{1}{3}}) \subseteq \mathbb{Q}(p^{\frac{1}{2}} \cdot q^{\frac{1}{3}})$. Therefore $\mathbb{Q}(p^{\frac{1}{2}} \cdot q^{\frac{1}{3}}) = \mathbb{Q}(p^{\frac{1}{2}}, q^{\frac{1}{3}})$. Let $\alpha = p^{\frac{1}{2}} + q^{\frac{1}{3}}$. Then

$$p^{\frac{1}{2}} = \frac{\alpha^3 + 3\alpha p - q}{3\alpha^2 + p}, \quad \text{and} \quad q^{\frac{1}{3}} = \frac{2\alpha^3 - 2\alpha p + q}{3\alpha^2 + p}$$

This shows that $\mathbb{Q}(p^{\frac{1}{2}}, q^{\frac{1}{3}}) \subseteq) \subseteq \mathbb{Q}(p^{\frac{1}{2}} + q^{\frac{1}{3}})$. Therefore $\mathbb{Q}(p^{\frac{1}{2}}, q^{\frac{1}{3}}) = \mathbb{Q}(p^{\frac{1}{2}} + q^{\frac{1}{3}})$. Cubing $\alpha - \sqrt{p} = q^{\frac{1}{3}}$ and simplifying we get $\alpha^3 + 3\alpha p - q = \sqrt{p}(3\alpha^2 + p)$. Finally squaring and further simplifying we get

$$\alpha^6 - 3p\alpha^4 - 2q\alpha^3 + 3p^2\alpha^2 - 6pq\alpha + q^2 - p^3 = 0$$

Therefore $\alpha = p^{\frac{1}{2}} + q^{\frac{1}{3}}$ is a root of the polynomial $f(x) = x^6 - 3px^4 - 2qx^3 + 3p^2x^2 - 6pqx + q^2 - p^3$. Since

$$f(x) \equiv x^6 - 2qx^3 + q^2 = (x^3 - q)^2 \text{ (modulo } p\text{), and}$$

$$f(x) \equiv x^6 - 3x^4 + 3p^2x^2 - p^3 = (x^2 - p)^3 \text{ (modulo } q\text{)}$$

it follows that $f(x) = x^6 - 3px^4 - 2qx^3 + 3p^2x^2 - 6pqx + q^2 - p^3$ is irreducible and so it is the minimal polynomial of $\alpha = \sqrt{p} + q^{\frac{1}{3}}$.

(31) The statement to be proved is an immediate consequence of the fact that the six roots of the equation $x^6 - 2x^3 - 1 = 0$are

$$(1+\sqrt{2})^{\frac{1}{3}},\ \ \omega(1+\sqrt{2})^{\frac{1}{3}},\ \ \omega^2(1+\sqrt{2})^{\frac{1}{3}},\ \ (1-\sqrt{2})^{\frac{1}{3}},\ \ \omega(1-\sqrt{2})^{\frac{1}{3}},\ \ \omega^2(1-\sqrt{2})^{\frac{1}{3}}$$

(32) Evidently $f(\alpha) \neq 0$ and is an algebraic number. As $f(\alpha) \in \mathbb{Q}[\alpha]$ its degree must be 1 or p. As the degree is certainly not one it must be p and so $\mathbb{Q}[\alpha] = \mathbb{Q}[f(\alpha)]$.

(33) Irreducibility of $x^p - x - 1 \in \mathbb{F}_p[x]$ is a consequence of Artin-Schreier theorem discussed in the text. The p roots of $x^p - x - 1$ in $\mathbb{K} = \mathbb{F}_p(\alpha)$ are

$$\alpha, \alpha^p, \alpha^{p^2}, \cdots \alpha^{p^{p-1}}$$

So we have

$$\alpha \cdot \alpha^p \cdot \alpha^{p^2} \cdots \alpha^{p^{p-1}} = (-1)^{p-1}, \qquad \text{or} \qquad \alpha^n = 1$$

(34) Let $f(x) = x^4 + 8x + 12$. Since $f(x)$ has no rational roots and modulo 5 it factors as $(x+1)(x^3 + 4x^2 + x + 2)$ it is irreducible in $\mathbb{Q}[x]$. Alternately it can be seen that in $\mathbb{Z}[x]$, $f(x)$ cannot factor as $(x^2 + ax + b)(x^2 + cx + d)$. Let G be the Galois group of $f(x)$ over $\mathbb{Q}$. As discriminant of $f(x) = 331776 = 2^{12} \cdot 3^4$, $G \subseteq A_4$. Since $f(x)$ is irreducible 4 divides order of G. Also since modulo 5 the polynomial $f(x)$ has an irreducible factor of degree 3 the order of G is divisible by 3. So the order of G is divisible by 12 and therefore G is isomorphic to A_4.

(35) Let $f(x) = x^4 + 3x^3 - 3x - 2$. Since $f(x)$ has no linear factor and modulo 2 it factors as $x(x^3 + x^2 + 1)$, $f(x)$ is irreducible in $\mathbb{Q}[x]$. Let G be the Galois group of $f(x)$ over $\mathbb{Q}$. The fact that $f(x)$ is irreducible in $\mathbb{Q}[x]$ and that modulo 2 it has an irreducible factor of degree 3 implies that the order of G is divisible by 12. The discriminant of $f(x)$ is -2183, therefore G is not a subgroup of A_4 hence G has order 24 and is isomorphic to the symmetric group S_4.

(36) Let $\mathbb{K}$ be the splitting field of $f(x) = x^4 - 3x^2 + 4$. Evidently $\mathbb{K} = \mathbb{Q}\left(\sqrt{\frac{3+\sqrt{-7}}{2}}\right)$ and so the Galois group G of $f(x)$ over $\mathbb{Q}$ has order 4. The discriminant of $f(x)$ is $2^6 \cdot 7^2$, and therefore $G \subseteq A_4$. Since G has order 4 and A_4 has no element of order 4, G is isomorphic to Klein-Four group.

(37) Let $f(x) = x^3 + x + 1$. Evidently $f(x)$ has no linear factor in $\mathbb{Q}$ and so it is irreducible in $\mathbb{Q}[x]$. Since the derivative $3x^2 + 1$ of $f(x)$ is positive the graph of $y = f(x)$ intersect the x-axis just at one point. Therefore $f(x)$ has one real root and two nonreal roots. It follows that the order of the Galois group of $f(x)$ is divisible by 6 and it is isomorphic to S_3, the symmetric group on three letters.

(38) The polynomial $f(x) = x^4 + 3x^3 - 3x - 2$ is an irreducible polynomial of degree 4 with Galois group isomorphic to S_4 as shown in one of the earlier problem.

We first note that any subgroup of order 6 of the symmetric group S_4 is isomorphic to S_3 and it **does not** lie inside the subgroup A_4 of S_4. Let $\mathbb{K}$ be the splitting field of $f(x)$. Since $\mathbb{Q}(\alpha)$ has degree 4 over $\mathbb{Q}$ the subgroup of S_4 (say H) which corresponds to $\mathbb{Q}(\alpha)$ (in Galois correspondence) is of index 4 in S_4 and is therefore isomorphic to S_3. If the algebraic number α were a constructible number then there would be a field $\mathbb{L}$ of degree 2 over $\mathbb{Q}$ such that $\mathbb{Q} \subseteq L \subseteq \mathbb{Q}(\alpha)$. According to the Fundamental Theorem of Galois Theory this would imply that S_4 has a subgroup of index 2 (that is of order 12) having S_3 as a subgroup. But the only subgroup of S_4 or order 12 is A_4 which does not have any subgroup of order 6. Hence α is not a constructible number.

(39) If $f(x)$ has a root $\mathbb{F}_p$ then we have nothing to prove. Suppose that $f(x)$ has no root in $\mathbb{F}_p$. We shall show that it factors as a product of two quadratic polynomials over $\mathbb{F}_p$. Since in $x^4 + 1 \equiv (x+1)^4$ (modulo 2) and $x^4 + 1 \equiv (x^2 + 2x + 2) \times (x^2 - 2x + 2)$ (modulo 3), we may assume that $p > 3$. Clearly $p^2 \equiv 1$ (modulo 8). Therefore we can find $\alpha \in \mathbb{F}_{p^2}$ such that as an element of the multiplicative group $\mathbb{F}_{p^2}^*$ of the non zero elements of $\mathbb{F}_{p^2}$ it has order 8. So α is a root of $x^8 - 1 \in \mathbb{F}_p[x]$. As $\alpha \in \mathbb{F}_{p^2} - \mathbb{F}_p$ its minimal polynomial over $\mathbb{F}_p$ has degree two. As $x^8 - 1 = (x^4 - 1)(x^4 + 1)$, we have $0 = \alpha^8 - 1 = (\alpha^4 - 1)(\alpha^4 + 1)$. Since and $\alpha^4 - 1 \neq 0$ we must have $\alpha^4 + 1 = 0$. But the minimal polynomial of α over $\mathbb{F}_p$ has degree two and it divides $x^4 + 1$. Therefore $x^4 + 1$ is **not** irreducible over $\mathbb{F}_p[x]$.

(40) Let $g(x) = x^m + p$, which is irreducible by Eisenstein Criterion. Let $\alpha \in \mathbb{C}$ be a root of $g(x)$. Therefore α is of degree m and we have $\alpha^m + p = 0$, $\alpha^m = -p$. Let $\beta = \alpha^n$. It will be enough to show that $f(\beta) = 0$ and $\mathbb{Q}(\alpha) = \mathbb{Q}(\beta)$ as this would imply that both α and β have degree m over $\mathbb{Q}$.

We have $f(\beta) = \beta^m + p^n = \alpha^{mn} + p^n = (\alpha^m)^n + p^n = (-p)^n + p^n = 0$. So β is a roots of $f(x)$. Evidently $\mathbb{Q}(\beta) \subseteq \mathbb{Q}(\alpha)$. Since $gcd(m, n) = 1$ we can find $a, b \in \mathbb{Z}$ such that $am + bn = 1$, or $bn = 1 - am$. Now

$$\beta = \alpha^n \Longrightarrow \beta^b = \alpha^{bn} = \alpha^{1-am} = \frac{\alpha}{\alpha^{am}} = \frac{\alpha}{(-p)^a}$$

Therefore $\alpha = (-p)^a \cdot \beta^b \in \mathbb{Q}(\beta)$. Hence $\mathbb{Q}(\alpha) = \mathbb{Q}(\beta)$. Likewise it can be shown that $x^m - p^n$ is irreducible if $gcd(m, n) = 1$.

(41) Let $f(x) = x^5 - kpx + p$. The polynomial $f(x)$ is irreducible by Eisenstein Criterion. Clearly $f(x)$ has a zero in each of the three intervals $(\infty, 0)$, $(0, 1)$, and $(1, \infty)$. Therefore $f(x)$ has at least three real zeros. In the text it has been shown that the discriminant of the trinomial $x^n + ax + b$ is

$$(-1)^{\frac{n(n-1)}{2}} \left[n^n \cdot b^{n-1} + (-1)^{n-1} \cdot a^n \cdot (n-1)^{n-1}\right]$$

—— Substituting $n = 5$, $a = -kp$, and $b = p$, we get the discriminant of $f(x)$ as $p^4(5^5 - pk^5 4^4)$. Now $k \geq 2$ implies that this discriminant is negative. It follows that not all zeros of $f(x)$ can be real and so we conclude that the irreducible quintic $f(x)$ has three real zeros and two nonreal zeros. So the Galois group G of $f(x)$ has a 5-cycle and a transposition and is therefore isomorphic to S_5 - the symmetric group on five letters.

Note. The same conclusion holds if $k = 1$, and $p \geq 13$.

(42) Evidently $\mathbb{Q}(\sqrt{p} + q^{\frac{1}{p}}) \subseteq \mathbb{Q}(\sqrt{p}, q^{\frac{1}{p}})$. Let $\alpha = \sqrt{p} + q^{\frac{1}{p}}$. Then $\alpha - \sqrt{p} = q^{\frac{1}{p}}$ and so $(\alpha - \sqrt{p})^p = q$. By expanding the left hand side by binomial theorem and separating the terms containing the even and odd powers of $\sqrt{p}$ we can write $(\alpha - \sqrt{p})^p = A + B\sqrt{p} = q$ where $A, B \in \mathbb{Q}(\alpha)$. This implies that $\sqrt{p} \in \mathbb{Q}(\alpha)$ which in turn implies that $q^{\frac{1}{p}} \in \mathbb{Q}(\alpha)$.

For the second part we let $\beta = \sqrt{p} \cdot q^{\frac{1}{p}}$. Then $\alpha^p = p^{\frac{p-1}{2}} \sqrt{p} \cdot q$. As before this implies that $\sqrt{p} \in \mathbb{Q}(\beta)$, which in turn implies that $q^{\frac{1}{p}} \in \mathbb{Q}(\beta)$.

(43) It will be enough to show that $\beta^3 = 27\alpha^3$, or $\beta = 3\alpha$. First we note that $\beta = 3^{\frac{1}{3}}(1 - 2^{\frac{1}{3}} + 2^{\frac{2}{3}})$. Substituting $x = -2^{\frac{1}{3}}$ in $x^3 - 1 = (x-1)(x^2+x+1)$, we get $3 = (2^{\frac{1}{3}}+1)(2^{\frac{2}{3}} - 2^{\frac{1}{3}} + 1)$. Therefore

$$\beta = 3^{\frac{1}{3}}(1 - 2^{\frac{1}{3}} + 2^{\frac{2}{3}}) = \frac{3 \cdot 3^{\frac{1}{3}}}{2^{\frac{1}{3}}+1} \Longrightarrow \beta^3 = \frac{81}{(2^{\frac{1}{3}}+1)^3}$$

$$= \frac{81}{3(1+2^{\frac{1}{3}}+2^{\frac{2}{3}})} = \frac{27}{1+2^{\frac{1}{3}}+2^{\frac{2}{3}}}$$

$$\beta^3 = \frac{27}{1+2^{\frac{1}{3}}+2^{\frac{2}{3}}} = \frac{27\cdot(2^{\frac{1}{3}}-1)}{(1+2^{\frac{1}{3}}+2^{\frac{2}{3}})(2^{\frac{1}{3}}-1)} = \frac{27\cdot(2^{\frac{1}{3}}-1)}{2-1}$$

$$= 27\cdot(2^{\frac{1}{3}}-1) = (3\alpha)^3$$

Hence $\mathbb{Q}(\alpha) = \mathbb{Q}(\beta)$. The minimal polynomial of α over $\mathbb{Q}$ is $x^9 + 3x^6 + 3x^2 - 1$.

(44) We first note that $\sqrt[3]{10+6\sqrt{3}} + \sqrt[3]{10-6\sqrt{3}}$ is a real number. Let $x = \sqrt[3]{10+6\sqrt{3}} + \sqrt[3]{10-6\sqrt{3}}$. Cubing both sides we get $x^3 + 6x - 20 = 0$. Let $f(x) = x^3 + 6x - 20$. Since $f'(x) = 3x^2 + 6$, $f(x)$ is an increasing function. So $f(x)$ has just one real root, namely 2. Therefore $\sqrt[3]{10+6\sqrt{3}} + \sqrt[3]{10-6\sqrt{3}} = 2$.

(45) This problem requires some results from the theory of permutation groups which have not been discussed in the text. Specifically among other things we shall need the fact that a normal subgroup of a primitive permutation group is transitive and familiarity with Sylow theorem, Frobenius groups etc. So the readers not familiar with these may skip/ignore this problem.

Since $f(x) \in \mathbb{Q}[x]$ is an irreducible polynomial of degree p, where p is an odd prime, it has at least one real root. We need to prove that if $f(x)$ has more than one real root then all of its roots are real. Let $\Omega = \{\alpha_1, \alpha_2, \cdots, \alpha_p\}$ be the p roots of $f(x)$ and G be its Galois group over $\mathbb{Q}$. Let $\mathbb{K} = \mathbb{Q}(\alpha_1, \alpha_2, \cdots, \alpha_p)$ be the splitting field of $f(x)$ over $\mathbb{K}$. The group G is transitive on Ω and so p divides the order of G. Suppose that the order of G is p and $\alpha \in \Omega$ be a real root of $f(x)$. The $\mathbb{K} = \mathbb{Q}(\alpha)$ and so $\mathbb{Q}(\alpha) \subseteq \mathbb{R}$, and so all the p roots of $f(x)$ are real.

Now suppose that $|G| > p$. As a transitive group of prime degree p the group G is a primitive group. So G is a primitive solvable group. As the prime p divides the order of the group G and G is a subgroup of the symmetric group S_p, the Sylow-p subgroup P of the group G has order p. We claim that P is the only subgroup of

G of order p and therefore is a normal subgroup of G. Suppose it is not so. Then G has more than one Sylow-p subgroup. As G is a solvable group of composite order it has a proper normal subgroup, say H. Since H is a normal subgroup of a primitive group it is transitive (of prime degree p), in particular its order is divisible by p. Therefore H contains all the Sylow-p subgroups of G as these happen to be conjugate to each other by Sylow Theorem. So H is a solvable primitive group of order greater than p having more than one Sylow-p subgroups. So the group H in turn will have a proper normal subgroup, say H_1 of order greater than p containing all the Sylow-p subgroups of H (and therefore of the group G). This will lead to an un-ending chain of subgroups of G each being a proper normal subgroup of the previous subgroup. Since G is finite, this is not possible. So we conclude that the Sylow-p subgroup P of G is a normal subgroup of G of order p.

It is known that the automorphism group of a cyclic group of prime order p is cyclic of order $p-1$ and so G/P is isomorphic to a subgroup of a cyclic group of order $p-1$. Let H be a subgroup of G of order $|G|/p$, which exists by Hall's generalization of Sylow's Theorem for solvable group. Then $G = PH$. In fact H is nothing but a stabilizer of one of the elements of Ω and G is a Frobenius group of degree p and order pt where t divides $p-1$, and is a subgroup of the normalizer of the P in the symmetric group S_p. Let $\alpha, \beta \in \Omega$ be two real roots of $f(x)$. Then in the Galois correspondence the subgroup G_α correspondence to the subfield $\mathbb{Q}(\alpha)$ and the subgroup G_β corresponds to the subfield $\mathbb{Q}(\beta)$. It follows that the subgroup $G_\alpha \cap G_\beta$ corresponds to the subfield $\mathbb{Q}(\alpha)\mathbb{Q}(\beta)$ - the smallest subfield of $\mathbb{K}$ containing the subfields $\mathbb{Q}(\alpha)$ and $\mathbb{Q}(\beta)$. But as G is a Frobenius group $G_\alpha \cap G_\beta = G_{\alpha\beta} = \{e\}$. But in the Galois correspondence the identity subgroup corresponds to the entire splitting field $\mathbb{K}$. Hence $\mathbb{K} = \mathbb{Q}(\alpha, \beta)$. Since both α and β are real numbers it follows that $\mathbb{K} \subseteq \mathbb{R}$. Hence all the roots of the $f(x)$ are real numbers.

Now suppose that $p \equiv 3$ (modulo 4). Suppose that $Disc(f) < 0$. Then not all the roots of $f(x)$ can be real. Since $f(x)$ does have at least one real root therefore $f(x)$ has just one real root.

Now suppose that $f(x)$ has just one real root. Then $f(x)$ has $\frac{p-1}{2}$ pairs of non-real complex-conjugate roots. Let these roots be α, β_i, $\bar{\beta}_i$, $1 \leq i \leq k$, where $k = \frac{p-1}{2}$.

Then the discriminant of $f(x)$ is

$$\prod_{1\le i\le k}(\alpha-\beta_i)^2\prod_{1\le i\le k}(\alpha-\bar{\beta}_i)^2\prod_{1\le i<j\le k}(\beta_i-\beta_j)^2\prod_{1\le i<j\le k}(\bar{\beta}_i-\bar{\beta}_j)^2\prod_{1\le i\le k}(\beta_i-\bar{\beta}_i)^2$$

In the above expression the product of all terms excepting for the last k terms is a positive number. Each of the last k terms, being the square of a purely imaginary number is a negative real number. Since k is odd we conclude that discriminant of $f(x)$ is negative.

(46) We have

$$\begin{aligned} a\alpha^3+b\alpha^2+c\alpha+d &= q_1 \\ a\beta^3+b\beta^2+c\beta+d &= q_2 \\ a\gamma^3+b\gamma^2+c\gamma+d &= q_3 \\ a\delta^3+b\delta^2+c\delta+d &= q_4 \end{aligned}$$

where $q_i \in \mathbb{Q}$ for $1 \le i \le 4$. Subtracting the last equation from each of the previous three equation we get the following three equations:

$$\begin{aligned} a(\alpha^3-\delta^3)+b(\alpha^2-\delta^2)+c(\alpha-\delta) &= q_1-q_4 \\ a(\beta^3-\delta^3)+b(\beta^2-\delta^2)+c(\beta-\delta) &= q_2-q_4 \\ a(\gamma^3-\delta^3)+b(\gamma^2-\delta^2)+c(\gamma-\delta) &= q_3-q_4 \end{aligned}$$

The determinant of the 3×3 matrix

$$\begin{bmatrix} (\alpha^3-\delta^3) & (\alpha^2-\delta^2) & (\alpha-\delta) \\ (\beta^3-\delta^3) & (\beta^2-\delta^2) & (\beta-\delta) \\ (\gamma^3-\delta^3) & (\gamma^2-\delta^2) & (\gamma-\delta) \end{bmatrix}$$

is equal to $(\alpha-\beta)(\alpha-\gamma)(\alpha-\delta)(\beta-\gamma)(\beta-\delta)(\gamma-\delta)$ - which is nonzero as α, β, γ and δ are distinct. So by Cramer's Rule the above system of three equations in the three unknowns a, b and c has (rational) solution. This in turn implies that d is also rational

(47) This is a generalization of the previous problem. In this case we let $f(\alpha_i) = \beta_i$ for $0 \le i \le n$, where β_i for $0 \le i \le n$ are algebraic numbers. Subtracting the equation

corresponding to $i = 0$ from each of the n equations corresponding to $1 \leq i \leq n$ we get the n equations $f(\alpha_i) - f(\alpha_0) = \beta_i - \beta_0 = \gamma_i($ say), that is

$$a_0(\alpha_i^n - \alpha_0^n) + a_1(\alpha_i^{n-1} - \alpha_0^{n-1}) + \cdots + a_{n-1}(\alpha_i - \alpha_0) = \gamma_i \text{ (say)},$$
$$i = 1, 2, 3, \cdots, n.$$

Or

$$\sum_{j=0}^{j=n-1} a_j(\alpha_i^{n-j} - \alpha_0^{n-j}) = \gamma_i, \quad i = 1, 2, \cdots, n,$$

Let $\alpha_i^{n-j} - \alpha_0^{n-j} = c_{ij}$, $1 \leq i \leq n$, $0 \leq j \leq n-1$. Then the $n \times n$ matrix $C = [c_{ij}]$ is nonsingular since its determinant is $\pm \prod_{0 \leq j < i \leq n}(\alpha_i - \alpha_j)$. So the above system of n equations (in the variables $a_0, a_1, \cdots a_{n-1}$) has unique solution and as the entries of the matrix C are algebraic numbers so are $a_0, a_1, \cdots a_{n-1}$. This in turn implies that a_n is also an algebraic number.

(48) This is very easy - in fact trivial.

(49) Let α, β, γ and δ be the roots of the polynomial $f(x) = x^4 + px^2 + r$ and let $S_i = \alpha^i + \beta^i + \gamma^i + \delta^i$. Then as we have seen in the text the discriminant $Disc(f(x))$ is the value of the following 4×4 determinant:

$$\begin{vmatrix} S_0 & S_1 & S_2 & S_3 \\ S_1 & S_2 & S_3 & S_4 \\ S_2 & S_3 & S_4 & S_5 \\ S_3 & S_4 & S_5 & S_6 \end{vmatrix}$$

It is easily seen that

$$S_{-1} = S_1 = S_3 = S_5 = 0, \quad S_2 = -2p, \quad S_4 = 2p^2 - 4r, \quad S_6 = -2p^3 + 6pr$$

Therefore

$$\begin{vmatrix} S_0 & S_1 & S_2 & S_3 \\ S_1 & S_2 & S_3 & S_4 \\ S_2 & S_3 & S_4 & S_5 \\ S_3 & S_4 & S_5 & S_6 \end{vmatrix} = \begin{vmatrix} 4 & 0 & -2p & 0 \\ 0 & -2p & 0 & 2p^2 - 4r \\ -2p & 0 & 2p^2 - 4r & 0 \\ 0 & 2p^2 - 4r & 0 & -2p^3 + 6pr \end{vmatrix}$$

This value of the last determinant is equal to

$$4\begin{vmatrix} -2p & 0 & 2p^2-4r \\ 0 & p^2-4r & 0 \\ 2p^2-4r & 0 & -2p^3+6pr \end{vmatrix} + (-2p)\begin{vmatrix} 0 & -2p & 0 \\ -2p & 0 & 2p^2-4r \\ 2p^2-4r & 0 & -2p^3+6pr \end{vmatrix}$$

which is equal to

$$4(2p^2-4r)\left[4p^2r-16r^2\right]-(2p)^2\left[4p^2r-16r^2\right]=16r(p^2-4r)^2$$

(50) Proceeding as in the previous problem we have

$$S_{-1}=-\frac{q}{r},\quad S_0=4,\quad S_1=S_2=S_5=0,\quad S_3=-3q,\quad S_4=-4r,\quad S_6=3q^2$$

The discriminant of x^4+qx+r is equal to

$$\begin{vmatrix} S_0 & S_1 & S_2 & S_3 \\ S_1 & S_2 & S_3 & S_4 \\ S_2 & S_3 & S_4 & S_5 \\ S_3 & S_4 & S_5 & S_6 \end{vmatrix} = \begin{vmatrix} 4 & 0 & 0 & -3q \\ 0 & 0 & -3q & -4r \\ 0 & -3q & -4r & 0 \\ -3q & -4r & 0 & 3q^2 \end{vmatrix}$$

This value of the last determinant is equal to

$$4\begin{vmatrix} 0 & -3q & -4r \\ -3q & -4r & 0 \\ -4r & 0 & 3q^2 \end{vmatrix} + 3q\begin{vmatrix} 0 & 0 & -3q \\ 0 & -3q & -4r \\ -q & -4r & 0 \end{vmatrix}$$

which is equal to

$$4\left[3q(-9q^3)-4r(-16r^2)\right]-9q^2(-9q^2)=-108q^4+256r^3+81q^4=256r^3-27q^4$$

Note.

The discriminant of x^4+qx+r can also be calculated by the formula for the discriminant of the trinomial x^n+ax+b discussed in the text. We had shown that the discriminant of the trinomial x^n+ax+b if given by

$$(-1)^{\frac{n(n-1)}{2}}\left[n^n\cdot b^{n-1}+(-1)^{n-1}\cdot a^n\cdot(n-1)^{n-1}\right]$$

In our case we have $n = 4$, $a = q$, $b = r$. Substituting these value in the above formula we get

$$Disc(x^4 + qx + r) = 4^4 \cdot r^3 + (-1)^3 \cdot q^4 \cdot 3^3 = 256r^3 - 27q^4$$

(51) **Hint.**

This has been discussed in the text somewhere and the answer is $y^3 + 6py^2 + 9p^2y + 4p^3 + 27q^2 = 0$.

(52) **Hint.**

This is an immediate consequence of the fact that A_4 is the only subgroup of order 12 in S_4 and A_4 has no subgroup of order 6 - and the FTGT (Fundamental Theorem of Galois Theory).

(53) The first four assertion are easily verified. Following the discussion of the solution of the cubic in an earlier chapter we can easily obtained the given expression for the roots of $f(x)$. For the rest we first observe that

$$\mathbb{K} = \mathbb{Q}(\alpha, \beta, \gamma, \omega) = \mathbb{Q}(A, \sqrt{-3})$$

We also note that as $(\alpha - \beta)^2(\beta - \gamma)^2(\gamma - \alpha)^2 = 2^2 \times 37$, so $\sqrt{37} \in \mathbb{K}$ but $\sqrt{-37} \notin \mathbb{K}$. Evidently $[\mathbb{K} : \mathbb{Q}] = 12$. So the group $G = Aut(\mathbb{K}/\mathbb{Q})$ has order 12. The elements of G are easily described. We observe that $A = \dfrac{\alpha\omega - \beta}{1 - \omega}$, so $A \in \mathbb{K}$ and $B \in \mathbb{K}$. As A and B are algebraic numbers of degree 6 any element of G is specified by its effect A and ω. We also note that $\sqrt{-111} = \sqrt{-3}\sqrt{37} = (2\omega + 1) \cdot \sqrt{37}$ and also $\sqrt{-111} = 9(A^3 + 1)$. Therefore $\sqrt{37} = \dfrac{9(A^3 + 1)}{2\omega + 1}$.

The minimal polynomial of A (or B) is $x^6 + 2x^3 + \frac{64}{27}$. It six roots are $A, B, \omega A$, $\omega B, \omega^2 A, \omega^2 B$. Let σ and η be the following elements of G.

$$\sigma : \begin{cases} A \longrightarrow \omega B \\ \omega \longrightarrow \omega^2 \end{cases} \qquad\qquad \eta : \begin{cases} A \longrightarrow B \\ \omega \longrightarrow \omega \end{cases}$$

The effect on A of the successive powers of σ is as follows:

$$A \longrightarrow \omega B \longrightarrow \omega A \longrightarrow B \longrightarrow \omega^2 A \longrightarrow \omega^2 B \longrightarrow A$$

So the order of σ is 6 and the order of η is 2 and $\eta\sigma = \sigma^5\eta$. Therefore

$$G = \langle \sigma, \eta, \sigma^6 = \eta^2 = e, \eta\sigma = \sigma^5\eta \rangle = \left\{ e, \sigma^i\, \eta^j, 0 \le i \le 5, \quad 0 \le j \le 1 \right\}$$

We remind the readers that the elements of G are considered as permutation and we multiply the permutations from left to right. So $\sigma\eta$ means that σ followed by η.

The following table gives the effect of various element of G on A, B, ω and $\sqrt{37}$

	e	σ	σ^2	σ^3	σ^4	σ^5	η	$\sigma\eta$	$\sigma^2\eta$	$\sigma^3\eta$	$\sigma^4\eta$	$\sigma^5\eta$
A	A	ωB	ωA	B	$\omega^2 A$	$\omega^2 B$	B	ωA	ωB	A	$\omega^2 B$	$\omega^2 A$
B	B	$\omega^2 A$	$\omega^2 B$	A	ωB	ωA	A	$\omega^2 B$	$\omega^2 A$	B	ωA	ωB
ω	ω	ω^2	ω	ω^2	ω	ω^2	ω	ω^2	ω	ω^2	ω	ω^2
$\sqrt{\mathbf{37}}$	$\sqrt{37}$	$-\sqrt{37}$	$\sqrt{37}$	$-\sqrt{37}$	$\sqrt{37}$	$-\sqrt{37}$	$-\sqrt{37}$	$-\sqrt{37}$	$-\sqrt{37}$	$-\sqrt{37}$	$-\sqrt{37}$	$-\sqrt{37}$

G is isomorphic to the $S_3 \times C_2$ as well as to the dihedral group of order 12.

The following table gives the subgroups of G and their corresponding fixed subfield.

Subgroup of G	Order	Type	Fixed Field
G	12	Dihedral	$\mathbb{K}$
$\langle\sigma\rangle$	6	Cyclic	$\mathbb{Q}(\sqrt{37})$
$\langle\sigma^2, \eta\rangle$	6	Dihedral	$\mathbb{Q}(\omega)$
$\langle\sigma^2, \sigma\eta\rangle$	6	Dihedral	$\mathbb{Q}(\sqrt{-111})$
$\langle\sigma^3, \eta\rangle$	4	Klein-Four	$\mathbb{Q}(\alpha)$
$\langle\sigma^3, \sigma\eta\rangle$	4	Klein-Four	$\mathbb{Q}(\gamma)$
$\langle\sigma^3, \sigma^2\eta\rangle$	4	Klein-Four	$\mathbb{Q}(\beta)$
$\langle\sigma^2\rangle$	3		$\mathbb{Q}(\omega, \sqrt{-37})$
$\langle\sigma^3\rangle$	2		$\mathbb{Q}(\alpha, \beta, \gamma)$
$\langle\eta\rangle$	2		$\mathbb{Q}(\alpha, \omega)$
$\langle\sigma\eta\rangle$	2		$\mathbb{Q}(\gamma, \sqrt{-111}) = \mathbb{Q}(\omega^2 A)$
$\langle\sigma^2\eta\rangle$	2		$\mathbb{Q}(\beta, \omega)$
$\langle\sigma^3\eta\rangle$	2		$\mathbb{Q}(\alpha, \sqrt{-111}) = \mathbb{Q}(A)$
$\langle\sigma^4\eta\rangle$	2		$\mathbb{Q}(\gamma, \omega)$
$\langle\sigma^5\eta\rangle$	2		$\mathbb{Q}(\beta, \sqrt{-111}) = \mathbb{Q}(\omega A)$

(54) **Hint.** Let $y = \alpha^2 + \alpha + 1$. We know that $\alpha^4 - 4\alpha + 2 = 0$. Eliminating α from these two equation we get $y^3 - 11y^2 + 37y - 37 = 0$.

(55) **Answer.** $x^4 - 28x^3 - 106x^2 + 4228x - 9199$.

(56) **Hint.** First show that

$$\sqrt{2} = \frac{\alpha^4 - 12\alpha^2 + 36}{6(\alpha^2 - 4)}, \qquad \sqrt{3} = \frac{\alpha^4 - 12\alpha^2 + 24}{4(\alpha^2 - 3)}$$

We note that $f(x) = (x^6 - 20x^4 + 64x^2 - 32)(x^2 - 4) + 16$. Therefore

$$\frac{1}{\alpha^2 - 4} = -\frac{1}{16}\left(\alpha^6 - 20\alpha^4 + 64\alpha^2 - 32\right).$$

So

$$\begin{aligned}\frac{\alpha^4 - 12\alpha^2 + 36}{6(\alpha^2 - 4)} &= \frac{1}{6}\left[\alpha^2 - 8 + \frac{4}{\alpha^2 - 4}\right] \\ &= \frac{1}{6}\left[\alpha^2 - 8 - \frac{1}{4}\left(x^6 - 20x^4 + 64x^2 - 32\right)\right]\end{aligned}$$

Therefore

$$\sqrt{2} = -\frac{1}{24}\left(\alpha^6 - 20\alpha^4 + 60\alpha^2\right)$$

Likewise $f(x) = (x^6 - 21x^4 + 81x^2 - 45)(x^2 - 3) + 9$. So

$$\frac{1}{\alpha^2 - 3} = -\frac{1}{9}\left(\alpha^6 - 21\alpha^4 + 81\alpha^2 - 45\right).$$

Proceeding as in the case of $\sqrt{2}$ we get

$$\sqrt{3} = \frac{1}{12}\left(\alpha^6 - 21\alpha^4 + 84\alpha^2 - 72\right)$$

(57) Let

$$f(x) = a(x - \alpha_1)(x - \alpha_2)\cdots(x - \alpha_n), \text{ and}$$

$$g(x) = b(x - \beta_1)(x - \beta_2)\cdots(x - \beta_m)$$

As the given assertion evidently holds if $Disc(f(x)g(x) = 0$, we may assume that $Disc(f(x)g(x) \neq 0$. Writing $f(x)g(x)$, $f(x)$ and $g(x)$ respectively as fg, f and g we see that the quotient

$$\frac{Disc(fg)}{Disc(f)\,Disc(g)}$$

is same as

$$= \frac{(ab)^{2m+2n-2}\prod_{1\le i<j\le n}(\alpha_i-\alpha_j)^2\cdot\prod_{1\le i<j\le m}(\beta_i-\beta_j)^2\cdot\prod_{\substack{1\le i\le n\\1\le j\le m}}(\alpha_i-\beta_j)^2}{a^{2n-2}\prod_{1\le i<j\le n}(\alpha_i-\alpha_j)^2\cdot b^{2m-2}\prod_{1\le i<j\le m}(\beta_i-\beta_j)^2}$$

$$= a^{2m}\prod_{\substack{1\le i\le n\\1\le j\le m}}(\alpha_i-\beta_j)^2\cdot b^{2n}$$

$$= \left(a^m b^n\prod_{\substack{1\le i\le n\\1\le j\le m}}(\alpha_i-\beta_j)^2\right)^2$$

$$= (Res(f,g))^2$$

(58) Evidently we may assume that $Disc(f_1f_2\cdots f_n)\neq 0$.

We shall prove given statement by induction on n. By the previous problem the statement holds for $n=2$. Assume that it holds for $n=r$, where $r\ge 2$. So we have

$$Disc(f_1f_2\cdots f_r)$$
$$=\prod_{i=1}^{i=r}Disc(f_i)\times(Res(f_1,f_2)\cdot Res(f_1f_2,f_3)\cdots\cdots Res(f_1f_2\cdots f_{r-1},f_r))^2$$

From the previous problem the statement is true for the product of two polynomials, therefore

$$Disc(f_1f_2\cdots f_rf_{r+1}) = Disc\left(\prod_{i=1}^{i=r}f_i\right)Disc(f_{r+1})\times(Res(f_1f_2\cdots f_r,f_{r+1}))^2.$$

Since the assertion is assumed to be true for $n=r$ this is equal to

$$\prod_{i=1}^{i=r+1}Disc(f_i)$$
$$\times(Res(f_1,f_2)\cdot Res(f_1f_2,f_3)Res(f_1f_2f_3,f_4)\cdots Res(f_1f_2\cdots f_r,f_{r+1}))^2.$$

(59) We first note that the polynomial $x^{12}-x^{10}+x^8-x^6+x^4-x^2+1$ is the 28-th cyclotomic polynomial Φ_{28}. It is known that the Galois group of x^n-1 over $\mathbb{Q}$ is isomorphic to $\mathbb{Z}_n^*$ - the group of units in the ring $\mathbb{Z}_n=\mathbb{Z}/n\mathbb{Z}$. Therefore the

Galois group of $x^{28}-1$ is isomorphic to $C_6 \times C_2$. However to exhibit the Galois correspondence we will need some information about the subfields of the splitting field $\mathbb{K} = \mathbb{Q}\left(e^{\frac{2\pi i}{28}}\right)$ of the polynomial $x^{28}-1$. We will first show that

$$\begin{aligned}\mathbb{K} &= \mathbb{Q}\left(e^{\frac{2\pi i}{28}}\right) = \mathbb{Q}\left(e^{\frac{2\pi i}{7}}, i\right) = \mathbb{Q}\left(\cos\frac{2\pi}{7}, \sqrt{7}, i\right) \\ &= \mathbb{Q}\left(\sin\frac{2\pi}{7}, i\right) = \mathbb{Q}\left(i\sin\frac{2\pi}{7}, i\right)\end{aligned}$$

The first part of the assertion is obvious as the product of a primitive 7th root of unity with a primitive 4th root of unit is a primitive 28th root of unity. For the rest we proceed as follows. Let $\alpha = e^{\frac{2\pi i}{7}}$. It is enough to show that

$$\mathbb{Q}(\alpha) = \mathbb{Q}\left(\cos\frac{2\pi}{7} + i\sin\frac{2\pi}{7}\right) = \mathbb{Q}\left(\cos\frac{2\pi}{7}, i\sin\frac{2\pi}{7}\right) = \mathbb{Q}\left(i\sin\frac{2\pi}{7}\right).$$

Now $\mathbb{Q}(\cos\frac{2\pi}{7}) = \mathbb{Q}(\alpha+\alpha^{-1}) \subseteq \mathbb{Q}(\alpha) \Longrightarrow \mathbb{Q}(i\sin\frac{2\pi}{7}) \subseteq \mathbb{Q}(\alpha)$, we only need to show that $\mathbb{Q}(\alpha) \subseteq \mathbb{Q}(i\sin\frac{2\pi}{7})$ which will follow if we could show that $\cos\frac{2\pi}{7} \in \mathbb{Q}(i\sin\frac{2\pi}{7})$. We shall show that $\cos\frac{2\pi}{7}$ can be expressed as a polynomial in $i\sin\frac{2\pi}{7}$ with rational coefficients. Let $\frac{2\pi}{7} = \theta$. Since

$$\begin{aligned}\cos 6\theta &= 4\cos^3 2\theta - 3\cos 2\theta = 4(1-2\sin^2\theta)^3 - 3(1-2\sin^2\theta) \\ &= 4\left[1+2(i\sin\theta)^2\right]^3 - 3\left[1+2(i\sin\theta)^2\right]\end{aligned}$$

we have $\cos\theta = \cos 6\theta \in \mathbb{Q}(i\sin\theta)$ as well as $\cos\theta = \cos 6\theta \in \mathbb{Q}(\sin\theta)$. As $\mathbb{Q}(\cos\frac{2\pi}{7})$ is algebraic of degree 3 we have $\mathbb{Q}(\cos\frac{2\pi}{7}) = \mathbb{Q}(\sin\frac{2\pi}{7}) \cap \mathbb{Q}(i\sin\frac{2\pi}{7})$. We also note that $\alpha+\alpha^2+\alpha^4$ and $\alpha^3+\alpha^6+\alpha^5$ are roots of the equation $x^2+x+2=0$ therefore

$$\sin\frac{2\pi}{7} + \sin\frac{4\pi}{7} + \sin\frac{8\pi}{7} = \frac{\sqrt{7}}{2}, \quad \text{and} \quad \sin\frac{6\pi}{7} + \sin\frac{12\pi}{7} + \sin\frac{10\pi}{7} = -\frac{\sqrt{7}}{2}.$$

The following table gives a list of subfields of $\mathbb{K} = \mathbb{Q}(e^{\frac{2\pi i}{28}}) = \mathbb{Q}(e^{\frac{2\pi i}{7}}, i)$ along with their degree over $\mathbb{Q}$.

Subfield of $\mathbb{K}$	Degree over $\mathbb{Q}$
$\mathbb{Q}(\alpha, i)$	12
$\mathbb{Q}(\sin\frac{2\pi}{7})$	6
$\mathbb{Q}(\alpha, i)$	6
$\mathbb{Q}(e^{\frac{2\pi i}{7}}) = \mathbb{Q}(\alpha) = \mathbb{Q}(i\sin\frac{2\pi}{7})$	6
$\mathbb{Q}(i, \sqrt{7})$	4
$\mathbb{Q}(\alpha + \alpha^{-1}) = \mathbb{Q}(\cos\frac{2\pi}{7})$	3
$\mathbb{Q}(i\sqrt{7})$	2
$\mathbb{Q}(\sqrt{7})$	2
$\mathbb{Q}(i)$	2
$\mathbb{Q}$	1

Let $\varepsilon = e^{\frac{2\pi i}{28}}$ and σ and η be the following automorphism of the field $\mathbb{K} = \mathbb{Q}(\varepsilon)$.

$$\sigma : \varepsilon \longrightarrow \varepsilon^5, \qquad\qquad \eta : \varepsilon \longrightarrow \varepsilon^{27}$$

Then $G = Aut(\mathbb{K}/\mathbb{Q}) = \langle \sigma, \eta, \sigma^6 = \eta^2 = e \;\; ; \;\; \sigma\eta = \eta\sigma \rangle$ and we have

$$G = \left\{e, \sigma, \sigma^2, \sigma^3, \sigma^4, \sigma^5, \eta, \sigma\eta, \sigma^2\eta, \sigma^3\eta, \sigma^4\eta, \sigma^5\eta\right\}$$

Although an automorphism of $\mathbb{K}$ is uniquely specified its effect on $\varepsilon = e^{\frac{2\pi i}{28}}$ but in the following table we describe the effect of various automorphisms of $\mathbb{K}$ on the elements $\varepsilon = e^{\frac{2\pi i}{28}}$, $\varepsilon^4 = \alpha = e^{\frac{2\pi i}{7}}$, and $\varepsilon^7 = i$ of $\mathbb{K}$. We also note that $\alpha - \alpha^6 = 2i\sin\frac{2\pi}{7}$. As can be easily seen the group $G = Aut(\mathbb{K}/\mathbb{Q})$ is an Abelian group and is isomorphic to the direct product of a cyclic group of order *six* with a group of order *two*. It has 3 (cyclic) subgroups of order six, one subgroup of order four (a Klein-Four group), one subgroup of order three and three subgroups of order two.

Automorphism	ε	$\alpha = \varepsilon^4$	$i = \varepsilon^7$	Order
e	ε	α	i	1
σ	ε^5	α^5	i	6
σ^2	ε^{25}	α^4	i	3
σ^3	ε^{13}	α^6	i	2
σ^4	ε^9	α^2	i	3
σ^5	ε^{17}	α^3	i	6
η	ε^{27}	α^6	$-i$	2
$\sigma\eta$	ε^{23}	α^2	$-i$	6
$\sigma^2\eta$	ε^3	α^3	$-i$	6
$\sigma^3\eta$	ε^{15}	α	$-i$	2
$\sigma^4\eta$	ε^{19}	α^5	$-i$	6
$\sigma^5\eta$	ε^{11}	α^4	$-i$	6

The following table describes the subgroups of $G = Aut(\mathbb{K}/\mathbb{Q})$ and the corresponding (fixed) subfields of $\mathbb{K}$.

Subgroup	Order	Fixed Subfield	Degree
$G = \langle \sigma, \eta \rangle$	12	$\mathbb{Q}$	1
$\langle \sigma \rangle$	6	$\mathbb{Q}(i)$	2
$\langle \sigma\eta \rangle$	6	$\mathbb{Q}(i\sqrt{7})$	2
$\langle \sigma^2\eta \rangle$	6	$\mathbb{Q}(\sqrt{7})$	2
$\langle \sigma^3, \eta \rangle$	4	$\mathbb{Q}(\alpha + \alpha^6) = \mathbb{Q}(\cos \frac{2\pi}{7})$	3
$\langle \sigma^2 \rangle$	3	$\mathbb{Q}(\sqrt{7}, i)$	4
$\langle \sigma^3 \rangle$	2	$\mathbb{Q}(\cos \frac{2\pi}{7}, i)$	6
$\langle \eta \rangle$	2	$\mathbb{Q}(\sin \frac{2\pi}{7})$	6
$\langle \sigma^3\eta \rangle$	2	$\mathbb{Q}(\alpha) = \mathbb{Q}(i \sin \frac{2\pi}{7})$	6

(60) To prove that $f(x) = x^4 - 5x + 5$ is the minimal polynomial of $\alpha - \alpha^2$, where α is a root of $x^4 + x^3 + x^2 + x + 1 = 0$ we proceed as follows:

We have $\alpha = e^{\frac{2\pi i}{5}} = \cos\frac{2\pi}{5} + i\sin\frac{2\pi}{5}$. We also note that $\alpha \longrightarrow \alpha^2$ is a generating automorphism of the splitting field $\mathbb{Q}(\alpha)$ of the polynomial $x^4+x^3+x^2+x+1$. Let

$$\beta_1 = \alpha - \alpha^2, \quad \beta_2 = \alpha^2 - \alpha^4, \quad \beta_3 = \alpha^4 - \alpha^3, \quad \beta_4 = \alpha^3 - \alpha$$

It is not too hard to show that

$$\begin{aligned}
\beta_1 + \beta_2 + \beta_3 + \beta_4 &= 0\\
\beta_1\beta_2 + \beta_1\beta_3 + \beta_1\beta_4 + \beta_2\beta_3 + \beta_2\beta_4 + \beta_3\beta_4 &= 0\\
\beta_1\beta_2\beta_3 + \beta_1\beta_2\beta_4 + \beta_1\beta_3\beta_4 + \beta_2\beta_3\beta_4 &= 5\\
\beta_1\beta_2\beta_3\beta_4 &= 5
\end{aligned}$$

Therefore $(x-\beta_1)(x-\beta_2)(x-\beta_3)(x-\beta_4) = x^4-5x+5$ is the minimal polynomial of $\alpha - \alpha^2$.

Alternately we notice that $\alpha+\alpha^4 = 2\cos\frac{2\pi}{5} = \dfrac{-1+\sqrt{5}}{2}$ and $\alpha^2+\alpha^3 = 2\cos\frac{4\pi}{5} = \dfrac{-1-\sqrt{5}}{2}$ Therefore

$$\beta_1 + \beta_3 = \sqrt{5}, \quad \text{and} \quad \beta_1\beta_3 = (\alpha - \alpha^2)(\alpha^4 - \alpha^3) = \frac{5-\sqrt{5}}{2}$$

So

$$(x-\beta_1)(x-\beta_3) = x^2 - \sqrt{5}x + \frac{5-\sqrt{5}}{2}$$

Similarly

$$(x-\beta_2)(x-\beta_4) = x^2 + \sqrt{5}x + \frac{5+\sqrt{5}}{2}$$

Therefore the minimal polynomial of $\alpha - \alpha^2$ is

$$\begin{aligned}
(x-\beta_1)(x-\beta_2)(x-\beta_3)(x-\beta_4) &= \left(x^2 - \sqrt{5}x + \frac{5-\sqrt{5}}{2}\right)\left(x^2 + \sqrt{5}x + \frac{5+\sqrt{5}}{2}\right)\\
&= x^4 - 5x + 5
\end{aligned}$$

The minimal polynomial of $\alpha+\alpha^2$ can likewise shown to be as $x^4+2x^3+4x^2+3x+1$. For another way of doing this we proceed as follows:

$$\alpha = \cos\frac{2\pi}{5} + i\sin\frac{2\pi}{5} = \frac{-1+\sqrt{5}}{4} + i\sqrt{\frac{5+\sqrt{5}}{8}}$$

$$\alpha^2 = \cos\frac{4\pi}{5} + i\sin\frac{4\pi}{5} = -\frac{1-\sqrt{5}}{4} + i\sqrt{\frac{5-\sqrt{5}}{8}}$$

Let

$$x = \alpha+\alpha^2 = -\frac{1}{2} + \frac{i}{\sqrt{8}}\left(\sqrt{5+\sqrt{5}} + \sqrt{5-\sqrt{5}}\right)$$

That is

$$\frac{1}{2}(2x+1) = \frac{i}{\sqrt{8}}\left(\sqrt{5+\sqrt{5}} + \sqrt{5-\sqrt{5}}\right)$$

Squaring this we get $2(4x^2+4x+1) = -(10+2\sqrt{20})$, or $2x^2+2x+3 = -\sqrt{5}$. Squaring again and simplifying we get $x^4+2x^3+4x^2+3x+1=0$. The polynomial $x^4+2x^3+4x^2+3x+1$ is easily seen to be irreducible (it is irreducible modulo 2) so it is the minimal polynomial of $\alpha+\alpha^2$.

(61) The statement is evidently true for $p=2$ so we assume that $p \geq 3$. We shall show that if $f(x)$ is reducible then it has a root in $\mathbb{K}$. Let $g(x)$ be a nontrivial factor of $f(x)$ of degree m where $1 < m < p$. Without loss of generality we may assume that m is even and $g(x)$ is monic. Let $\alpha = e^{\frac{2\pi i}{p}}$. Then the p roots of $f(x)$ in $\mathbb{C}$ are $a^{\frac{1}{p}}\alpha^i$, for $0 \leq i \leq p-1$ - some m of these are the roots of $g(x)$. Let these m roots of $g(x)$ be $a^{\frac{1}{p}}\alpha^{i_j}$, for $1 \leq j \leq m$. Then if c is the constant term of $g(x)$ then

$$c = \prod_{j=1}^{j=m}\left(a^{\frac{1}{p}}\alpha^{i_j}\right) = a^{m/p}\alpha^{\alpha_1^i+\alpha_2^i+\cdots+\alpha_m^i} = a^{m/p}\alpha^k, \text{ where } k = i_1+i_2+\cdots+i_m.$$

So $c^p = a^m$. Since $gcd(m,p) = 1$ there exist $s, t \in \mathbb{Z}$ such that $sm + pt = 1$. Now $c^{ps} = a^{ms} = a^{1-pt} = \frac{a}{a^{pt}}$. Therefore $a = (a^t c^s)^p$. That is a is a p-th power of $a^t c^s \in \mathbb{K}$, hence $f(x)$ has a root in $\mathbb{K}$.

(62) Let $\mathbb{K} = \mathbb{Q}[\alpha]$. Clearly $(x-\beta)(x-\gamma) = x^2 - (\beta+\gamma)x + \beta\gamma \in \mathbb{K}[x]$. Therefore $(\alpha-\beta)(\alpha-\gamma) \in \mathbb{K}$. Let $(\alpha-\beta)(\beta-\gamma)(\gamma-\alpha) = \pm\sqrt{\Delta}$. Then

$$\beta - \gamma = \pm\frac{\sqrt{\Delta}}{(\alpha-\beta)(\alpha-\gamma)} \in \mathbb{K}(\sqrt{\Delta}) = \mathbb{Q}(\alpha, \sqrt{\Delta}).$$

Since $\beta+\gamma \in \mathbb{K}$, it follows that $\beta, \gamma \in \mathbb{K}(\sqrt{\Delta}) = \mathbb{Q}(\alpha, \sqrt{\Delta})$.

(63) $\mathbf{A} \Longrightarrow \mathbf{B}$

Let $\alpha_1 \in \mathbb{L} - \mathbb{K}$ and $f_1(x) \in \mathbb{K}[x]$ be its minimal polynomial. By hypothesis the polynomial $f_1(x)$ splits completely in $\mathbb{L}$. Let $\mathbb{K}_1$ denote the subfield of $\mathbb{L}$ generated by all the roots of $f_1(x)$ lying in $\mathbb{L}$. So $\mathbb{K}_1 \subseteq \mathbb{L}$. If $\mathbb{K}_1 = \mathbb{L}$ then we are done. Otherwise let $\alpha_2 \in \mathbb{L} - \mathbb{K}_1$ and repeat this process. As $\mathbb{L}/\mathbb{K}$ is a finite extension this process must come to an end in finitely many (say r) steps. That is there are irreducible polynomials $f_i(x) \in \mathbb{K}[x]$, $1 \leq i \leq r$ such that the field $\mathbb{L}$ is the splitting field of the polynomial $f(x) = f_1(x)f_2(x)\cdots f_r(x)$. This completes the proof that $\mathbf{A} \Longrightarrow \mathbf{B}$.

Before we start the proof $\mathbf{B} \Longrightarrow \mathbf{A}$ we shall make some observations. We know that if $f(x) \in \mathbb{K}[x]$ is an irreducible polynomial and $\alpha, \beta \in \mathbb{C}$ are two roots of $f(x)$ then the fields $\mathbb{K}(\alpha)$ and $\mathbb{K}(\beta)$ are isomorphic. Also if $\mathbb{K}_1$ and $\mathbb{K}_2$ are isomorphic fields then an isomorphism between $\mathbb{K}_1$ to $\mathbb{K}_2$ extends to an isomorphism between the polynomial rings $\mathbb{K}_1[x]$ and $\mathbb{K}_2[x]$. So if $f(x) \in \mathbb{K}_1[x]$ is an irreducible polynomial then so is its image $f'(x) \in \mathbb{K}_2[x]$. Thus if α is a root of $f(x) \in \mathbb{K}_1[x]$ and β is a root of $f'(x) \in \mathbb{K}_2[x]$ then the extension fields $\mathbb{K}_1(\alpha) = \mathbb{K}_1[\alpha]$ and $\mathbb{K}_2(\beta) = \mathbb{K}_2[\beta]$ are also isomorphic. By extension of this argument it follows that if $\mathbb{L}_1$ is the splitting field of the polynomial $f(x) \in \mathbb{K}_1[x]$ and $\mathbb{L}_2$ is the splitting field of the polynomial $f'(x) \in \mathbb{K}_2[x]$ then the fields $\mathbb{L}_1$ and $\mathbb{L}_2$ are isomorphic. Now we proceed with the proof that $\mathbf{B} \Longrightarrow \mathbf{A}$.

We now assume that the field $\mathbb{L}$ is a splitting field of the polynomial $f(x) \in \mathbb{K}[x]$. Let $p(x) \in \mathbb{K}[x]$ is an irreducible polynomial such that one of its roots (say α) lies in $\mathbb{L}$. Suppose that the polynomial $p(x) \in \mathbb{K}[x]$ has another root $\beta \notin \mathbb{L}$. We shall show this assumption leads to a contradiction.

Since $\beta \notin \mathbb{L}$, the field $\mathbb{L}$ is a proper subfield of the field $\mathbb{L}(\beta)$. As α, β are two roots of the same irreducible polynomial $p(x) \in \mathbb{K}[x]$, the fields $\mathbb{K}(\alpha)$ and $\mathbb{K}(\beta)$ are isomorphic. We also note that $f(x) \in \mathbb{K}[x] \subseteq \mathbb{K}(\beta)[x]$. Since $\mathbb{L}$ is the splitting field of $f(x) \in \mathbb{K}[x]$ it can also be regarded as the splitting field of the polynomial $f(x) \in \mathbb{K}(\alpha)[x]$. Likewise the field $\mathbb{L}(\beta)$ can be regarded as the splitting field of the polynomial $f(x) \in \mathbb{K}(\beta)[x]$. So as the fields $\mathbb{L}$ and $\mathbb{L}(\beta)$ are the splitting fields of the polynomials $f(x) \in \mathbb{K}(\alpha)[x]$ and $f(x) \in \mathbb{K}(\beta)[x]$ respectively, and as the fields $\mathbb{K}(\alpha)$ and $\mathbb{K}(\beta)$ are isomorphic, we conclude that the fields $\mathbb{L}$ and $\mathbb{L}(\beta)$ are also isomorphic. That means that these have same dimension over the base field $\mathbb{K}$. But this contradicts the fact that the field $\mathbb{L}$ is a proper subfield of the field $\mathbb{L}(\beta)$.

(64) Before we start proving the stated assertion we would like to discuss its precise meaning. As discussed in the text the polynomial x^3+x^2-2x-1 is irreducible and has three real roots. In fact its three roots are $2\cos\frac{2\pi}{7}$, $2\cos\frac{4\pi}{7}$, and $2\cos\frac{6\pi}{7}$. In the text we obtained the following expressions for the three roots of the irreducible cubic equation $x^3+x^2-2x-1=0$ using Cardano's Formula

$$\begin{aligned}
2\cos\frac{2\pi}{7} &= \frac{\sqrt{7}}{3}\left(\frac{1+3\sqrt{-3}}{2\sqrt{7}}\right)^{\frac{1}{3}} + \frac{\sqrt{7}}{3}\left(\frac{1-3\sqrt{-3}}{2\sqrt{7}}\right)^{\frac{1}{3}} - \frac{1}{3} &\approx 1.2469796\\
2\cos\frac{4\pi}{7} &= \frac{\sqrt{7}}{3}\left(\frac{1+3\sqrt{-3}}{2\sqrt{7}}\right)^{\frac{1}{3}}\omega^2 + \frac{\sqrt{7}}{3}\left(\frac{1-3\sqrt{-3}}{2\sqrt{7}}\right)^{\frac{1}{3}}\omega - \frac{1}{3} &\approx -0.4450418\\
2\cos\frac{6\pi}{7} &= \frac{\sqrt{7}}{3}\left(\frac{1+3\sqrt{-3}}{2\sqrt{7}}\right)^{\frac{1}{3}}\omega + \frac{\sqrt{7}}{3}\left(\frac{1-3\sqrt{-3}}{2\sqrt{7}}\right)^{\frac{1}{3}}\omega^2 - \frac{1}{3} &\approx -1.8019377
\end{aligned}$$

where $\omega = e^{\frac{2\pi i}{3}}$ is a primitive cube root of unit. Notice that although the roots are real but to evaluate these from their expression given above we are forced to take cube root of (nonreal) complex numbers. The point of this problem is that this situation is unavoidable in case of irreducible cubic with three real roots.

Let $f(x) \in \mathbb{Q}[x]$ be an irreducible cubic with three real roots α, β and γ. Let $\Delta = (\alpha-\beta)^2(\beta-\gamma)^2(\gamma-\alpha)^2 \in \mathbb{Q}$ be the discriminant of $f(x)$. We note that since the three roots are real the discriminant $\Delta > 0$, so $\sqrt{\Delta}$ is a real number. In an earlier exercise we have seen that $\mathbb{Q}(\alpha, \sqrt{\Delta}) = \mathbb{Q}(\beta, \sqrt{\Delta}) = \mathbb{Q}(\gamma, \sqrt{\Delta})$, is the splitting field of $f(x) \in \mathbb{Q}[x]$. Suppose that a root of $f(x)$ has an expression using real radicals alone. This implies that there is a radical extension of $\mathbb{Q}$ (an extension obtained from $\mathbb{Q}$ by successively adding n-th roots of some element in the preceding field) containing a root of the given cubic. If a natural number n is a product of r prime numbers $n = p_1 p_2 \cdots p_r$ then n-th root of any number can be obtained by taking p_i-th root $1 \leq i \leq r$ in succession. For instance $7^{1/30} = \sqrt{\sqrt[3]{\sqrt[5]{7}}}$. So our assumption that a root of $f(x)$ can be expressed in terms of real radicals implies that there is a field $\mathbb{F} = \mathbb{Q}(a_1^{1/p_1}, a_2^{1/p_2}, \cdots, a_n^{1/p_r}) \subseteq \mathbb{R}$ containing a root of the polynomial $f(x)$. We define the fields $\mathbb{K}_i$, for $1 \leq i \leq r+1$ as follows. $\mathbb{K}_1 = \mathbb{Q}(\sqrt{\Delta})$, $\mathbb{K}_{i+1} = \mathbb{K}_i(a_i^{1/p_i})$, where $a_i \in \mathbb{K}_i$ for $1 \leq i \leq r$. Now $a_i^{1/p_i} \notin \mathbb{K}_i$ implies that $x^{p_i} - a_i$ is irreducible in $\mathbb{K}_i$ and so $[\mathbb{K}_{i+1} : \mathbb{K}_i] = p_i$ for $1 \leq i \leq r$. The roots of the cubic do not lie in $\mathbb{Q}$ or in $\mathbb{K}_1 = \mathbb{Q}(\sqrt{\Delta})$ as the given cubic is irreducible and so its roots are algebraic of degree three and $\mathbb{K}_0 = \mathbb{Q}(\sqrt{\Delta})$ has degree two over $\mathbb{Q}$. In the chain of fields

$$\mathbb{Q} \subseteq \mathbb{Q}(\sqrt{\Delta}) = \mathbb{K}_1 \subseteq \mathbb{K}_1 \subseteq \mathbb{K}_2 \subseteq \cdots \subseteq \mathbb{K}_{r-1} \subseteq \mathbb{K}_r \subseteq \mathbb{K}_{r+1} = \mathbb{L}$$

we have implicitly assumed that $\mathbb{K}_{r+1} = \mathbb{L}$ is the first field in the above chain of fields containing a root of the given cubic polynomial. It follows that the field L contains all the three roots of the cubic. The cubic $f(x)$ must be irreducible over $\mathbb{K}_r$ as otherwise the field $\mathbb{K}_r$ would have a root of the cubic. Since each of the root of the cubic has degree three it follows that $[\mathbb{L} : \mathbb{K}_r] = 3$ and the polynomial $x^3 - a_r \in \mathbb{K}_r[x]$ has a root (namely a_k^{1/p_r}) in $\mathbb{L}$. Also since $\mathbb{L}$ is the first field in the above chain to contain a root of the cubic it is the splitting field of the cubic as it already contains $\sqrt{\Delta}$. Therefore the extension $\mathbb{L}/\mathbb{K}_r$ is normal. But then as $\mathbb{L}$ has one root of the irreducible polynomial $x^3 - a_r$ it has all the three roots of this polynomial. These three roots of $x^3 - a_r$ are $a_r^{1/3}$, $\omega a_r^{1/3}$, and $\omega^2 a_r^{1/3}$, where $\omega = e^{\frac{2\pi i}{3}}$. It follows that $\omega \in \mathbb{L}$. This contradicts the fact that $\mathbb{L}$ is a subfield of the field $\mathbb{R}$ or real numbers. This contradiction completes the proof of the stated assertion.

(65) The minimal polynomial of $\alpha = (4 + \sqrt{15})^{\frac{1}{3}} + (4 - \sqrt{15})^{\frac{1}{3}}$ is $x^3 - 3x - 8$ which has just one real root.

(66)

$$\begin{aligned}
\alpha &= \frac{1+\sqrt{5}}{4} + \frac{\sqrt{3}}{2}\sqrt{\frac{5-\sqrt{5}}{2}} \\
4\alpha - 1 &= \sqrt{5} + 2\sqrt{3}\sqrt{\frac{5-\sqrt{5}}{2}} \\
\frac{4\alpha-1}{\sqrt{5}} &= 1 + \frac{2\sqrt{3}}{\sqrt{5}}\sqrt{\frac{5-\sqrt{5}}{2}} \\
\frac{4\alpha-1}{\sqrt{5}} - 1 &= \frac{2\sqrt{3}}{\sqrt{5}}\sqrt{\frac{5-\sqrt{5}}{2}}, \qquad \text{Squaring both sides we get} \\
\frac{16\alpha^2-8\alpha+1}{5} + 1 - \frac{8\alpha-2}{\sqrt{5}} &= \frac{12}{5}\left(\frac{5-\sqrt{5}}{2}\right) \\
\frac{16\alpha^2-8\alpha+6}{5} - \frac{8\alpha-2}{\sqrt{5}} &= \frac{6}{5}(5-\sqrt{5}) = 6 - \frac{6}{\sqrt{5}} \\
\frac{16\alpha^2-8\alpha+6}{5} - 6 &= \frac{8\alpha-2}{\sqrt{5}} - \frac{6}{\sqrt{5}} \\
\frac{16\alpha^2-8\alpha-24}{5} &= \frac{1}{\sqrt{5}}(8\alpha-8) \\
\frac{2\alpha^2-\alpha-3}{5} &= \frac{\alpha-1}{\sqrt{5}} \\
\frac{(\alpha+1)(2\alpha-3)}{5} &= \frac{\alpha-1}{\sqrt{5}}, \qquad \text{Squaring both sides we get} \\
(\alpha^2+2\alpha+1)(4\alpha^2-12\alpha+9) &= 5(\alpha^2-2\alpha+1) \\
4\alpha^4-4\alpha^3-11\alpha^2+6\alpha+9 &= 5\alpha^2-10\alpha+5 \\
4\alpha^4-4\alpha^3-16\alpha^2+16\alpha+4 &= 0 \\
\alpha^4-\alpha^3-4\alpha^2+4\alpha+1 &= 0
\end{aligned}$$

Therefore the minimal polynomial of $\alpha = 2\cos\frac{2\pi}{15}$ is $x^4 - x^3 - 4x^2 + 4x + 1$.

(67) The minimal polynomial of $\alpha = \sqrt{(p+\sqrt{p})(q+\sqrt{q})}$ is

$$x^8 - 4x^6pq + 2x^4pq(3pq-p-q-1) - 4x^2p^2q^2(pq-p-q+1)$$
$$+ p^2q^2\left[p^2q^2 - 2pq(p+q) + p^2 + q^2 + 4pq - 2(p+q) + 1.\right]$$

In particular the minimal polynomial of $\sqrt{(3+\sqrt{3})(2+\sqrt{2})}$ is $x^8 - 24x^6 + 144x^4 - 288x^2 + 144$.

(68) Clearly $\mathbb{Q}\left((2+\sqrt{2})(-1+i)\right) \subseteq \mathbb{Q}(\sqrt{2}, i)$.

We need to show that $\mathbb{Q}(\sqrt{2}, i) \subseteq \mathbb{Q}\left((2+\sqrt{2})(-1+i)\right)$.

Let $a = (2+\sqrt{2})(-1+i)$. We have $a = -2 - \sqrt{2} + 2i + i\sqrt{2} \Longrightarrow a + 2 - 2i = \sqrt{2}(-1+i)$. Squaring we get $a^2 + 4a = 4i(a+1)$ and therefore $i = \dfrac{a^2+4a}{4(a+1)}$. So $i \in \mathbb{Q}\left((2+\sqrt{2})(-1+i)\right)$ which in turn implies that $\sqrt{2} \in \mathbb{Q}\left((2+\sqrt{2})(-1+i)\right)$. This proves that $\mathbb{Q}(\sqrt{2}, i) \subseteq \mathbb{Q}\left((2+\sqrt{2})(-1+i)\right)$.

(69) Let $\alpha = \sqrt{(2+\sqrt{2})(-1+i)}$. In the previous problem it was proved that

$$\mathbb{Q}\left((2+\sqrt{2})(-1+i)\right) = \mathbb{Q}(\sqrt{2}, i).$$

Therefore α has degree 8 over $\mathbb{Q}$. By squaring the expression for α and by further simplifying we find that α is a zero of an irreducible polynomial of degree eight. This can be seen as follows:

$$\begin{aligned}
\alpha &= \sqrt{(2+\sqrt{2})(-1+i)} \quad \text{squaring this we get} \\
\alpha^2 &= -2 + 2i + \sqrt{2}(-1+i) \quad \text{or} \\
\alpha^2 + 2 - 2i &= \sqrt{2}(-1+i) \quad \text{squaring this we get} \\
\alpha^4 + 4\alpha^2 &= 4i(\alpha^2+1) \quad \text{finally squaring this we get}
\end{aligned}$$

$$\alpha^8 + 8\alpha^6 + 32\alpha^4 + 32\alpha^2 + 16 = 0$$

Therefore α is a zero of the polynomial $f(x) = x^8 + 8x^6 + 32x^4 + 32x^2 + 128$ and α^2 is a zero of the polynomial $g(x) = x^4 + 8x^3 + 32x^2 + 32x + 16$. We will show that $f(x)$ is irreducible over $\mathbb{Q}$. It will be enough to show that α is an algebraic number of degree 8 over $\mathbb{Q}$. As the field

$$\mathbb{Q}(\alpha^2) = \mathbb{Q}\left((2+\sqrt{2})(1+i)\right) = \mathbb{Q}(\sqrt{2}, i)$$

is of degree four over $\mathbb{Q}$, it follows that $g(x)$ is irreducible over $\mathbb{Q}$ and so α^2 is of degree four over $\mathbb{Q}$ and α is of degree eight over $\mathbb{Q}$. The four roots of $g(x)$ (which are squares of the roots of $f(x)$) are as follows

$$\begin{aligned}\alpha^2 &= (2+\sqrt{2})(-1+i)\\ \beta^2 &= (2+\sqrt{2})(-1-i)\\ \gamma^2 &= (2-\sqrt{2})(-1+i)\\ \delta^2 &= (2-\sqrt{2})(-1-i).\end{aligned}$$

So we have

$$\mathbb{Q}(\sqrt{2},i)=\mathbb{Q}(\alpha^2)=\mathbb{Q}(\beta^2)=\mathbb{Q}(\gamma^2)=\mathbb{Q}(\delta^2).$$

The eight zeros of $f(x)$ are: $\pm\alpha, \pm\beta, \pm\gamma, \pm\delta$. Consider the field diagram:

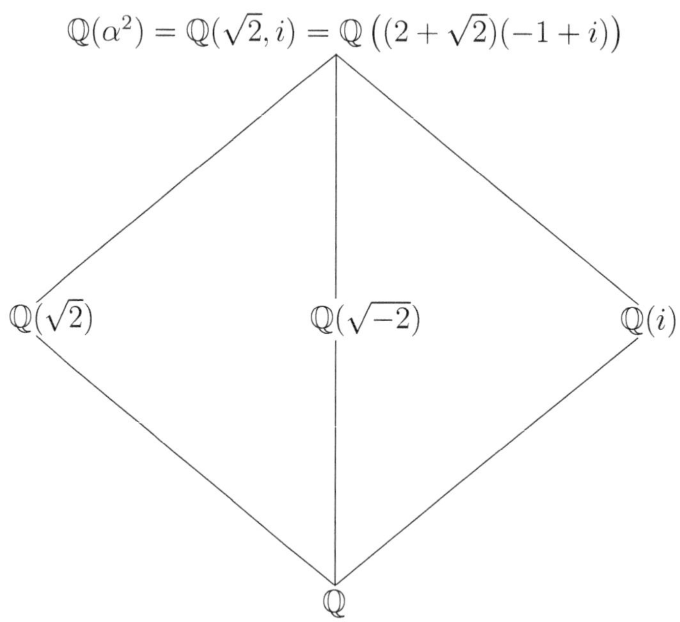

We claim that $\alpha \notin \mathbb{Q}(\alpha^2)$. This would imply that α is of degree two over $\mathbb{Q}(\alpha^2)$, so α is an algebraic number of degree 8 and $f(x)$ is its minimal polynomial over $\mathbb{Q}$. From this we would conclude that $f(x)$ is irreducible over $\mathbb{Q}$.

Suppose that $\alpha \in \mathbb{Q}(\alpha^2)=\mathbb{Q}(\sqrt{2},i)$. The group $Aut\left(\mathbb{Q}(\sqrt{2},i)/\mathbb{Q}(i)\right)=\langle\sigma\rangle$ has order two and its nontrivial automorphism σ maps $\sqrt{2}$ to $-\sqrt{2}$ and $\mathbb{Q}(i)$ is its fixed field. Therefore σ maps $\alpha\sigma(\alpha)$ to itself and so $\alpha\sigma(\alpha)\in\mathbb{Q}(i)$. We have

$$(\alpha\cdot\sigma(\alpha))^2=\alpha^2\cdot\sigma(\alpha^2)=(2+\sqrt{2})(-1+i)(2-\sqrt{2})(-1+i)=2(-1+i)^2$$

Therefore $\alpha\sigma(\alpha) = \pm\sqrt{2}(-1+i) \in \mathbb{Q}(i)$ - which is not true.

Now the eight roots of the equation $f(x) = x^8 + 8x^6 + 32x^4 + 32x^2 + 128 = 0$ are $\pm\alpha, \pm\beta, \pm\gamma$ and $\pm\delta$, where

$$\begin{aligned}
\alpha &= \sqrt{(2+\sqrt{2})(-1+i)} \\
&= \sqrt{2+\sqrt{2}} \cdot \sqrt{-1+i} = \sqrt{2+\sqrt{2}} \cdot \sqrt{\sqrt{2}\left(\frac{-1+i}{\sqrt{2}}\right)} \\
&= \sqrt{2+\sqrt{2}} \cdot 2^{\frac{1}{4}} \cdot \sqrt{\frac{-1+i}{\sqrt{2}}} \\
&= \sqrt{2+\sqrt{2}} \cdot 2^{\frac{1}{4}} \cdot \sqrt{e^{\frac{3\pi i}{4}}} \\
&= \sqrt{2+\sqrt{2}} \cdot 2^{\frac{1}{4}} \cdot e^{\frac{3\pi i}{8}} \\
&= \sqrt{2+\sqrt{2}} \cdot 2^{\frac{1}{4}} \cdot \left(\cos\tfrac{3\pi}{8} + i\sin\tfrac{3\pi}{8}\right) \\
&= \sqrt{2+\sqrt{2}} \cdot 2^{\frac{1}{4}} \cdot (\cos 67.5^\circ + i\sin 67.5^\circ) \\
&= \sqrt{2+\sqrt{2}} \cdot 2^{\frac{1}{4}} \cdot \left(\sqrt{\frac{\sqrt{2}-1}{2\sqrt{2}}} + i\sqrt{\frac{\sqrt{2}+1}{2\sqrt{2}}}\right)
\end{aligned}$$

likewise

$$\beta = \sqrt{2+\sqrt{2}} \cdot 2^{\frac{1}{4}} \cdot \left(-\sqrt{\frac{\sqrt{2}-1}{2\sqrt{2}}} + i\sqrt{\frac{\sqrt{2}+1}{2\sqrt{2}}}\right)$$

$$\gamma = \sqrt{2-\sqrt{2}} \cdot 2^{\frac{1}{4}} \cdot \left(\sqrt{\frac{\sqrt{2}-1}{2\sqrt{2}}} + i\sqrt{\frac{\sqrt{2}+1}{2\sqrt{2}}}\right)$$

$$\delta = \sqrt{2-\sqrt{2}} \cdot 2^{\frac{1}{4}} \cdot \left(-\sqrt{\frac{\sqrt{2}-1}{2\sqrt{2}}} + i\sqrt{\frac{\sqrt{2}+1}{2\sqrt{2}}}\right)$$

So if we denote its splitting field by $\mathbb{K}$ then

$$\mathbb{K} = \mathbb{Q}(\alpha, \beta, \gamma, \delta) = \mathbb{Q}(\alpha) = \mathbb{Q}(\beta) = \mathbb{Q}(\gamma) = \mathbb{Q}(\delta)$$

In fact as $\alpha - \beta = 2^{\frac{3}{4}}$ it turns out that $\mathbb{K} = \mathbb{Q}\left(2^{\frac{1}{4}}, i\right)$ and therefore $Gal(f)$ is isomorphic to the dihedral group of order 8. We omit the details.

Note. The polynomial $f(x) = x^8 - 24x^6 + 144x^4 - 288x^2 + 144$ could also have been obtained by multiplying $(x^2-\alpha^2)(x^2-\beta^2)(x^2-\gamma^2)(x^2-\delta^2)$ as follows:

$$\begin{aligned}
&(x^2-\alpha^2)(x^2-\beta^2)(x^2-\gamma^2)(x^2-\delta^2)\\
&= \left[x^2-(2+\sqrt{2})(3+\sqrt{3})\right]\left[x^2-(2+\sqrt{2})(3-\sqrt{3})\right]\left[x^2-(2-\sqrt{2})(3+\sqrt{3})\right]\\
&\quad \times \left[x^2-(2-\sqrt{2})(3-\sqrt{3})\right]\\
&= \left[x^4-6x^2(2+\sqrt{2})+6(2+\sqrt{2})^2\right]\left[x^4-6x^2(2-\sqrt{2})+6(2-\sqrt{2})^2\right]\\
&= \left[x^4-12x^2-6\sqrt{2}x^2+6(6+4\sqrt{2})\right]\left[x^4-12x^2+6\sqrt{2}x^2+6(6-4\sqrt{2})\right]\\
&= \left[x^4-12x^2+36-6\sqrt{2}(x^2-4)\right]\left[x^4-12x^2+36+6\sqrt{2}(x^2-4)\right]\\
&= (x^4-12x^2+36)^2-72(x^2-4)^2\\
&= x^8+144x^2+1296-24x^6+72x^4-864x^2-72(x^4-8x^2+16)\\
&= \mathbf{x^8-24x^6+144x^4-288x^2+144}.
\end{aligned}$$

(70) **Hint.** Observe that

- $f(x)$ is irreducible modulo 3.
- $f(x) \equiv (x+4)(x+5)(x^3+5x^2+5x+2)$ (modulo 7).
- $f(x) \equiv (x^2+5x+1)(x^3+6x^2+2x+6)$ (modulo 11).

Note that the discriminant of $f(x)$ is $2^8 \times 799,999$ which in particular implies that its Galois group is not a subgroup of A_5 - the alternating group on five letters.

(71) Let $\alpha = \frac{1}{3}\cos^{-1}\frac{11}{6}$. So $3\alpha = \cos^{-1}\frac{11}{6}$. Therfore $\cos 3\alpha = \frac{11}{6}$. Let $\cos\alpha = x$. Since

$$\cos 3\alpha = 4\cos^3\alpha - 3\cos\alpha$$

we have $\frac{11}{16} = 4x^3 - 3x$ or $64x^3 - 48x - 11 = 0$. Since

$$64x^3 - 48x - 11 = (4x+1)(16x^2-4x-11)$$

and $16x^2 - 4x - 11 = 0 \Longrightarrow x = \frac{1\pm\sqrt{3}}{8}$, it follows that $\cos\alpha$ is a constructible number.

(72) Evidently $f(x)$ has at least one real root. If $f(x)$ had just one real root then its Galois group over $\mathbb{Q}$ will be isomorphic to S_3, the symmetric group on three letters,

(73) Let $x^3+ax^2+bx+c \in \mathbb{Q}[x]$ be the minimal polynomial of α. Evidently $\alpha^2+b \neq 0$. We have

$$\alpha^3+a\alpha^2+b\alpha+c=\alpha(\alpha^2+b)+a\alpha^2+c=0 \Longrightarrow \alpha=-\frac{a\alpha^2+c}{\alpha^2+b}$$

Therefore $\mathbb{Q}(\alpha) \subseteq \mathbb{Q}(\alpha^2)$.

(74) The first part is easy. Since $x^2-\alpha^2 \in \mathbb{Q}(\alpha^2)[x]$, $[\mathbb{Q}(\alpha):\mathbb{Q}(\alpha^2)] \leq 2$. Since $[\mathbb{Q}(\alpha):\mathbb{Q}]$ is odd, $[\mathbb{Q}(\alpha):\mathbb{Q}(\alpha^2)] \neq 2$. Therefore $[\mathbb{Q}(\alpha):\mathbb{Q}(\alpha^2)]=1$ and $\mathbb{Q}(\alpha)=\mathbb{Q}(\alpha^2)$.

For the second part we proceed as follows. Assume that α is algebraic of degree $2n+1$, where $n \geq 1$. Let

$$f(x)=\sum_{i=0}^{2n+1} a_i x^i \in \mathbb{Q}[x], \qquad a_{2n+1}=1$$

be the minimal polynomial of α. So we have $f(\alpha)=0$. We write

$$\begin{aligned} f(x) &= a_0+a_1x+a_2x^2+a_3x^3+\cdots\cdots+a_{2n}x^{2n}+a_{2n+1}x^{2n+1} \\ &= \left(a_1x+a_3x^3+\cdots+a_{2n+1}x^{2n+1}\right)+\left(a_0+a_2x+\cdots+a_{2n}x^{2n}\right) \\ &= x\left(a_1+a_3x^2+\cdots+a_{2n+1}x^{2n}\right)+\left(a_0+a_2x+\cdots+a_{2n}x^{2n}\right) \\ &= x\sum_{i=0}^{n} a_{2i+1}x^{2i}+\sum_{i=0}^{n} a_{2i}x^{2i} \end{aligned}$$

Therefore $0=f(\alpha)=\alpha\sum_{i=0}^{n} a_{2i+1}\alpha^{2i}+\sum_{i=0}^{n} a_{2i}\alpha^{2i}$ and so

$$\alpha=-\frac{\sum_{i=0}^{n} a_{2i}\alpha^{2i}}{\sum_{i=0}^{n} a_{2i+1}\alpha^{2i}}.$$

Since α is algebraic of degree $2n+1$, $\sum_{i=0}^{n} a_{2i+1}\alpha^{2i} \neq 0$.

(75) It is easily checked that σ and η are members of $G = Aut\,(\mathbb{K}/\mathbb{Q})$ and σ has order p and η has order $p-1$. Since every element of $G = Aut\,(\mathbb{K}/\mathbb{Q})$ is completely specified by its effect on ε and $2^{\frac{1}{p}}$, we have

$$G=Aut\,(\mathbb{K}/\mathbb{Q})=\langle\sigma,\eta\rangle=\left\{\sigma^i\eta^j : 0 \leq i \leq p-1,\ \ 0 \leq j \leq p-2\right\}$$

Also if we compose $\sigma\eta$ the usual way then

$$\sigma\eta\left(2^{\frac{1}{p}}\right) = \sigma\left(2^{\frac{1}{p}}\right) = \varepsilon 2^{\frac{1}{p}}, \quad \text{and} \quad \eta\sigma\left(2^{\frac{1}{p}}\right) = \eta\left(\varepsilon 2^{\frac{1}{p}}\right) = \varepsilon^r 2^{\frac{1}{p}},$$

or as

$$2^{\frac{1}{p}} \xrightarrow{\sigma} \varepsilon 2^{\frac{1}{p}} \xrightarrow{\eta} \varepsilon^r 2^{\frac{1}{p}}, \qquad 2^{\frac{1}{p}} \xrightarrow{\eta} 2^{\frac{1}{p}} \xrightarrow{\sigma} \varepsilon 2^{\frac{1}{p}}$$

Since $r \neq 1$, $\sigma\eta \neq \eta\sigma$ and the group G is nonabelian.

This group is so-called a Frobenius group. It is isomorphic to the normalizer of the Sylow-p subgroup of the group S_p, the symmetric group on p-letters.

(76) First we note that since $f(x) \in \mathbb{Q}[x]$ is a monic irreducible polynomial it is the minimal polynomial of each of its root. Also if two monic irreducible polynomials of the same degree n have a root in common then they are identical.

(a) Suppose that $\alpha - \beta = q \in \mathbb{Q}$. Then $\alpha = \beta + q$. Then $0 = f(\alpha) = f(\beta + q)$. Let $g(x) = f(x + q)$. Then $g(x) \in \mathbb{Q}[x]$ is a monic irreducible polynomial of degree n and as $g(\beta) = f(\beta + q) = f(\alpha) = 0$ it is the minimal polynomial of β. Therefore $f(x) = g(x) = f(x + q)$. Since $\alpha_i + q$, $1 \leq i \leq n$ are the n roots of $f(x)$ as well we have $\Omega = \{\alpha_1, \alpha_2, \cdots, \alpha_n\} = \{\alpha_1 + q, \alpha_2 + q, \cdots, \alpha_n + q\}$. Therefore

$$\alpha_1 + \alpha_2 + \cdots + \alpha_n = (\alpha_1 + q) + (\alpha_2 + q) + \cdots + (\alpha_n + q)$$

This implies that $q = 0$ which contradicts the assumption that $\alpha \neq \beta$.

(b)
- $\alpha + \beta \notin \mathbb{Q}$. Suppose $\alpha + \beta = q \in \mathbb{Q}$ or $\alpha = q - \beta$. So $0 = f(\alpha) = f(q - \beta)$. Let $g(x) = -f(q - x)$. Then as n is odd $g(x)$ is a monic irreducible polynomial of degree n having β as one of its roots. Therefore it is the minimal polynomial of β. Thus $f(x) = -f(q - x)$. If we substitute $x = \dfrac{q}{2}$ in $f(x) = -f(q - x)$, we get $f(q/2) = 0$ which is not possible.
- $\alpha\beta \notin \mathbb{Q}$. For this we first note that $\alpha\beta \neq 1$. Suppose that $\alpha\beta = 1$. So $\beta = \dfrac{1}{\alpha}$. Let $g(x) = \dfrac{x^n}{a_n} f\left(\dfrac{1}{x}\right)$. As $g(x)$ and $f(x)$ are irreducible, monic polynomial of the same degree having a root in common they are identical. Now $f(x) = g(x)$ implies that $\gamma \in \Omega$ if and only if $\dfrac{1}{\gamma} \in \Omega$.

Since n is odd one of the root of $f(x)$ is self-reciprocal namely ± 1. This contradicts the irreducibility of $f(x)$.

Next we argue that $\alpha\beta$ is not a perfect square of a rational number. For if $\alpha\beta = r^2$ then $g(x) = r^{-n} f(xr) \in \mathbb{Q}[x]$ is a monic irreducible polynomial of odd degree $n > 1$ with roots $\frac{\alpha_i}{r}$, $1 \le i \le n$. Since $\alpha\beta = r^2$, the product of some two roots of $g(x)$ is equal to 1 which is not possible.

Now suppose that $\alpha\beta = q$, $q > 0$, and is not a perfect square. So $\sqrt{q} \notin \mathbb{Q}$. Let $g(x) = \frac{x^n}{a_n} f\left(\frac{q}{x}\right)$. Therefore $g(x)$ is a monic polynomial in $\mathbb{Q}[x]$ of degree n having α as one of its roots. So $f(x) = g(x)$. Therefore

$$f(\sqrt{q}) = g(\sqrt{q}) = \frac{q^{\frac{n-1}{2}}\sqrt{q}}{a_n} f(\sqrt{q}) \Longrightarrow f(\sqrt{q}) \left(1 - \frac{q^{\frac{n-1}{2}}\sqrt{q}}{a_n}\right)$$

$$\Longrightarrow f(\sqrt{q}) = 0$$

Now $f(\sqrt{q}) = 0 \Longrightarrow f(-\sqrt{q}) = 0$, which implies that $x^2 + q \in \mathbb{Q}[x]$ is a factor of $f(x)$ contradicting irreducibility of $f(x)$.

Finally let $\alpha\beta = q$, $q < 0$. We substitute $x = i\sqrt{-q}$ in $f(x) = \frac{x^n}{a_n} f\left(\frac{q}{x}\right)$. We have

$$f(i\sqrt{-q}) = \frac{(i\sqrt{-q})^n}{a_n} f(i\sqrt{-q}) \Longrightarrow f(i\sqrt{-q}) = 0.$$

Now $f(i\sqrt{-q}) = 0 \Longrightarrow f(-i\sqrt{-q}) = 0$ which implies that $x^2 + q \in \mathbb{Q}[x]$ is a factor of $f(x)$ contradicting its irreducibility.

- $\frac{\alpha}{\beta} \notin \mathbb{Q}$. Let $\frac{\alpha}{\beta} = q$. In this case we let $g(x) = \frac{1}{q^n} f(qx)$. Then $g(x)$ is a monic irreducible polynomial of degree n having β as one of its roots. Therefore $g(x) = f(x)$. Since $\frac{\alpha_i}{q}$, $1 \le i \le n$ are roots of $g(x)$ we have

$$\Omega = \{\alpha_1, \alpha_2, \cdots, \alpha_n\} = \left\{\frac{\alpha_1}{q}, \frac{\alpha_2}{q}, \cdots, \frac{\alpha_n}{q}\right\}.$$

Therefore

$$\sum_{i=1}^{n} |\alpha_i| = \sum_{i=1}^{n} \left|\frac{\alpha_i}{q}\right| \Longrightarrow |q| = 1$$

which implies that either $\alpha - \beta$ or $\alpha + \beta$ is rational. These two cases have been dealt with earlier.

(77) Since $[\mathbb{K} : \mathbb{Q}] = 4$, α_0 has four conjugates in $\mathbb{K}$. Let these be $\alpha_0 = \alpha, \beta, \gamma$ and δ, where

$$\begin{aligned} \alpha &= a + b\sqrt{2} + c\sqrt{3} + d\sqrt{6} \\ \beta &= a - b\sqrt{2} + c\sqrt{3} - d\sqrt{6} \\ \gamma &= a + b\sqrt{2} - c\sqrt{3} - d\sqrt{6} \\ \delta &= a - b\sqrt{2} - c\sqrt{3} + d\sqrt{6}. \end{aligned}$$

Then

$$\begin{aligned} \beta\delta &= (a - b\sqrt{2} + c\sqrt{3} - d\sqrt{6})(a - b\sqrt{2} - c\sqrt{3} + d\sqrt{6}) \\ &= a^2 + 2b^2 - 3c^2 - 6d^2 - 2\sqrt{2}(ab - 3cd) \\ \alpha\gamma &= (a + b\sqrt{2} + c\sqrt{3} + d\sqrt{6})(a + b\sqrt{2} - c\sqrt{3} - d\sqrt{6}) \\ &= a^2 + 2b^2 - 3c^2 - 6d^2 + 2\sqrt{2}(ab - 3cd). \end{aligned}$$

Let $N(\alpha) = \alpha\beta\gamma\delta$. Then we have

$$\begin{aligned} N(\alpha) = a^4 &+ 4b^4 + 9c^4 + 36d^4 - 4a^2b^2 - 6a^2c^2 - 12b^2c^2 - 12a^2d^2 - 24b^2d^2 \\ &- 36c^2d^2 + 48abcd \in \mathbb{Q}. \end{aligned}$$

Also as

$$\beta\delta\gamma = \left(a^2 + 2b^2 - 3c^2 - 6d^2 - 2\sqrt{2}(ab - 3cd)\right)(a + b\sqrt{2} - c\sqrt{3} - d\sqrt{6})$$

we can write $\beta\delta\gamma = x + y\sqrt{2} + z\sqrt{3} + t\sqrt{6}$, where

$$\begin{aligned} x &= a^3 - 2ab^2 - 3ac^2 - 6ad^2 + 12bcd \\ y &= -(a^2b - 2b^3 + 3bc^2 - 6acd + 6bd^2) \\ z &= -(a^2c + 2b^2c - 3c^3 + 6cd^2 - 4abd) \\ t &= (2abc - a^2d - 2b^2d - 3c^2d + 6d^3). \end{aligned}$$

Therefore

$$\alpha^{-1} = \frac{1}{a + b\sqrt{2} + c\sqrt{3} + d\sqrt{6}} = \frac{\beta\gamma\delta}{\alpha\beta\gamma\delta}\frac{\beta\gamma\delta}{N(\alpha)} = A + B\sqrt{2} + C\sqrt{3} + D\sqrt{6},$$

where A, B, C, D are as follows:

$$A = \frac{x}{N(\alpha)} = \frac{(a^3 - 2ab^2 - 3ac^2 - 6ad^2 + 12bcd)}{N(\alpha)}$$

$$B = \frac{y}{N(\alpha)} = -\frac{(a^2b - 2b^3 + 3bc^2 - 6acd + 6bd^2)}{N(\alpha)}$$

$$C = \frac{z}{N(\alpha)} = -\frac{(a^2c + 2b^2c - 3c^3 + 6cd^2 - 4abd)}{N(\alpha)}$$

$$D = \frac{t}{N(\alpha)} = -\frac{(a^2d + 2b^2d + 3c^2d - 6d^3 - 2abc)}{N(\alpha)}.$$

If $\alpha = 1+\sqrt{2}+\sqrt{3}+\sqrt{6}$ then substituting $a = b = c = d = 1$ in this problem we get $N(\alpha) = 4$ and so $\alpha^{-1} = \frac{1}{2}\left(1 - \sqrt{2} - \sqrt{3} + \sqrt{6}\right)$ which can be easily verified.

Another way of finding the value of the inverse of $\alpha = 1 + \sqrt{2} + \sqrt{3} + \sqrt{6}$ is to first calculate the minimal polynomial of α which turns out to be $f(x) = x^4 - 4x^3 - 16x^2 - 8x + 4$. Substituting $x = \alpha$ in it we get $\alpha^{-1} = -\frac{1}{4}\left(\alpha^3 - 4\alpha^2 - 16\alpha - 8\right)$. Since $\alpha^2 = 12+8\sqrt{2}+6\sqrt{3}+4\sqrt{6}$ and $\alpha^3 = 70+50\sqrt{2}+43\sqrt{3}+30\sqrt{6}$, substituting the values of α, α^2 and α^3 in

$$\alpha^{-1} = -\frac{1}{4}\left(\alpha^3 - 4\alpha^2 - 16\alpha - 8\right).$$

we get

$$\alpha^{-1} = \frac{1}{1+\sqrt{2}+\sqrt{3}+\sqrt{6}} = \frac{1}{2}\left(1 - \sqrt{2} - \sqrt{3} + \sqrt{6}\right)$$

(78) Let

$$\begin{aligned}
\alpha &= a + b\sqrt{p} + c\sqrt{q} + d\sqrt{pq} \\
\beta &= a - b\sqrt{p} + c\sqrt{q} - d\sqrt{pq} \\
\gamma &= a + b\sqrt{p} - c\sqrt{q} - d\sqrt{pq} \\
\delta &= a - b\sqrt{p} - c\sqrt{q} + d\sqrt{pq}.
\end{aligned}$$

Proceeding as in the previous problem we find that $N(\alpha) = \alpha\beta\gamma\delta$ is equal to

$$a^4 - 2a^2b^2p + b^4p^2 - 2a^2d^2pq - 2b^2d^2p^2q + d^4p^2q^2 + 8abcdpq$$
$$- 2a^2c^2q - 2b^2c^2pq - 2c^2d^2pq^2 + c^4q^2$$

and

$$\beta\gamma\delta = x + y\sqrt{p} + z\sqrt{q} + t\sqrt{pq}$$

where

$$\begin{aligned} x &= a^3 - ab^2p - ad^2pq + 2bcdpq - ac^2q \\ y &= -a^2b + b^3p - bd^2pq + 2acdq - bc^2q \\ z &= 2abdp - a^2c - b^2cp - cd^2pq + c^3q \\ t &= -a^2d - b^2dp + 2abc - c^2dq + d^3pq. \end{aligned}$$

Then the values of $A, B, C, D \in \mathbb{Q}$ are as follows:

$$\begin{aligned} A = \frac{x}{N(\alpha)} &= \frac{a^3 - ab^2p - ad^2pq + 2bcdpq - ac^2q}{N(\alpha)} \\ B = \frac{y}{N(\alpha)} &= \frac{-a^2b + b^3p - bd^2pq + 2acdq - bc^2q}{N(\alpha)} \\ C = \frac{z}{N(\alpha)} &= \frac{2abdp - a^2c - b^2cp - cd^2pq + c^3q}{N(\alpha)} \\ D = \frac{t}{N(\alpha)} &= \frac{-a^2d - b^2dp + 2abc - c^2dq + d^3pq}{N(\alpha)}. \end{aligned}$$

Substituting $p = 2$, $q = 3$, we get the answer of previous problem.

(79) Irreducibility of $p(x) \in \mathbb{Z}[x] \subset \mathbb{Q}[x]$ is easily verified. Note that $G = Aut(\mathbb{K}/\mathbb{Q})$, the Galois group of $p(x)$ is isomorphic to $C_2 \times C_2$, the Klein-Four Group. By Dedekind's Theorem proved in the appendix it follows that if $p(x)$ were irreducible modulo p then G would be cyclic of order four.

(80) For this problem we extend the definition of constructible numbers to include complex numbers as well. That is $\alpha \in \mathbb{C}$ is considered to be a constructible number if

- α is algebraic of degree 2^n.
- There is a chain of fields $\mathbb{K}_i$, $0 \le i \le n$ such that

$$\mathbb{Q} = \mathbb{K}_0 \subset \mathbb{K}_1 \subset \mathbb{K}_2 \subset \cdots \subset \mathbb{K}_{n-1} \subset \mathbb{K}_n = \mathbb{Q}(\alpha)$$

satisfying $[\mathbb{K}_{i+1} : \mathbb{K}_i] = 2$ for $0 \le i \le n-1$.

Let K denote the splitting field of $f(x) \in \mathbb{Q}[x]$ and $G = Aut(\mathbb{K}/\mathbb{Q})$.

Therefore if α of the problem were a constructible number then there would be a field $L \subset \mathbb{K}$ such that $\mathbb{Q} \subset \mathbb{L} \subset \mathbb{Q}(\alpha) \subset \mathbb{K}$ satisfying

$$[\mathbb{Q}(\alpha) : \mathbb{L}] = [\mathbb{L} : \mathbb{Q}] = 2$$

By Fundamental Theorem of Galois Theory it would imply that the group A_4 (which is the only subgroup of S_4 of index two) will have a subgroup of index two. Since the group A_4 has no subgroup of order six we conclude that the splitting field $\mathbb{K}$ of the polynomial $f(x)$ cannot have a subfield like $\mathbb{L}$ which is such that $[\mathbb{Q}(\alpha) : \mathbb{L}] = [\mathbb{L} : \mathbb{Q}] = 2$

(81) **First Method**

We note that $\alpha^4 = -(2\alpha^2 + \alpha)$ and $\alpha^{-1} = -(\alpha^2 + 2)$. We have

$$\begin{aligned} \frac{1}{\alpha^2 - \alpha + 1} &= \frac{\alpha + 1}{(\alpha^2 - \alpha + 1)(\alpha + 1)} = \frac{\alpha + 1}{\alpha^3 + 1} = \frac{\alpha + 1}{-2\alpha} \\ &= -\frac{1}{2}\left(1 + \frac{1}{\alpha}\right) = -\frac{1}{2}\left(-\alpha^2 - 1\right) = \frac{\alpha^2 + 1}{2}. \end{aligned}$$

Second Method

Let $g(x) = x^2 - x + 1$. As $f(x) = x^3 + 2x + 1$ is irreducible $GCD(f(x), g(x)) = 1$. So there exist polynomials $A(x), B(x) \in \mathbb{Q}[x]$ such that $f(x)A(x) + g(x)B(x) = 1$. By using Euclidean Algorithm we find one such pair of polynomials $A(x), B(x)$ as

$$A(x) = \frac{x^2 + 1}{2}, \qquad B(x) = -\frac{x - 1}{2}.$$

So we have

$$(x^2 - x + 1)\left(\frac{x^2 + 1}{2}\right) + (x^3 + 2x + 1)\left(-\frac{x - 1}{2}\right) = 1$$

Substituting $x = \alpha$ in above and using the fact that $f(\alpha) = \alpha^3 + 2\alpha + 1 = 0$ we get

$$(\alpha^2 - \alpha + 1)\left(\frac{\alpha^2 + 1}{2}\right) = 1. \qquad \text{Therefore} \qquad \frac{1}{\alpha^2 - \alpha + 1} = \frac{\alpha^2 + 1}{2}.$$

(82) It is an immediate consequence of the following:

- $f(x)$ is irreducible modulo 3 and so $|G|$ is divisible by 5.
- The discriminant of $f(x) = 1024000000 = 2^{16}5^6$, therefore $G \subseteq A_5$.

- $f^{'}(x) = 5x^4 + 20 > 0$ for all x and therefore $f(x)$ has just one real root. It follows that the complex conjugation is an element of $G = Aut(\mathbb{K}/\mathbb{Q})$ therefore $|G|$ is even.
- $f(x) \equiv (x+2)(x+3)(x^3+2x^2+5x+5)$ (modulo 7). Therfore $|G|$ is divisible by 3.
- $|G|$ is divisible by 30 and A_5 does not have any subgroup of order 30, hence G is isomorphic to A_5.

(83) We recall that an angle θ can be drawn with ruler and compass if and only if $\cos\theta$ is a constructible number.

First suppose that 3 divides m and $m = 3n$. We shall show that the angle $\alpha = \frac{m\pi}{180} = \frac{3n\pi}{180} = \frac{n\pi}{60}$ can be drawn by using ruler and compass. It is enough to show that we can draw the angle $\frac{\pi}{60}$ by ruler and compass. We know that

$$\cos\frac{\pi}{5} = \cos 36^o = \frac{1+\sqrt{5}}{4}, \qquad \cos\frac{\pi}{3} = \cos 60^o = \frac{1}{2}, \qquad \cos\frac{\pi}{4} = \cos 45^o = \frac{\sqrt{2}}{2}$$

Since $\cos\frac{\pi}{3}$, $\cos\frac{\pi}{4}$, and $\cos\frac{\pi}{5}$ are constructible numbers it is possible to draw the angles $\frac{\pi}{3}$, $\frac{\pi}{4}$, and $\frac{\pi}{5}$ using ruler and compass. Since $\frac{1}{60} = \frac{3}{5} - \frac{1}{3} - \frac{1}{4}$, we have $\frac{\pi}{60} = \frac{3\pi}{5} - \frac{\pi}{3} - \frac{\pi}{4}$ It is possible to draw the angle $\frac{\pi}{60}$ by ruler and compass and therefore same is true for the angle $\frac{n\pi}{60} = \frac{3n\pi}{180} = \frac{m\pi}{180} = m$ degrees.

Another way of showing that the angle $\frac{\pi}{60}$ can be drawn by ruler and compass is to prove that $\cos\frac{\pi}{60}$ is a constructible number. This can be done by finding an expression for $\cos 24^o$ involving rational operations and square root radicals alone. We note that

$$\cos 36^o = \frac{1+\sqrt{5}}{4} \Longrightarrow \sin 36^o = \frac{\sqrt{10-2\sqrt{5}}}{4},$$

and so we have

$$\begin{aligned}\cos 24^o &= \cos(60^o - 36^o) = \cos 60^o \cos 36^o + \sin 60^o \sin 36^o \\ &= \frac{1}{2}\left(\frac{1+\sqrt{5}}{4}\right) + \frac{\sqrt{3}}{2}\frac{\sqrt{10-2\sqrt{5}}}{4}.\end{aligned}$$

Now

$$\cos 24^o \text{ is constructible } \Longrightarrow \cos 12^o \text{ is costructible } \Longrightarrow \cos 6^o \text{ is constructible}$$

$$\cos 6^o \text{ is contructible } \Longrightarrow \cos 3^o \text{ is constructible } \Longrightarrow \cos\frac{\pi}{60} \text{ is constructible.}$$

Now suppose that m is a natural number not divisible by 3. Assume that it is possible to draw an angle measuring m degrees using a ruler and compass alone. Since $GCD(3, m) = 1$ we can find integers a, b such that $3a + mb = 1$. We have

$$\frac{3a\pi}{180} + \frac{mb\pi}{180} = \frac{\pi}{180}.$$

Since by assumption the angle $\frac{m\pi}{180}$ can be drawn by ruler and compass, and as proved above the angle $\frac{3a\pi}{180}$ can be drawn by ruler and compass it follows that the angle $\frac{\pi}{180}$ can be drawn using ruler and compass. In particular the angle $\frac{20\pi}{180} = 20^o$ can be drawn by ruler and compass. This would imply that $\cos 20^o$ is a constructible number which is not true as shown in the text.

(84) Let $G = Aut(\mathbb{K}/\mathbb{Q})$ be the Galois group of $f(x)$. We note that the discriminant of $f(x) = 24217$, is not a square of a rational number. Therefore $G \not\subset A_5$.

  (a) $f(x)$ is irreducible modulo 2 therefore it is irreducible over $\mathbb{Q}$. Therefore $|G|$ is divisible by 5.

  (b) $f(x) \equiv (x^2 + 5x + 1)(x^3 + 6x^2 + 8x + 10)$ (modulo 11). So $|G|$ is divisible by 6 and therefore by 30. Since G contains an odd permutation G is isomorphic to S_5.

  (c) The polynomial $f(x)$ has five real roots, one in each of the following intervals: $(-3, -2)$, $(-1, 0)$, $(0, \frac{1}{2})$, $(\frac{1}{2}, 1)$, and $(1, 2)$.

(85) This problem is similar to the previous problem.

(86) Let $G = Aut(\mathbb{K}/\mathbb{Q})$ be the Galois group of $f(x)$. We note that the discriminant of $f(x) = 33489 = 183^2$. Therefore $G \subseteq A_4$. As $f(x) \equiv (x+1)(x^3+x^2+1)$ (modulo 2) and $f(x) \equiv (x^2 + 1)^2$ (modulo 3), G has order divisible by 12 and therefore G is isomorphic to A_4. The polynomial $f(x)$ has four real roots, one in each of the following intervals: $(-3, -2)$, $(-1, 0)$, $(0, 1)$, and $(2, 3)$.

(87) This problem is similar to the previous problem.

(88) Let $G = Aut(\mathbb{K}/\mathbb{Q})$ be the Galois group of $f(x)$. We note that the discriminant of $f(x) = 1957 = 19 \times 103$. Therefore $G \not\subseteq A_4$. As $f(x)$ irreducible modulo 2 and $f(x) \equiv (x + 2)(x^3 + x^2 + 2)$ (modulo 3), G has order divisible by 12 and therefore

G is isomorphic to S_4. The polynomial $f(x)$ has four real roots, one in each of the following intervals: $(-2,-1)$, $(-1,0)$, $(0,1)$, and $(2,3)$.

(89) This problem is similar to the previous problem.

(90) We have seen in the text (in the chapter on '**A Radical Expression for** $\cos\frac{2\pi}{11}$') that the minimal polynomial of $2\cos\frac{2\pi}{11}$ over $\mathbb{Q}$ is

$$f(y) = y^5 + y^4 - 4y^3 - 3y^2 + 3y + 1$$

The five roots of the above polynomial are

$$2\cos\frac{2\pi}{11},\quad 2\cos\frac{4\pi}{11},\quad 2\cos\frac{6\pi}{11},\quad 2\cos\frac{8\pi}{11},\quad 2\cos\frac{10\pi}{11}$$

and $f(y)$ is the minimal polynomial of each of these.

Let $x = 2i\sin\frac{2\pi}{11}$ then $x^2 = -4\sin^2\frac{2\pi}{11}$. Now

$$x^2 = -4\sin^2\frac{2\pi}{11} = 2\left(-2\sin^2\frac{2\pi}{11}\right) = 2\left(\cos\frac{4\pi}{11} - 2\right) = 2\cos\frac{4\pi}{11} - 2 = y - 2.$$

Substituting $y = x^2 + 2$ in $f(y)$ we have

$$\begin{aligned}&(x^2+2)^5 + (x^2+2)^4 - 4(x^2+2)^3 - 3(x^2+2)^2 + 3(x^2+2) + 1\\&= x^{10} + 11x^8 + 44x^6 + 77x^4 + 55x^2 + 11.\end{aligned}$$

The polynomial $x^{10} + 11x^8 + 44x^6 + 77x^4 + 55x^2 + 11$ which is easily seen to be irreducible is the minimal polynomial of $2i\sin\frac{2\pi}{11}$. Since $2i\sin\frac{2\pi}{11} \in \mathbb{Q}(\alpha)$ and is an algebraic number (in fact an algebraic integer) of degree 10 it follows that $\mathbb{Q}(\alpha) = \mathbb{Q}\left(2i\sin\frac{2\pi}{11}\right) = \mathbb{Q}\left(i\sin\frac{2\pi}{11}\right)$.

Let $z = 2\sin\frac{2\pi}{11}$. Therefore $zi = 2i\sin\frac{2\pi}{11}$ or $-z^2 = x^2$. Substituting $x^2 = -z^2$ in the equation $x^{10} + 11x^8 + 44x^6 + 77x^4 + 55x^2 + 11 = 0$ we get $z^{10} - 11z^8 + 44z^6 - 77z^4 + 55z^2 - 11 = 0$.

The above polynomial is irreducible and its roots are

$$\pm 2\sin\frac{2\pi}{11},\quad \pm 2\sin\frac{2\pi}{11},\quad \pm 2\sin\frac{2\pi}{11},\quad \pm 2\sin\frac{2\pi}{11},\quad \pm 2\sin\frac{2\pi}{11}.$$

So it is the minimal polynomial of each of them.

For the last part let

$$A = \alpha + \alpha^4 + \alpha^5 + \alpha^9 + \alpha^3, \qquad B = \alpha^2 + \alpha^8 + \alpha^{10} + \alpha^7 + \alpha^6$$

We have $A + B = -1$ and $A \cdot B = 3$. Therefore A, B are roots of the equation $x^2 + x + 3 = 0$. So

$$A = \frac{-1 + \sqrt{-11}}{2}, \qquad B = \frac{-1 - \sqrt{-11}}{2}$$

Since $A, B \in \mathbb{Q}(\alpha)$ it follows that $\sqrt{-11} \in \mathbb{Q}(\alpha)$. The following diagram shows the lattice of subfields of the field $\mathbb{K} = \mathbb{Q}(\alpha)$.

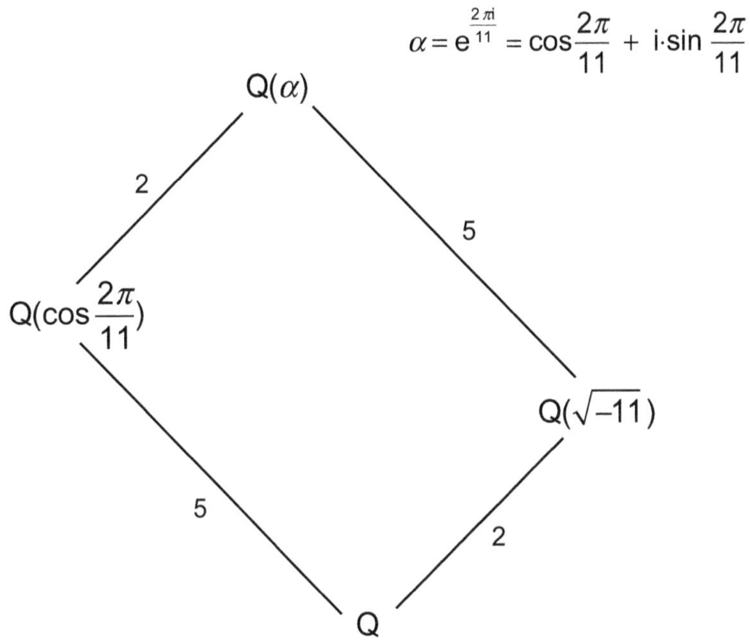

It is clear from this diagram that $\mathbb{Q}(\alpha) = \mathbb{Q}\left(\cos \frac{2\pi}{11}, \sqrt{-11}\right)$.

(91) Although we do not need to explicitly compute the polynomial $\Phi_{55}(x)$ but

$$\Phi_{55}(x) = \frac{x^{55} - 1}{(x^{11} - 1)(x^4 + x^3 + x^2 + x + 1)}$$

which is equal to

$$x^{40} - x^{39} + x^{35} - x^{34} + x^{30} - x^{28} + x^{25} - x^{23} + x^{20} - x^{17} + x^{15} - x^{12}$$
$$+ x^{10} - x^6 + x^5 - x + 1.$$

Let $\alpha = e^{\frac{2\pi i}{11}} = \cos \frac{2\pi}{11} + i \sin \frac{2\pi}{11}$, $\quad \varepsilon = e^{\frac{2\pi i}{5}} = \cos \frac{2\pi}{5} + i \sin \frac{2\pi}{5}$. Then Since $\phi(55) = 40$, the order of the group $G = Aut(\mathbb{K}/\mathbb{Q})$ is 40 and $|\mathbb{K} : \mathbb{Q}| = 40$. $\mathbb{K} = \mathbb{Q}(\alpha, \varepsilon)$. Let $\sigma, \eta \in Aut(\mathbb{K}/\mathbb{Q})$ be the following two automorphisms:

$$\sigma(\alpha) = \alpha^2, \ \ \sigma(\varepsilon) = \varepsilon, \qquad \eta(\alpha) = \alpha, \ \ \eta(\varepsilon) = \varepsilon^2.$$

As can be easily checked that $Order(\alpha) = 10,\quad Order(\eta) = 4$ and $\sigma\eta = \eta\sigma$. We have

$$G = \langle \sigma, \eta : \sigma^{10} = \eta^4 = e, \sigma\eta = \eta\sigma \rangle = \left\{ \sigma^i \eta^j : 0 \le i \le 9,\ \ 0 \le j \le 3 \right\}.$$

The orders of elements of G

Element	Order	Element	Order	Element	Order	Element	Order
e	1	η	4	η^2	2	η^3	4
σ	10	$\sigma\eta$	20	$\sigma\eta^2$	10	$\sigma\eta^3$	20
σ^2	5	$\sigma^2\eta$	20	$\sigma^2\eta^2$	10	$\sigma^2\eta^3$	20
σ^3	10	$\sigma^3\eta$	20	$\sigma^3\eta^2$	10	$\sigma^3\eta^3$	20
σ^4	5	$\sigma^4\eta$	20	$\sigma^4\eta^2$	10	$\sigma^4\eta^3$	20
σ^5	2	$\sigma^5\eta$	4	$\sigma^5\eta^2$	2	$\sigma^5\eta^3$	4
σ^6	5	$\sigma^6\eta$	20	$\sigma^6\eta^2$	10	$\sigma^6\eta^3$	20
σ^7	10	$\sigma^7\eta$	20	$\sigma^7\eta^2$	10	$\sigma^7\eta^3$	20
σ^8	5	$\sigma^8\eta$	20	$\sigma^8\eta^2$	10	$\sigma^8\eta^3$	20
σ^9	10	$\sigma^9\eta$	20	$\sigma^9\eta^2$	10	$\sigma^9\eta^3$	20

Proper Subgroups of G

- Two cyclic subgroups of order 20.

$$\langle \sigma\eta \rangle = \left\{ e, \sigma\eta, \sigma^2\eta^2, \sigma^3\eta^3, \sigma^4, \sigma^5\eta, \sigma^6\eta^2, \sigma^7\eta^3, \sigma^8, \sigma^9\eta, \eta^2, \sigma\eta^3, \sigma^2, \sigma^3\eta, \sigma^4\eta^2, \sigma^5\eta^3, \sigma^6, \sigma^7\eta, \sigma^8\eta^2, \sigma^9\eta^3 \right\}$$

$$\langle \sigma^2\eta \rangle = \left\{ e, \sigma^2\eta, \sigma^4\eta^2, \sigma^6\eta^3, \sigma^8, \eta, \sigma^2\eta^2, \sigma^4\eta^3, \sigma^6, \sigma^8\eta, \eta^2, \sigma^2\eta^3, \sigma^4, \sigma^6\eta, \sigma^8\eta^2, \eta^3, \sigma^2, \sigma^4\eta, \sigma^6\eta^2, \sigma^8\eta^3 \right\}$$

- One subgroup $\langle \sigma, \eta^2 \rangle$ of order 20 isomorphic to $C_{10} \times C_2$.

$$\langle \sigma, \eta^2 \rangle = \left\{ e, \sigma, \sigma^2, \sigma^3, \sigma^4, \sigma^5, \sigma^6, \sigma^7, \sigma^8, \sigma^9, \eta^2, \sigma\eta^2, \sigma^2\eta^2, \sigma^3\eta^2, \sigma^4\eta^2, \sigma^5\eta^2, \sigma^6\eta^2, \sigma^7\eta^2, \sigma^8\eta^2, \sigma^9\eta^2 \right\}$$

- Three (cyclic) subgroups of order 10 each.

$$\langle\sigma\rangle = \{e, \sigma, \sigma^2, \sigma^3, \sigma^4, \sigma^5, \sigma^6, \sigma^7, \sigma^8, \sigma^9\}$$

$$\langle\sigma\eta^2\rangle = \{e, \sigma\eta^2, \sigma^2, \sigma^3\eta^2, \sigma^4, \sigma^5\eta^2, \sigma^6, \sigma^7\eta^2, \sigma^8, \sigma^9\eta^2\}$$

$$\langle\sigma^2\eta^2\rangle = \{e, \sigma^2\eta^2, \sigma^4, \sigma^6\eta^2, \sigma^8, \eta^2, \sigma^2, \sigma^4\eta^2, \sigma^6, \sigma^8\eta^2\}$$

- One subgroup $\langle\sigma^5, \eta\rangle$ of order eight isomorphic to $C_4 \times C_2$.

$$\langle\sigma^5, \eta\rangle = \{e, \sigma^5, \eta, \sigma^5\eta, \eta^2, \sigma^5\eta^2, \eta^3, \sigma^5\eta^3\}$$

- One subgroup of order five $\langle\sigma^2\rangle = \{e, \sigma^2, \sigma^4, \sigma^6, \sigma^8\}$
- Two cyclic subgroups of order four. $\langle\eta\rangle = \{e, \eta, \eta^2, \eta^3\}$, $\langle\sigma^5\eta\rangle = \{e, \sigma^5\eta, \eta^2, \sigma^5\eta^3\}$
- One subgroup $\langle\sigma^5, \eta^2\rangle$ of order four isomorphic to Klein-Four group. $\langle\sigma^5, \eta^2\rangle = \{e, \sigma^5, \eta^2, \sigma^5\eta^2\}$
- Three subgroups of order two. $\langle\sigma^5\rangle$, $\langle\eta^2\rangle$, and $\langle\sigma^5\eta^2\rangle$.

The following table gives a gives a description of the nonidentity subgroups of $G = Aut(\mathbb{K}/\mathbb{Q})$.

Order of Subgroup	Number of Subgroups	Isomorphic Type of Subgroup
40	1	$C_{10} \times C_4 \approx C_5 \times C_2 \times C_4 \approx C_{20} \times C_2$
20	2	Cyclic
20	1	$C_{10} \times C_2$
10	3	Cyclic
8	1	$C_4 \times C_2$
4	2	Cyclic
4	1	Klein-Four
2	3	C_2

Galois Correspondence

Subgroup	Order	Fixed Subfield	Degree
$\langle \sigma\eta \rangle$	20	$\mathbb{Q}(\sqrt{-55})$	2
$\langle \sigma^2\eta \rangle$	20	$\mathbb{Q}(\sqrt{-11})$	2
$\langle \sigma\eta^2 \rangle$	20	$\mathbb{Q}(\sqrt{5})$	2
$\langle \sigma \rangle$	10	$\mathbb{Q}(\varepsilon)$	4
$\langle \sigma\eta^2 \rangle$	10	$\mathbb{Q}\left((\varepsilon - \varepsilon^4)\sqrt{-11}\right)$	4
$\langle \sigma^2\eta^2 \rangle$	10	$\mathbb{Q}(\sqrt{5}, \sqrt{-11})$	4
$\langle \sigma^5, \eta \rangle$	8	$\mathbb{Q}\left(\alpha + \dfrac{1}{\alpha}\right) = \mathbb{Q}\left(\cos\frac{2\pi}{11}\right)$	5
$\langle \sigma^2 \rangle$	5	$\mathbb{Q}(\varepsilon, \sqrt{-11})$	8
$\langle \eta \rangle$	4	$\mathbb{Q}(\alpha)$	10
$\langle \sigma^5\eta \rangle$	4	$\mathbb{Q}(\cos\frac{2\pi}{11}, \sqrt{-55})$	10
$\langle \sigma^5, \eta^2 \rangle$	4	$\mathbb{Q}(\cos\frac{2\pi}{11}, \sqrt{5})$	10
$\langle \sigma^5 \rangle$	2	$\mathbb{Q}(\cos\frac{2\pi}{11}, \varepsilon)$	20
$\langle \eta^2 \rangle$	2	$\mathbb{Q}(\alpha, \sqrt{5})$	20
$\langle \sigma^5\eta^2 \rangle$	2	$\mathbb{Q}(\cos\frac{2\pi}{11}, (\varepsilon - \varepsilon^4)\sqrt{-11})$ $= \mathbb{Q}\left(\cos\frac{2\pi}{11}, \sqrt{11(10+2\sqrt{5})}\right)$	20

(92) This problem is similar to the previous problem. In this case the splitting field is:

$$\mathbb{K} = \mathbb{Q}(\alpha, \omega), \quad \text{where} \quad \alpha = e^{\frac{2\pi i}{11}} = \cos\frac{2\pi}{11} + i\sin\frac{2\pi}{11},$$

$$\omega = e^{\frac{2\pi i}{3}} = \cos\frac{2\pi}{3} + i\sin\frac{2\pi}{3} = \frac{-1+\sqrt{-3}}{2}.$$

Let $\sigma, \eta \in G = Aut(\mathbb{K}/\mathbb{Q})$ be the following two automorphisms:

$$\sigma(\alpha) = \alpha^2, \quad \sigma(\omega) = \omega, \qquad \eta(\alpha) = \alpha, \quad \eta(\omega) = \omega^2.$$

As can be easily checked that $Order(\sigma) = 10$, $Order(\eta) = 2$ and $\sigma\eta = \eta\sigma$. We have

$$G = \langle \sigma, \eta : \sigma^{10} = \eta^2 = e, \sigma\eta = \eta\sigma \rangle$$
$$= \left\{\sigma^i\eta^j : 0 \le i \le 9, \;\; 0 \le j \le 1\right\} G \approx C_{10} \times C_2.$$

The proper subgroups of G are as follows:

- Three subgroups of order 10. These are $\langle\sigma\rangle$, $\langle\sigma\eta\rangle$, and $\langle\sigma^2\eta\rangle$.
- One subgroup of order five, namely $\langle\sigma^2\rangle$.
- One subgroup of order four, namely $\langle\sigma^5, \eta\rangle = \{e, \sigma^5, \eta, \sigma^5\eta\}$.
- Three subgroups of order two. These are $\langle\sigma^5\rangle$, $\langle\eta\rangle$, and $\langle\sigma^5\eta\rangle$.

The subfields of $\mathbb{K} = \mathbb{Q}(\alpha, \omega)$ are as follows:

- Three subfields of degree two over $\mathbb{Q}$. These are $\mathbb{Q}(\omega) = \mathbb{Q}(\sqrt{-3})$, $\mathbb{Q}(\sqrt{-11})$, and $\mathbb{Q}(\sqrt{33})$.
- One subfield of degree four over $\mathbb{Q}$, $\mathbb{Q}(\sqrt{-3}, \sqrt{-11})$.
- One subfield of degree five over $\mathbb{Q}$, $\mathbb{Q}(\cos\frac{2\pi}{11})$.
- Three subfields of degree ten over $\mathbb{Q}$. These are: $\mathbb{Q}(\cos\frac{2\pi}{11}, \sqrt{-3})$, $\mathbb{Q}(\alpha)$, and $\mathbb{Q}(\cos\frac{2\pi}{11}, \sqrt{33})$.

Galois Correspondence

Subgroup	Order	Fixed Subfield	Degree Over $\mathbb{Q}$
$\langle\sigma^5\rangle$	2	$\mathbb{Q}(\cos\frac{2\pi}{11}, \sqrt{-3})$	10
$\langle\eta\rangle$	2	$\mathbb{Q}(\alpha)$	10
$\langle\sigma^2\eta\rangle$	2	$\mathbb{Q}(\cos\frac{2\pi}{11}, \sqrt{33})$	10
$\langle\sigma^5\eta\rangle$	4	$\mathbb{Q}(\cos\frac{2\pi}{11})$	5
$\langle\sigma^2\rangle$	5	$\mathbb{Q}(\sqrt{-3}, \sqrt{-11})$	4
$\langle\sigma\rangle$	10	$\mathbb{Q}(\sqrt{-3})$	2
$\langle\sigma\eta\rangle$	10	$\mathbb{Q}(\sqrt{33})$	2
$\langle\sigma^2\eta\rangle$	10	$\mathbb{Q}(\sqrt{-11})$	2

The two diagrams on the following pages give the lattice of subfields of $\mathbb{K}$ and the lattice of subgroups of the Galois group G respectively.

Lattice of Subfields of $K = Q(\alpha, \omega)$, $\alpha = e^{\frac{2\pi i}{11}}$, $\omega = e^{\frac{2\pi i}{3}}$

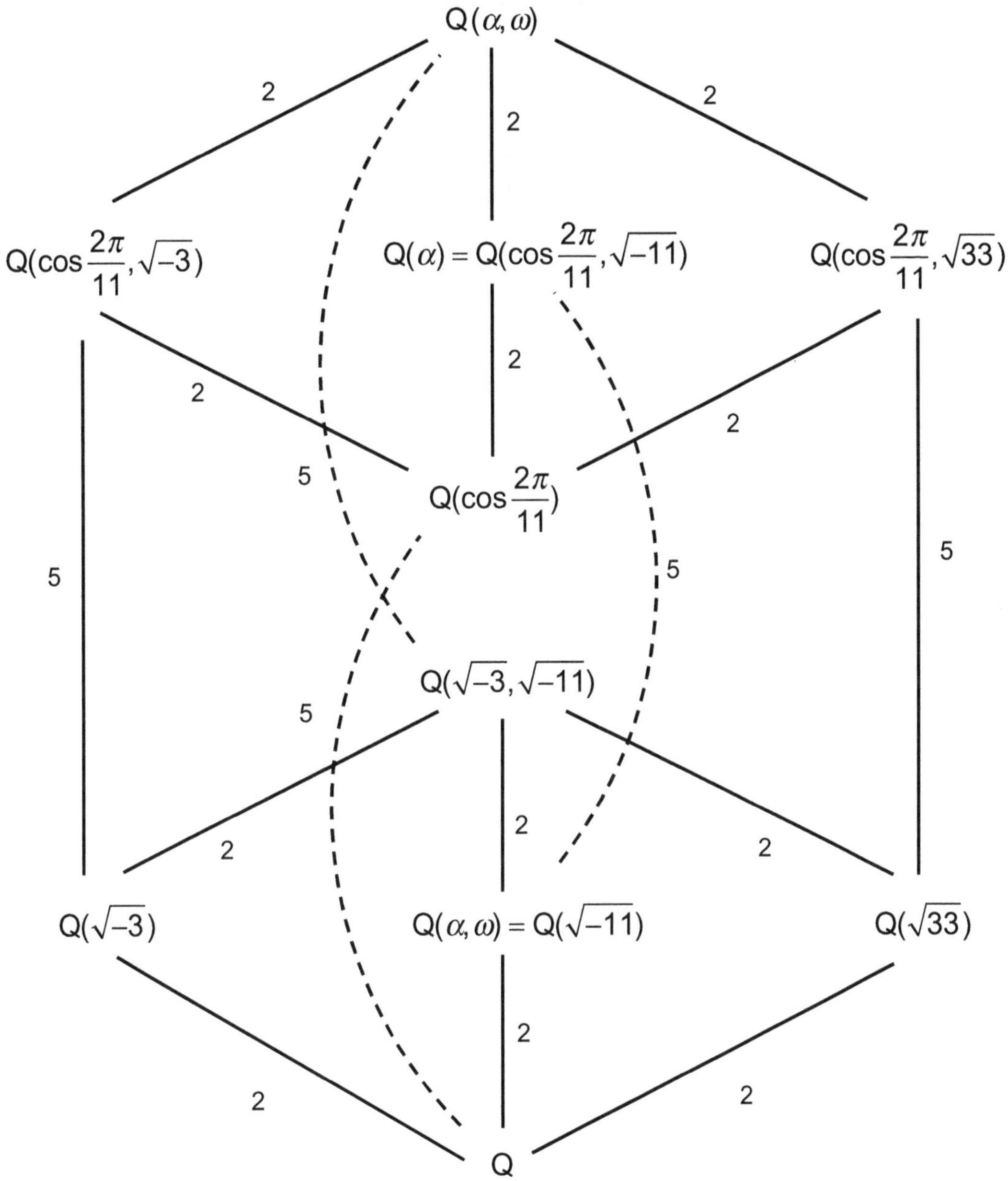

Lattice of Subgroups of the Group $G = \mathrm{Aut}(K/Q)$

$K = Q(\alpha,\omega),\ \alpha = e^{\frac{2\pi i}{11}},\ \omega = e^{\frac{2\pi i}{3}}$

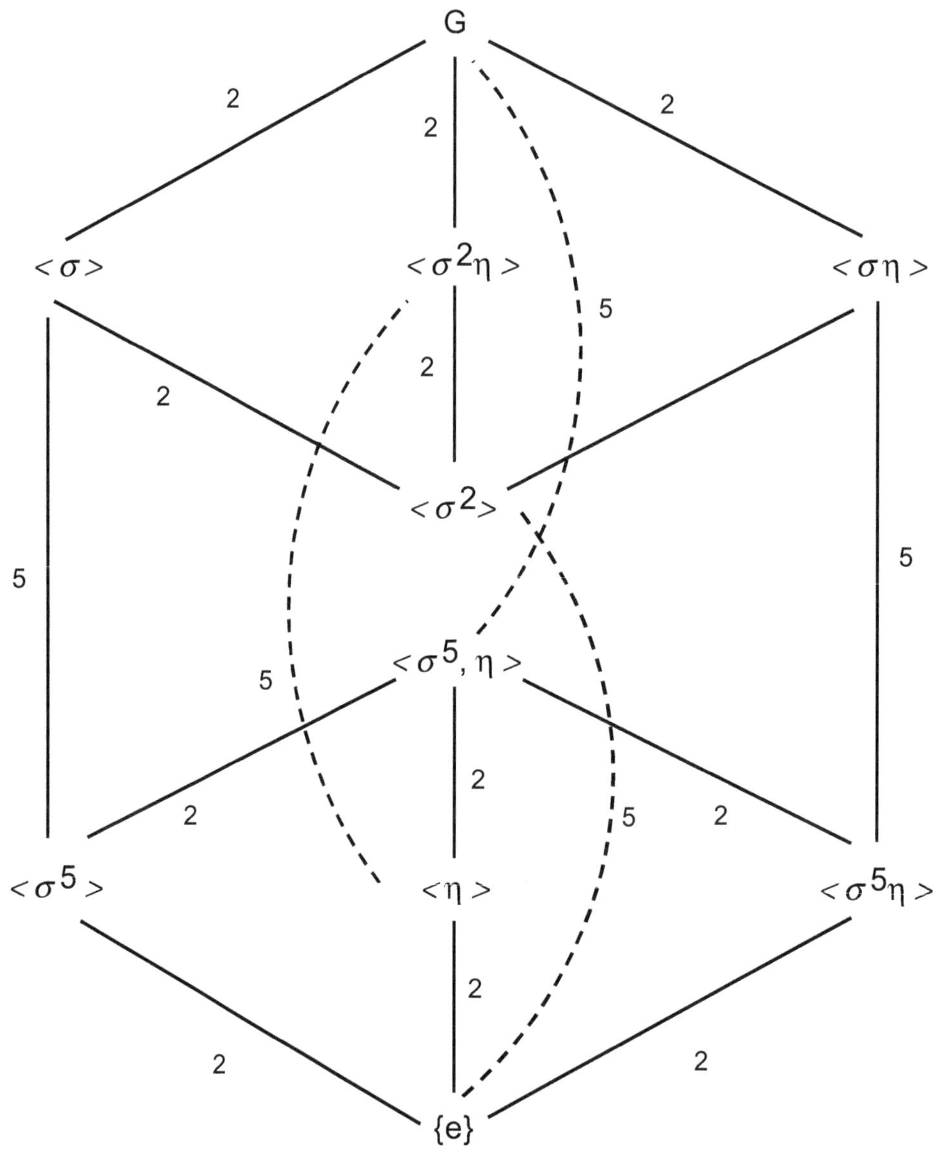

(93) Remember the requirement that $A \cdot B = -\dfrac{p}{3} = 1$ in this case. We have the following three possible values for A

$$\frac{\sqrt{3}}{2} + \frac{i}{2}, \quad \omega\left(\frac{\sqrt{3}}{2} + \frac{i}{2}\right) \quad \omega^2\left(\frac{\sqrt{3}}{2} + \frac{i}{2}\right)$$

where

$$\omega = e^{\frac{2\pi}{3}} = \cos\frac{2\pi}{3} + i\sin\frac{2\pi}{3} = \frac{-1 + i\sqrt{3}}{2}$$

Since $A \cdot B = 1$, once we pick A then we have to have in this case $B = \dfrac{1}{A}$. The following table gives the possible matchings of A and B and the corresponding values of $A + B$ giving the three roots of $x^3 - 3x = 0$ as before

A	B	$A + B$
$\frac{\sqrt{3}}{2} + \frac{i}{2}$	$\omega^2 i$	$\sqrt{3}$
$\omega\left(\frac{\sqrt{3}}{2} + \frac{i}{2}\right)$	ωi	$-\sqrt{3}$
$\omega^2\left(\frac{\sqrt{3}}{2} + \frac{i}{2}\right)$	i	0

(94) Let $f(x) = a_0 + a_1x + a_2x^2 + \cdots + a_{n-1}x^{n-1} + a_nx^n \in \mathbb{C}[x]$, where $a_k \in \mathbb{C}$, $0 \le k \le n$. Let $a_k = b_k + ic_k$, where $i = \sqrt{-1}$. Then $\bar{a_k} = b_k - c_k$. We have

$$\begin{aligned} f(x) &= a_0 + a_1x + a_2x^2 + \cdots + a_{n-1}x^{n-1} + a_nx^n \\ &= (b_0 + ic_0) + (b_1 + ic_1)x + (b_2 + ic_2)x^2 + \cdots + (b_{n-1} + ic_{n-1})x^{n-1} + (b_n + ic_n)x^n \\ &= g(x) + ih(x) \end{aligned}$$

where

$$\begin{aligned} g(x) &= b_0 + b_1x + b_2x^2 + \cdots + b_{n-1}x^{n-1} + b_nx^n \\ h(x) &= c_0 + c_1x + c_2x^2 + \cdots + c_{n-1}x^{n-1} + c_nx^n. \end{aligned}$$

Now $f(x) = g(x) + ih(x) \implies f(x)\bar{f}(x) = g(x)^2 + h(x)^2 \in \mathbb{R}[x]$. Since $f(x)\bar{f}(x)$ has real coefficients there exists $\alpha \in \mathbb{C}$ such that $f(\alpha)\bar{f}(\alpha) = 0$. So either $f(\alpha) = 0$ or $\bar{f}(\alpha) = 0$, which implies that $f(\bar{\alpha}) = 0$.

(95) We have $\alpha^4 = -p$ and therefore

$$\frac{1}{\alpha^3 + \alpha^2 + \alpha + 1} = \frac{\alpha - 1}{x^4 - 1} = \frac{\alpha - 1}{-(p+1)}$$

Since α is a root of the equation, $\alpha - 1$ is a root of the equation $(y+1)^4 + p = 0$ or $y^4 + 4y^3 + 6y^2 + 4y + (p+1) = 0$. Let $z = -\dfrac{y}{p+1}$ or $y = -(p+1)z$. So $\dfrac{\alpha - 1}{-(p+1)}$ is a root of the equation

$$(p+1)^4z^4 - 4(p+1)^3z^3 + 6(p+1)^2z^2 - 4(p+1)z + (p+1) = 0$$

So the polynomial with a root $\alpha^3 + \alpha^2 + \alpha + 1$ is

$$z^4 - 4z^3 + 6(p+1)z^2 - 4(p+1)^2z + (p+1)^3 = 0$$

(96) Since $\alpha\beta$ is a primitive mn-th root of unity it follows that

$$[\mathbb{Q}(\alpha\beta) : \mathbb{Q}] = \phi(mn) = \phi(m)\phi(n), \qquad \text{therefore} \qquad \mathbb{Q}(\alpha\beta) = \mathbb{Q}(\alpha, \beta) = \mathbb{K}(\text{say}).$$

To prove that

$$\Omega = \left\{\alpha^i\beta^j : 0 \le i \le \phi(m) - 1, \ \ 0 \le j \le \phi(n) - 1\right\}$$

is a basis of $\mathbb{K} = \mathbb{Q}(\alpha\beta)$ over $\mathbb{Q}$ we need to show that every element of $\mathbb{Q}(\alpha\beta)$ can be expressed as a linear combination of elements of Ω with rational coefficients. Since $\alpha\beta$ is a primitive mn-th root of unity the set

$$\left\{(\alpha\beta)^k : 0 \le k \le \phi(mn) - 1\right\}$$

is a basis of $\mathbb{Q}(\alpha\beta) = \mathbb{Q}(\alpha, \beta)$ over $\mathbb{Q}$. So we can write every element of $\mathbb{Q}(\alpha, \beta)$ as a linear combination of the set $\{(\alpha\beta)^k : 0 \leq k \leq \phi(mn) - 1\}$ with rational coefficients like

$$\sum_{k=0}^{\phi(mn)-1} a_k(\alpha\beta)^k = \sum_{k=0}^{\phi(mn)-1} a_k\alpha^k\beta^k.$$

Since $degree(\alpha) = \phi(m)$ and $degree(\beta) = \phi(n)$, we can express every α^i, $i \geq \phi(m)$ as a linear combination of $\{1, \alpha, \alpha^2, \cdots, \alpha^{\phi(m)-1}\}$ with rational coefficients and likewise we can write every β^j, $j \geq \phi(n)$ as a linear combination of $\{1, \beta, \beta^2, \cdots, \beta^{\phi(n)-1}\}$ with rational coefficients. It follows that

$$\sum_{k=0}^{\phi(mn)-1} a_k(\alpha\beta)^k \quad \text{can be expressed as} \quad \sum_{j=0}^{\phi(n)-1}\sum_{i=0}^{\phi(m)-1} a_{ij}\alpha^i\beta^j.$$

Therefore the set of $\phi(mn)$ elements

$$\Omega = \{\alpha^i\beta^j : 0 \leq i \leq \phi(m) - 1, \;\; 0 \leq j \leq \phi(n) - 1\}$$

spans $\mathbb{K}$ over $\mathbb{Q}$. Since $[\mathbb{K} : \mathbb{Q}] = \phi(mn)$ it follows that the elements of the set Ω are linearly independent over $\mathbb{Q}$. Now suppose that

$$\sum_{i=0}^{\phi(m)-1} a_i\alpha_i = \sum_{j=0}^{\phi(n)-1} b_j\beta_j$$

This implies that

$$(a_0 - b_0) + \sum_{i=1}^{\phi(m)-1} a_i\alpha_i - \sum_{j=1}^{\phi(n)-1} b_j\beta_j = 0.$$

Since the set $\{1, \alpha, \alpha^2, \cdots, \alpha^{\phi(m)-1}, \beta, \beta^2, \cdots, \beta^{\phi(n)-1}\}$ is a subset of a basis, we conclude that

$$a_0 = b_0, \;\; a_i = 0, \;\; 1 \leq i \leq \phi(m) - 1, \;\; b_j = 0, \;\; 1 \leq j \leq \phi(n) - 1.$$

This proves that $\mathbb{Q}(\alpha) \cap \mathbb{Q}(\beta) = \mathbb{Q}$.

(97) The first part of this problem has been discussed in problem # 53. The splitting field of $g(x) = f(x)(x^2+x+1) = (x^3-4x+2)(x^2+x+1)$ is

$$\mathbb{K}(\sqrt{-3}, \sqrt{37}, A), \quad \text{where } A = \left[-1 + \frac{1}{9}\sqrt{-111}\right]^{\frac{1}{3}}.$$

Since the roots of $f(x)$ are all real and its Galois group is isomorphic to S_3 its splitting field is not a radical extension of $\mathbb{Q}$. The splitting field of $g(x)$ is a radical extension of $\mathbb{Q}$ since

$$\mathbb{Q} \subset \mathbb{Q}(\sqrt{-3}) \subset \mathbb{Q}(\sqrt{-3}, \sqrt{37}) \subset \mathbb{Q}\left(\sqrt{-3}, \sqrt{37}, \left[-1 + \frac{1}{9}\sqrt{-111}\right]^{\frac{1}{3}}\right).$$

The main point of this problem is that the Galois group of $x^3 - 4x + 2$ is S_3 (a solvable group) but its splitting field is *not* a radical extension of $\mathbb{Q}$ - but the field obtained by adding ω (a primitive cube root of unity) to it makes the resulting field a radical extension of $\mathbb{Q}$.

(98) Let $\mathbb{F}_p = GF(p)$ be finite field of order p. Let $\mathbb{K}$ be a splitting field of the polynomial $f(x) = x^6 + x^3 + 1 \in \mathbb{F}_p[x]$ - which in some cases may turn out to the field $\mathbb{F}_p$ itself. Let $\zeta \in \mathbb{K}$ be a root of $f(x)$. The as $x^9 - 1 = (x^3-1)(x^6+x^3+1)$ we have $\zeta^9 - 1 = 0$ or $\zeta^9 = 1$. Therefore ζ as an element of the field $\mathbb{K}$ has order 1, 3 or 9. If the order were one or three then $3 = \zeta^6 + \zeta^3 + 1 = 0$. Since $p \neq 3$ this is not possible therefore ζ must have order 9. This observation will be used repeatedly in what follows.

- (a) Suppose that $p \equiv 1$ (modulo 9). So 9 divides $p-1$ and $\mathbb{F}_p^*$ has a (cyclic) subgroup of order 9. Each element of this group is a root of the polynomial $x^9 - 1$. Since $x^9 - 1 = (x^3-1)(x^6+x^3+1)$ the polynomial $f(x) = x^6 + x^3 + 1$ splits into linear factors modulo p if $p \equiv 1$ (modulo 9). For example for $p = 19$ we have

$$x^6 + x^3 + 1 \equiv (x-4)(x-5)(x-6)(x-9)(x-13)(x-17) \text{ (modulo 19)}$$

- (b) $p \equiv 4, 7$ (modulo 9).

  In this case since 9 does not divide $p-1$ the polynomial $f(x) = x^6+x^3+1$ does not have any linear factors modulo p. We note that $p \equiv 4, 7$ (modulo 9) imply that $\left(\dfrac{-3}{p}\right) = 1$, and so -3 is a square modulo p. We shall find $a, b \in \mathbb{F}_p$ such that $x^6 + x^3 + 1) = (x^3 + a)(x^3 + b)$. Such a and b can be found as the roots of the equation $z^2 - z + 1$, which are $\dfrac{1 \pm -3}{2}$. Since $\left(\dfrac{-3}{p}\right) = 1$ $\dfrac{1 \pm -3}{2} \in \mathbb{F}_p$. Since $p-1$ is not divisible by 9 the polynomial does not have any linear factor. Therefore both $x^3 + a$ and $x^3 + b$ are irreducible over $\mathbb{F}_p$. For example

$$f(x) \equiv (x^3 - 3)(x^3 - 9) \text{ (modulo 13) and } f(x) \equiv (x^3 - 2)(x^3 - 4) \text{ (modulo 7)}.$$

- (c) $p \equiv 2, 5$ (modulo 9)

  Since 9 does not divide $p-1$, the polynomial $f(x)$ cannot have any linear factor modulo p. Therefore it is enough to show that $f(x)$ does not any quadratic or a cubic factor. We note that

  $$p \equiv 2, 5 \text{ (modulo 9)} \Longrightarrow p^2 \equiv 4, 7 \text{ (modulo 9)}, \quad \text{and} \quad p^3 \equiv 8 \text{ (modulo 9)}$$

  In particular 9 does not divide $p^2 - 1$ or $p^3 - 1$.

  So if $f(x)$ had a quadratic factor then $\mathbb{F}_p$ will have an extension of degree two having a root of this quadratic factor. If ζ is a root of this quadratic factor then its order will be 9 which is not possible since 9 does not divide $p^2 - 1$. Likewise if $f(x)$ had a cubic factor then $\mathbb{F}_p$ will have an extension of degree three having a root of this cubic factor. If ζ is a root of this cubic factor then its order will be 9 which is not possible since 9 does not divide $p^3 - 1$. For example the polynomial $f(x) = x^6 + x^3 + 1$ is irreducible modulo 2 as well as modulo 5.

- (d) Let $p \equiv 8$ (modulo 9).

  We first note that $p \equiv 8$ (modulo 9) $\Longrightarrow p^2 \equiv 1$ (modulo 9). We shall show that then $f(x)$ splits into three quadratic factors modulo p. Let $\mathbb{K}$ splitting field of $f(x) \in \mathbb{F}_p[x]$. We shall show that if $\zeta \in \mathbb{K}$ is a root of $f(x)$ then $(x - \zeta)(x - \zeta^p) \in \mathbb{F}_p[x]$. We have seen that such a ζ has order 9 as an element of $\mathbb{K}^*$. Since

  $$(\alpha + \alpha^p)^p = \alpha^p + \alpha^{p^2} = \alpha^p + \alpha, \text{ and } \quad (\alpha\alpha^p)^p = \alpha^p\alpha^{p^2} = \alpha^p + \alpha$$

it follows that every element of $\mathbb{K}$ which is a root of $f(x)$ is of degree 2 over $\mathbb{F}_p$ $Aut(\mathbb{K}/\mathbb{F}_p)$ has order 2 and so $\alpha + \alpha^p$ and $\alpha\alpha^p$ are in $\mathbb{F}_p$. Thus if ζ is a root of $f(x)$ then $(x-\zeta)(x-\zeta^p) \in \mathbb{F}_p[x]$ is a quadratic factor of $f(x)$. It follows that $f(x)$ splits into three quadratic factors modulo p if $p \equiv 8$ (modulo 9).

The Polynomial $f(x) = \Phi_9(x) = x^6 + x^3 + 1$

- The elements 2 and 5 as elements of the group $\mathbb{Z}_9^*$ have order 6. So modulo primes $p \equiv 2, 5$ (modulo 9) $f(x)$ remains irreducible.
- The elements 4 and 7 as elements of the group $\mathbb{Z}_9^*$ have order 3. So modulo primes $p \equiv 4, 7$ (modulo 9) $f(x)$ factors as a product of two irreducible cubics.
- The element 8 as an element of the group $\mathbb{Z}_9^*$ has order 2. So modulo primes $p \equiv 8$ (modulo 9) $f(x)$ factors as a product of three irreducible quadratics.
- The elements 1 as an element of the group $\mathbb{Z}_9^*$ has order 1. So modulo primes $p \equiv 1$ (modulo 9) $f(x)$ factors as a product six linear factors.

Factoring $\Phi_9(x) = x^6 + x^3 + 1$ modulo various primes

p	Order of p (modulo 9)	Factors of $x^6 + x^3 + 1$ modulo p
2	6	$x^6 + x^3 + 1$
5	6	$x^6 + x^3 + 1$
13	3	$(x^3 - 3)(x^3 - 9)$
7	3	$(x^3 - 2)(x^3 - 4)$
17	2	$(x^2 + 3x + 1)(x^2 + 4x + 1)(x^2 + 10x + 1)$
19	1	$(x - 4)(x - 5)(x - 6)(x - 9)(x - 16)(x - 17)$

Note.

It is clear from the factorization of $\Phi_9(x) = x^6 + x^3 + 1$ modulo a prime $p \neq 3$ is determined by the order of the Frobenius automorphism $\zeta \longrightarrow \zeta^p$, which is the generator of the group $Aut(\mathbb{K}/\mathbb{F}_p)$, where $\mathbb{K}$ is the splitting field of $x^6+x^3+1 \in \mathbb{F}_p[x]$. The following table illustrates this.

p	Order of $\zeta \longrightarrow \zeta^p$	$\Phi_9(x)$ Factors as
$p \equiv 1$ (modulo 9)	1	Linear Factors
$p \equiv 8$ (modulo 9)	2	Three quadratic Factors
$p \equiv 2, 5$ (modulo 9)	6	Irreducible
$p \equiv 4, 7$ (modulo 9)	3	Two cubic Factors

(99) This is just like the previous problem. Let $\mathbb{K}$ be the splitting field of $f(x) = x^{18} + x^9 + 1 \in \mathbb{F}_p[x]$. If $\zeta \in \mathbb{K}$ is a root of $f(x)$ then order of ζ must be 27. This is so as $\zeta \neq 1$ and if $\zeta^3 = 1$ or $\zeta^9 = 1$ then we would have $\zeta^{18} + \zeta^9 + 1 = 3 \neq 0$ as $p \neq 3$. The orders of p modulo 27 in the above table can be easily verified. The following table gives the information about factors of $f(x) = x^{18} + x^9 + 1$ modulo p for various primes.

$p \equiv ?$ (modulo 27)	Prime	Factors of $x^{18}+x^9+1$ modulo p
$p \equiv 1$ (modulo 27)	109	All Linear
	163	All Linear
$p \equiv 2$ (modulo 27)	29	Irreducible
	83	Irreducible
$p \equiv 4$ (modulo 27)	31	$(x^9+6)(x^9+26)$
	139	$(x^9+43)(x^9+97)$
$p \equiv 5$ (modulo 27)	59	Irreducible
	113	Irreducible
$p \equiv 7$ (modulo 27)	61	$(x^9+14)(x^9+48)$
	223	$(x^9+40)(x^9+184)$
$p \equiv 8$ (modulo 27)	89	$(x^6+12x^3+1)(x^6+36x^3+1)(x^6+41x^3+1)$
	197	$(x^6+6x^3+1)(x^6+28x^3+1)(x^6+163x^3+1)$
$p \equiv 10$ (modulo 27)	37	$(x^3+3)(x^3+4)(x^3+21)(x^3+25)(x^3+28)(x^3+30)$
	199	$(x^3+19)(x^3+21)(x^3+24)(x^3+37)(x^3+141)(x^3+156)$
$p \equiv 11$ (modulo 27)	173	Irreducible
	227	Irreducible
$p \equiv 13$ (modulo 27)	67	$(x^9+30)(x^9+38)$
	229	$(x^9+95)(x^9+135)$

$p \equiv ?$ (modulo 27)	Prime	Factors of $x^{18}+x^9+1$ modulo p
$p \equiv 14$ (modulo 27)	41	Irreducible
	149	Irreducible
$p \equiv 16$ (modulo 27)	43	$(x^9+7)(x^9+37)$
	151	$(x^9+33)(x^9+119)$
$p \equiv 17$ (modulo 27)	71	$(x^6+16x^3+1)(x^6+25x^3+1)(x^6+30x^3+1)$
	179	$(x^6+34x^3+1)(x^6+46x^3+1)(x^6+99^3+1)$
$p \equiv 19$ (modulo 27)	73	$(x^3+18)(x^3+36)(x^3+41)(x^3+57)(x^3+69)(x^3+71)$
	181	$(x^3+101)(x^3+108)(x^3+116)(x^3+119)(x^3+138)(x^3+142)$
$p \equiv 20$ (modulo 27)	47	Irreducible
	101	Irreducible
$p \equiv 22$ (modulo 27)	103	$(x^9+47)(x^9+57)$
	157	$(x^9+13)(x^9+145)$
$p \equiv 23$ (modulo 27)	131	Irreducible
	239	Irreducible
$p \equiv 25$ (modulo 27)	79	$(x^9+24)(x^9+56)$
	241	$(x^9+16)(x^9+226)$
$p \equiv 26$ (modulo 27)	53	$(x^2+3x+1)(x^2+6x+1)(x^2+12x+1)(x^2+17x+1)(x^2+19x+1)$ $(x^2+30x+1)(x^2+31x+1)(x^2+46x+1)(x^2+48x+1)$
	107	$(x^2+8x+1)(x^2+9x+1)(x^2+10x+1)(x^2+23x+1)(x^2+28x+1)$ $(x^2+34x+1)(x^2+45x+1)(x^2+74x+1)(x^2+90x+1)$

(100) If $f(x) \in \mathbb{Z}[x]$ is reducible then it must have either a linear factor or a quadratic factor. We shall show that if it has a linear factor then $k = 0$ or $k = 2$ and if it does not have linear factor but has a quadratic factor then $a = -1$.

- Suppose that $f(x) = x^5 - kx - 1$ has a linear factor. By rational root theorem this linear factor is either $x - 1$ or $x + 1$. If $f(x)$ has $x - 1$ as linear factor then $\mathbf{k = 0}$ and if it has $x + 1$ as a linear factor then $\mathbf{k = 2}$.
- Suppose that $f(x) = x^5 - kx - 1 \in \mathbb{Z}[x]$ has an irreducible quadratic factor. Then it factors in $\mathbb{Z}[x]$ as

$$x^5 - kx - 1 = (x^3 + ax^2 + bx + c)(x^2 + ex + f).$$

Equating the coefficients we have

$$a + e = 0, \qquad f + ae + b = 0, \qquad af + be + c = 0, \qquad bf + ce = -k, \qquad cf = -1.$$

Replacing e by $-a$ we have

$$b + f = a^2, \qquad af - ab + c = 0, \qquad bf - ac = -k, \qquad cf = -1.$$

Since $cf = 1$, we have either $c = 1, \;\; f = -1$ or $c = -1, \;\; f = 1$.

Case I. $c = 1, \;\; f = -1$. Substituting these values we have

$$b - 1 = a^2, \qquad -a - ab + 1 = 0, \qquad -b - a = -k$$

that is

$$a^2 = b - 1, \qquad a(b + 1) = 1, \qquad a + b = k.$$

Now $a(b+1) = 1 \Longrightarrow a = \pm 1$. If $a = 1$ then $b = 0$, but this is not possible since $a^2 = b - 1$. If $a = -1$ then $b = -2$ and this too is not possible since $a^2 = b - 1$.

Case II. $c = -1, \;\; f = 1$. Substituting these values we have

$$a^2 = b + 1, \qquad a(1 - b) = 1, \qquad a + b = -k.$$

Now $a(1 - b) = 1 \Longrightarrow a = \pm 1$. If $a = 1$ then $b = 0$ and $\mathbf{k = -1}$. If $a = -1$ then $b = 2$ which is not possible since $a^2 = b + 1$.

(101) Let $\mathbb{K} = \mathbb{Q}(x)$, $\mathbb{E} = \mathbb{Q}(x^2)$, and $\mathbb{F} = \mathbb{Q}((x-1)^2)$.

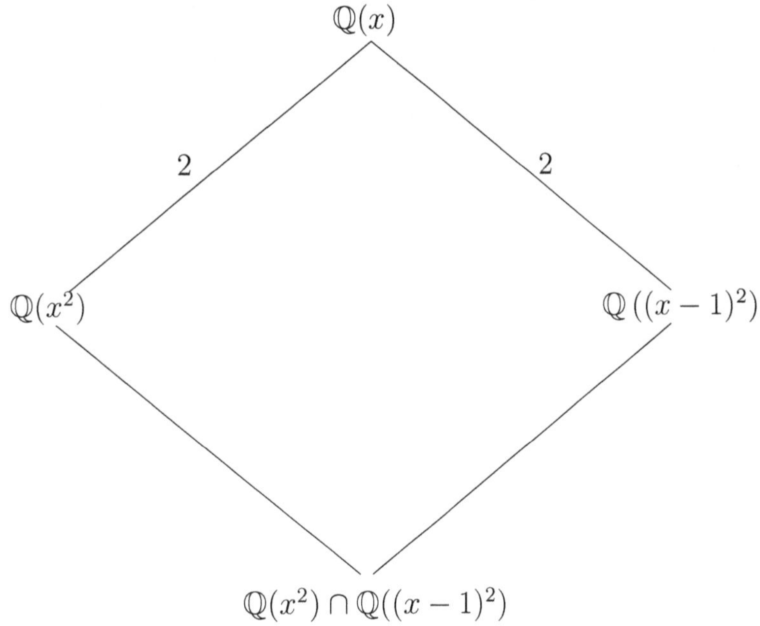

Then $\mathbb{Q}(x^2) \cap \mathbb{Q}((x-1)^2) = \mathbb{Q}$. This is an immediate consequence of the following:

Lemma. Let $f(x) \in \mathbb{Q}(x)$ be a rational function whose graph is symmetric about the lines $x = 0$ and $x = 1$, then $f(x)$ is a constant function.

Proof. It is easy to see that $f(2n) = f(0)$ for all $n \in \mathbb{Z}$. It is obvious that $f(2) = f(0)$ as $f(x)$ is symmetric about the line $x = 1$. Assume that $f(2n) = f(0)$ for some $n > 0$, then

$$\begin{aligned} f(2n) &= f(-2n) && \text{Since } f(x) \text{ is symmetric about } x = 0 \\ &= f(1 + (2n+1) && \text{Since } f(x) \text{ is symmetric about } x = 1 \\ &= f(2n+2). \end{aligned}$$

Since a nonconstant rational function cannot take any value infinitely often the function $f(x)$ must be a constant function.

(102)

Let $\mathbb{K} = \mathbb{C}(x)$, $\mathbb{L} = \mathbb{C}(x^m)$, and $\mathbb{M} = \mathbb{C}((x-1)^n)$.

Evidently

$$[\mathbb{C}(x) : \mathbb{L}] = m, \quad \text{and} \quad [\mathbb{C}(x) : \mathbb{M}] = n.$$

Let $\mathbb{F} = \mathbb{L} \cap \mathbb{M}$. We shall show that the field extension $\mathbb{C}(x)/\mathbb{F}$ is not algebraic. Suppose that the extension $\mathbb{C}(x)/\mathbb{F}$ is algebraic. Then x is algebraic over $\mathbb{F}$ and therefore $\mathbb{C}(x)/\mathbb{F}$ is a finite extension. In particular, the group $\mathrm{Aut}(\mathbb{C}(x)/\mathbb{F})$ is finite. Let $\omega = e^{2\pi/m}$ be primitive m-th root of unity and $\zeta = e^{2\pi/n}$ be primitive n-th root of unity.

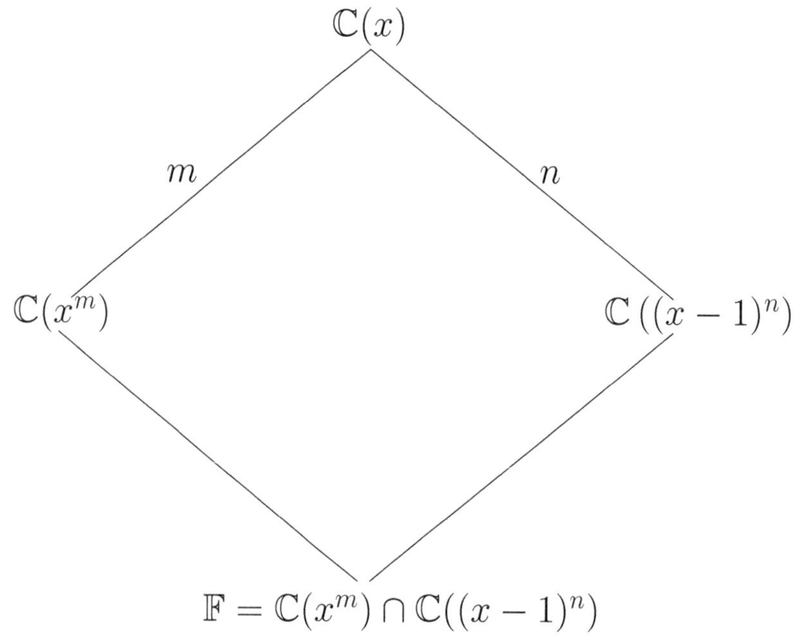

Consider σ and η following two automorphisms of the field $\mathbb{C}(x)$.

$$\sigma : x \longmapsto \omega x, \qquad \text{and } \eta : (x-1) \longmapsto \zeta(x-1).$$

Evidently

$$\sigma^{-1} : x \longmapsto \omega^{-1} x, \qquad \text{and} \qquad \eta^{-1} : (x-1) \longmapsto \zeta^{-1}(x-1).$$

We note that

$$\eta : x \longmapsto \zeta(x-1)+1 = \zeta x - \zeta + 1, \text{ and } \eta^{-1} : x \longmapsto \zeta^{-1}(x-1)+1 = \zeta^{-1}x - \zeta^{-1} + 1.$$

We have

$$\begin{aligned}
x &\xrightarrow{\quad\sigma\quad} \omega x \\
\omega x &\xrightarrow{\quad\eta\quad} \omega(\zeta x - \zeta + 1) = \omega\zeta x - \omega\zeta + \omega \\
\omega\zeta x - \omega\zeta + \omega &\xrightarrow{\quad\sigma^{-1}\quad} \omega\zeta(\omega^{-1}x) - \omega\zeta + \omega = \zeta x - \omega\zeta + \omega \\
\zeta x - \omega\zeta + \omega &\xrightarrow{\quad\eta^{-1}\quad} \zeta(\zeta^{-1}x - \zeta^{-1} + 1) - \omega\zeta + \omega = x - (\omega - 1)(\zeta - 1).
\end{aligned}$$

Therefore composing these mappings in the usual way we have

$$x \xrightarrow{\eta^{-1}\sigma^{-1}\eta\sigma} x - (\omega - 1)(\zeta - 1).$$

It is clear that this mapping fixes the field $\mathbb{F} = \mathbb{C}(x^m) \cap \mathbb{C}((x-1)^n)$ and as a nontrivial translation this mapping has infinite order. But this mapping is an element of the group $\mathrm{Aut}(\mathbb{C}(x)/\mathbb{F})$ so the extension $\mathbb{C}(x)/\mathbb{F}$ cannot be finite. As this contradiction results from our assumption that the extension $\mathbb{C}(x)/\mathbb{F}$ is algebraic it follows that $\mathbb{C}(x)/\mathbb{F}$ is not algebraic.

Alternately, if $\mathbb{F} = \mathbb{C}(x^m) \cap \mathbb{C}((x-1)^n) \neq \mathbb{C}$, then there would exist a non-constant rational function in $\mathbb{F}$ with infinitely many zeros. This is not possible.

Glossary

- **Algebraic Integer**

  A complex number α is said to be an algebraic integer if there exists a monic polynomial (that is with leading coefficients unity) $f(x) \in \mathbb{Z}[x]$ of degree $n > 0$ having α as one of its roots.

- **Algebraic Number**

  A complex number α is said to be an algebraic number if there exists a polynomial $f(x) \in \mathbb{Z}[x]$ of degree $n > 0$ having α as one of its roots.

- **Algebraic Number Field**

  A finite extension of the field of rational numbers $\mathbb{Q}$ is called an Algebraic Number Field.

- **Algebraic (Number) over a Field**

  Let $\mathbb{F} \subset \mathbb{K}$ be field. $\alpha \in \mathbb{K}$ is said to be algebraic over $\mathbb{F}$ is there exits a polynomial $f(x) \in \mathbb{F}[x]$ of degree $n > 0$ having α as one of its roots.

- **Algebraically Closed Field**

  A field $\mathbb{K}$ is said to be algebraically closed if for every polynomial $f(x) = a_0x^n + a_1x^{n-1} + \cdots\cdots a_{n-1}x + a_n \in \mathbb{K}[x]$, $n > 0$ there exist $\alpha_i \in \mathbb{K}$, $1 \leq i \leq n$ such that

$$f(x) = a_0x^n + a_1x^{n-1} + \cdots\cdots a_{n-1}x + a_n = a_0 \prod_{i=1}^{i=n} (x - \alpha_i)$$

- **Axiom of Choice**

  In theory of sets this axiom is an apparently intuitively obvious statement that Cartesian product of a number of nonempty sets is nonempty. In other words, if I is a nonempty index set and for each $i \in I$, S_i is a nonempty set then $\prod_{i \in I} S_i$ is nonempty.

- **Burnside's N/C Theorem**

Let G be a finite group and p be a prime number dividing the order $|G|$ of the group G. Let P be a Sylow p-subgroup of G and $N_G(P) = C_G(P)$. Then the group G has a normal subgroup of order $|G|/|P|$.

- **Centralizer $C_G(H)$ of a Subgroup H of a Group G**

$$C_G(H) = \{x \in G, \text{ such that } xg = gx \text{ for all } g \in H\}$$

- **Characteristic of a Field**

Let $\mathbb{F}$ be a field and e its (multiplicative) identity. The smallest nonnegative integer n such that $ne = 0$ is called the characteristic of the field $\mathbb{F}$. The characteristic of a field is either zero or a prime number.

- **Commutator Subgroup of a Group**

Let G be a group. The commutator subgroup $G^{'}$ of G is the subgroup of G generated by all the elements $x^{-1}y^{-1}xy \in G$. That is

$$G^{'} = \langle x^{-1}y^{-1}xy, x, y \in G\rangle$$

The group G is Abelian if and only its commutator subgroup is identity.

- **Cyclotomic Extension and Cyclotomic Field**

Let $n > 2$ be a natural number and $\alpha = e^{\frac{2\pi i}{n}}$. Then the field $\mathbb{K} = \mathbb{Q}(\alpha)$ is called a cyclotomic field and the extension $\mathbb{K}/\mathbb{Q}$ is called cyclotomic extension.

- **Cyclotomic Polynomial**

Let n be a natural number and $\alpha = e^{\frac{2\pi i}{n}}$ be primitive n-th root of unity. Then the polynomial

$$\Phi_n(x) = \prod_{(n,r)=1} (x - \alpha^r)$$

of degree $\phi(n)$ has integer coefficients. The polynomial $\Phi_n(x)$ is called n-th cyclotomic polynomial.

- **Cyclotomy**

Study of the problems arising from the arithmetical considerations of the divisions of the unit circle into a number of equal parts.

Tignol (2001) says the following about cyclotomy,

> ... the study of values of the cosine and sine functions at $\frac{2k\pi}{n}$ for integers k, n is called **cyclotomy**, meaning literally : division of the circle [into equal parts]

- **Dihedral Group**

The dihedral group D_n is the group of symmetries of a regular n-gon for $n \geq 3$. Equivalently, it is a group of order $2n$ defined by the following generators and relations:

$$D_n = \langle x, y : x^n = e, y^2 = e, y^{-1}xy = x^{-1} \rangle = \{x^i y^j, 0 \leq i \leq n-1, 0 \leq j \leq 1\}$$

- **Dirichlet's Theorem on Primes in Arithmetical Progression**

If a and d are natural numbers such that $gcd(a, d) = 1$, then the set

$$\{a + nd : n = 0, 1, 2, \cdots\cdots\}$$

contains infinitely many primes numbers.

- **Division Algorithm**

It is simply the statement that if we divide a nonzero integer by a positive integer then there is some quotient and some remainder. More precisely if a and b be nonzero integers then there exist unique integers q and r such that $a = bq + r$, such that $0 \leq r < |b|$.

- **Eisenstein's Criterion**

Let $f(x) = a_0x^n + a_1x^{n-1} + \cdots + a_{n-1}x + a_n \in \mathbb{Z}[x]$ be a polynomial of degree $n > 0$ and p be a prime such that

(1) p does not divide a_0,

(2) p divides a_i, for $1 \leq i \leq n$,

(3) p^2 does not divide a_n,

then the polynomial $f(x)$ is irreducible over the field $\mathbb{Q}$.

- **Elementary Abelian Group**

A finite Abelian group of order p^n, where p is a prime number is said to be an elementary Abelian group or an elementary Abelian p-Group, if each of its nonidentity element has order p. An elementary Abelian p-group of order p^n is isomorphic to direct product of n-copies of the group C_p, the cyclic group or order p.

- **Euclidean Algorithm**

  It is name given to the method of computing the greatest common divisor of two integers by the successive use of Division Algorithm. It is based upon the observation that $GCD(a,0) = a$ if $a \neq 0$ and if $a, b \neq 0$ and $a = bq + r$, $0 \leq r \leq |b|$ then $GCD(a,b) = GCD(b,r)$.

- **Euclidean Domain**

  An integral domain D is said to be an *Euclidean Domain* if there exists a mapping

  $$\phi :\longrightarrow \mathbb{N} \cup 0$$

  satisfying

  - Given $a, b \in D$, if a divides b then $\phi(a) \leq \phi(b)$
  - Given $a, b \in D$, $a \neq 0$, there exist $q, r \in D$ such that $b = aq+r$, $0 \leq \phi(r) < \phi(a)$

- **Euler's Phi-function, $\phi(n)$**

  For a positive integer n, $\phi(n)$ denotes the number of positive integers less than or equal to n which are coprime to n.

- **Fermat Prime**

  Primes of the form $2^n + 1$ are called Fermat primes. It is easy to see that if $2^n + 1$ is a prime then n must be a power of 2.

- **Fermat's Little Theorem**

  It is the statement that if p is a prime number then $a^p - a$ is divisible by p for all integers a.

- **Frobenius Automorphism**

  In a field $\mathbb{K}$ of characteristic p the mapping $\alpha \rightarrow \alpha^p$, for all $\alpha \in \mathbb{K}$ is known as Frobenius Automorphism.

- **Frobenius Group**

  A finite permutation group on a set Ω is said to be Frobenius group if it satisfies the following conditions:

  (1) G is transitive on Ω.

  (2) $G_\alpha \neq \{e\}$ for $\alpha \in \Omega$.

  (3) $G_{\alpha\beta} = \{e\}$ for $\alpha, \beta \in \Omega$, $\alpha \neq \beta$.

A finite group G having any of the following two properties has a (transitive) permutation representation satisfying the property of being a Frobenius group:

(1) G has a normal subgroup K such that $C_G(K) \subseteq K$.

(2) G has a subgroup H which is its own normalizer in G and is disjoint from its conjugates.

- Fundamental Theorem of Algebra

Given a nonconstant polynomial $f(x) \in \mathbb{C}[x]$, there exist $\alpha \in \mathbb{C}$ such that $f(\alpha) = 0$. Equivalently if $a_0x^n + a_1x^{n-1} + \cdots + a_{n-1}x + a_n \in \mathbb{C}[x]$, $a_0 \neq 0$, then there exist $\alpha_i \in \mathbb{C}$, $1 \leq i \leq n$ such that

$$a_0x^n + a_1x^{n-1} + \cdots + a_{n-1}x + a_n = a_0(x-\alpha_1)(x-\alpha_2)\cdot\cdots\cdots(x-\alpha_n)$$

- **Galois Extension (Finite)**

In the case of characteristic zero (or finite extension of finite fields) a Galois extension of a field $\mathbb{F}$ is same as normal extension of the field $\mathbb{F}$.

- **Gaussian Sums**

Certain types of finite sums generally involving roots of unity similar to the ones first discussed by Gauss in his discussion of the construction of regular 17-gon. For example, if $p > 2$ is a prime number then sums such as $\sum_{j-1}^{p-1} \left(\frac{j}{p}\right) \alpha^j$, where $\alpha = e^{\frac{2\pi}{p}}$ are termed *Gaussian sums* in number theory.

- **Leibniz's Rule for the n-th Derivative of Product of Two Functions**

Leibniz's Rule is a theorem about finding the n-th derivative of product of two functions. If f and g are differentiable functions of x then

$$\frac{d^n(fg)}{dx^n} = \sum_{r=0}^{n} \binom{n}{r} \frac{d^{n-r}f}{dx^{n-r}}\frac{d^r g}{dx^r}$$

where $\binom{n}{r}$ is the Binomial coefficient.

- **Mobius μ- Function**

Let n be a natural number. The Mobius μ- function is defined as

$$\mu(n) = \begin{cases} 1 & \text{if } n = 1 \\ 0 & \text{if } p^2 \text{ divides } n \text{ for some prime } p \\ (-1)^n & \text{if } n = p_1p_2p_3\cdots p_n \text{ where } p_i, 1 \leq i \leq n \text{ are distinct primes} \end{cases}$$

- **Multiplicative Arithmetical Function**

  A real or complex or valued function whose domain is the set of natural numbers $\mathbb{N}$ is called a Multiplicative Arithmetical function if $f(mn) = f(m)f(n)$.

- **NC-Theorem** See Burnside's N/C Theorem.

- **Normal Extension (Finite)**

  In the case of characteristic zero (or finite extension of finite fields) a normal extension of a field $\mathbb{F}$ is simply a splitting field of some polynomial $f(x) \in \mathbb{F}[x]$.

- **Normal Subgroup**

  A subgroup H of a group G is said to be normal if $x^{-1}Hx = H$ for all $x \in G$.

- **Normalizer of Subgroup of a Group**

  Let H be a subgroup of a group G. Then the subgroup $N_G(H) = \{x : x^{-1}Hx = H\}$ is called the normalizer of H in G.

- **Primitive Group**

  A finite permutation group G on a set Ω is said to be primitive if

  (1) G is transitive on Ω.

  (2) For every $\alpha \in \Omega$, the subgroup $G_\alpha = \{g \in G, g \text{ fixes } \alpha\}$ is a maximal subgroup of G.

  Being a primitive group is not an intrinsic property of a group but that of its representation as a permutation group.

- **Primitive n-th Root of Unity**

  Given a natural number n, $\alpha \in \mathbb{C}$ is said to be a primitive n-th root of unity if $\alpha^n = 1$ but $\alpha^r \neq 1$ for $1 \leq r < n$. For every natural number n there are $\phi(n)$ primitive roots of unity, where $\phi(n)$ is the Euler's Phi-function.

- **Primitive Root Modulo n**

  Let n be a natural number. An integer x is said to be primitive root modulo n if $x^{\phi(n)} \equiv 1 \pmod{n}$ but $x^r \not\equiv 1 \pmod{n}$ if $1 \leq r < \phi(n)$. A natural number n has a primitive root if and only if the group $\mathbb{Z}_n^*$ is cyclic (of order $\phi(n)$. The only natural numbers having primitive rotos are $1, 2, 4, p^k$, and $2p^k$ where p is an odd prime. If the natural number n has a primitive root then any generator of the cyclic group $\mathbb{Z}_n^*$ is called primitive root mod n.

- **Principal Ideal Domain**

  An integral domain whose every ideal is principal—that is all of its element consist of multiple of single element.

- **Quadratic Reciprocity - Law of**

  Let $p \neq q$ be two odd prime numbers. Consider the two congruence

  (I) $x^2 \equiv p \pmod{q}$, (II) $x^2 \equiv q \pmod{p}$.

  The Law of Quadratic Reciprocity states that

  (1) If $p \equiv q \equiv 3 \pmod{4}$ then congruence (I) has a solution if and only if the congruence (II) does not have a solution.

  (2) If p or q (or both) are $\equiv 1 \pmod{4}$ the congruence (I) has a solution if and only if congruence (II) has a solution.

  Using the notation of Legendre Symbol, The Law of Quadratic Reciprocity can be stated as

$$\left(\frac{p}{q}\right) \cdot \left(\frac{q}{p}\right) = (-1)^{\frac{(p-1)(q-1)}{4}}$$

- **Radical Extension**

  A field extension $\mathbb{K}/\mathbb{F}$ which is such that

$$\mathbb{F} = \mathbb{F}_1 \subseteq \mathbb{F}_2 \subseteq \mathbb{F}_3 \subseteq \cdots \subseteq \mathbb{F}_k \subseteq \mathbb{F}_{k+1} = \mathbb{K}$$

  where there are natural numbers $n_i > 1$ and $a_i \in \mathbb{F}_i$, $1 \leq i \leq k$ are such that $\mathbb{F}_{i+1} = \mathbb{F}_i(\sqrt[n_i]{a_i})$

- **Rational Root Theorem**

  If $f(x) = a_0x^n + a_1x^{n-1} + a_2x^{n-2} + \cdots + a_{n-1}x + a_n \in \mathbb{Z}[x]$ is a polynomial of degree $n \geq 1$ then the rational roots of the equation $f(x) = 0$, if there are any, are of the form $\pm\frac{p}{q}$ where p, q are natural numbers, such that p divides a_n and q divides a_0.

- **SemiDirect Product**

  Roughly speaking in group theory when elements of a group K act as automorphism of a group H then we can construct a group G which is such that $G = HK$, G contains H as a normal subgroup and $G/H \approx K$. In such cases we say that G

is a semidirect product of H by K—also known as extension of H by K. Direct product of groups is a special case of semidirect product. For example,

(i) S_3 is a semidirect product of C_3 by C_2.

(ii) A_4 is a semidirect product of the group $C_2 \times C_2$ by C_3.

(iii) Frobenius group of order 21 is semidirect product of C_7 by C_3.

(iv) The Galois group of $x^5 - 2$ over $\mathbb{Q}$ is semidirect product of C_5 by C_4.

- **Simple Extension**

  A field extension $\mathbb{K}/\mathbb{F}$ such that $\mathbb{K} = \mathbb{F}(\alpha)$ for some $\alpha \in \mathbb{K}$.

- **Simple Group**

  A group which does not have any proper normal subgroup.

- **Solvable Group**

  A group G is said to be solvable if there exists a chain of subgroups G_i, $0 \le i \le n$, such that

$$G = G_0 \supseteq G_1 \supseteq G_2 \supseteq \cdots\cdots \supseteq G_n = \{e\}$$

and G_{i+1} is a normal subgroup of G_i and G_i/G_{i+1} is Abelian for $0 \le i \le n-1$. (For finite groups G this amounts to saying that G_i/G_{i+1} is cyclic of prime order.)

- **Sylow p-Subgroup of a Group**

  Let G be a finite group of order $p^k m$, where p is a prime which does not divide m. Then A subgroup P of G of order p^k is called Sylow p-subgroup of G. Sylow p-subgroups exist for every prime divisor p of the order of the group G.

- **Sylow Theorem**

  Let G be a finite group whose order is divisible by a prime p, and $|G| = p^n \cdot m$ where m is not divisible by p. Then.

  (1) G has a subgroup of order p^n called a Sylow p-subgroup of G. The number k of such subgroups in G is $\equiv 1 \pmod{p}$.

  (2) Any two Sylow p-subgroups of a finite group are conjugate.

- **Sylow Theorem—Hall's Generalization to Solvable Groups**

  See Hall's Theorem.

- **Transcendental Number**

  A complex number which is not an algebraic number is called a transcendental number.

- **Transitive Group**

  A finite permutation acting on a set Ω is said to be transitive on Ω if given $\alpha, \beta \in \Omega$, there is an element $g \in G$ which maps α to β.

  Likewise a finite permutation group G acting on set Ω is said to be **k-transitive** on Ω if given two ordered k-tuples $(\alpha_1, \alpha_2, \cdots, \alpha_k)$ and $(\beta_1, \beta_2, \cdots, \beta_k)$ of distinct elements of Ω, there exists an element $g \in G$ which maps α_i to β_i, for $1 \leq i \leq k$. The symmetric group S_n is n-transitive and the alternative group A_n is $(n-2)$-transitive (for $n \geq 3$) on n letters.

- **Zorn's Lemma**

  This is an axiom in logic/set theory which is equivalent to Axiom of Choice.

Credits and Sources Acknowledged

> *As for credits-who discovered what?, who was the first?, whose solution is the best? ... It wouldn't be fair to give credit in some cases and not in others. We are all pretty sure that Pythagoras did not discover the Pythagorean theorem, and it doesn't really matter, does it? The beauty of the mathematics speaks for itself, and so be it.*
> *P. R. Halmos (1991)*

As mentioned in the introduction during the course of our long teaching career we had the opportunity to look at and benefit from the available books on Galois Theory and related mathematics literature. Some of the problems whose solutions are included in this book are generally somewhat modified form of the problems which also appear in one or more books listed in the bibliography. Some have been suggested by colleagues, some are folklore, and some we had created based on our interaction with the students. Below is only a partial list of the sources that we specifically remember and recognize:

- For the chapter on Transcendence on of e and π (Adamson 1982; Stewart 1989).
- Problem # 11 (Gaal 1971).
- Problem # 18 (Bromwich 1991; Williams and Hardy 1996).
- Problem # 19 (Williams and Hardy 1996).
- Problem # 53 (Morandi 1996).
- Problem # 56 (Dummit and Foote 2004).
- Problem # 57 and # 58 (Childs 1979).
- Problem # 76 (Yahaghi 2010).
- Problem # 84–87 (Gaal 2002).
- Problem # 101 (Lang 1993).

- Solutions to problems # 101 & 102 were provided by friends who prefer to remain anonymous.
- For converting our Power Point slides into eps files we thank our student Christopher Hegland.

Bibliography

Adamson, I. *Introduction to Field Theory.* Cambridge: Cambridge University Press, 1982.

Artin, M. *Algebra.* Englewood Cliff, NJ: Prentice-Hall, 1991.

Ayad, M. *Galois Theory and Applications.* Calais, France: World Scientific, 2018.

Bewersdorff, J. *Galois Theory for Beginners.* Providence, RI: American Mathematical Society, 2006.

Birkhoff, G. D., and McLane, S. *A Survey of Modern Algebra.* 3rd ed. Wellesley, MA: A K Peters, 1965.

Brandl, R. "Integer Polynomials That are Reducible Modulo All Primes." *American Mathematical Monthly* 93 (1986): 286–8.

Bromwich, T. J. I'A. *An Introduction to the Theory of Infinite Series.* New York: Chelsea Publishing Company, 1991.

Burnside, W. S., and Panton, A. W. *The Theory of Equations Volume I.* Delhi, India: S. Chand and Company, 1957.

Childs, L. *A Concrete Introduction to Higher Algebra.* New York: Springer-Verlag, 1979.

Coleman, Robert F. "On the Galois Group of the Exceptional Taylor Polynomials." *L'Enseignement Mathematique* t33 (1987): 183–9.

Cox, David A. *Galois Theory.* Hoboken, NJ: John Wiley & Sons, 2012.

Dean, R. "A Rational Polynomial Whose Group is the Quaternion." *American Mathematical Monthly* 88 (1981): 42–45.

Dixon, John D. *Problems in Group Theory.* Waltham, MA: Blaisdell, 1967.

Dummit, D. S. "Solving Solvable Quintics." *Mathematics of Computation* 57, no. 195 (1991): 387–401.

Dummit, D. S., and Foote, R. M. *Abstract Algebra.* Hoboken, NJ: John Wiley and Sons, Inc., 2004.

Edwards, H. M. *Galois Theory.* New York: Springer-Verlag, 1984.

Erbach, D. W., Fisher, J., and McKay, J. "Polynomials with PSL(2,7) as Galois Group." *Journal of Number Theory* 11 (1979): 69–75.

Escofier, Jean-Pierre. *Galois Theory.* New York: Springer-Verlag, 2000.

Fenrick, Maureen H. *Introduction to Galois Correspondence.* Boston, MA: Birkhäuser, 1992.

Fraleigh, J. B. *A First Course in Abstract Algebra.* Reading, MA: Addison Wesley, 2003.

Gaal, I. *Diophantine Equations and Integral Bases.* Boston, MA: Birkhäuser, 2002.

Gaal, L. *Classical Galois Theory.* New York: Chelsea Publishing Company, 1971.

Garling, D. J. H. *A Course in Galois Theory.* Cambridge: Cambridge University Press, 1986.

Guralnick, R., Schacher, Murray M., and Sonn, J. "Irreducible Polynomials Which are Locally Reducible Everywhere." *Proceedings of the American Mathematical Society* 133 (2005): 3171–7.

Hadlock, C. R. *Field Theory and Its Classical Problems.* Washington, DC: Mathematical Association of America, 1978.

Halmos, P. R. *Problems for Mathematicians, Young and Old.* Washington, DC: The Mathematical Association of America, 1991.

Hamburg, Robin R. "The Theory of Equations in the 18th Century. The Work of Joseph Lagrange." *Archive for the History of Exact Sciences* 16 (1976–77): 17–36.

Isaacs, I. M. "Solutions of Polynomials by Real Radicals." *The American Mathematical Monthly* 92 (1985): 571–5.

Jacobson, N. *Basic Algebra I.* 2nd ed. Mineola, NY: Dover Publications, Inc., 2014.

Jones, A. Morris, Sidney A., and Pearson, Kenneth R. *Abstract Algebra and Famous Impossibilities.* New York: Springer-Verlag, 1991.

Kappe, L., and Warren, B. "An Elementary Test for the Galois Group of a Quartic Equation." *The American Mathematical Monthly* 96, no. 2 (1989): 133–7.

Kiernan, B. M. "The Development of Galois Theory from Lagrange to Artin." *Archive for History of Exact Sciences* 8 (1971): 40–154.

Kurosh, A. *Higher Algebra.* Moscow, Russia: MIR Publishers, 1980.

Lagrange, J. *Oeuvres*, 354–5. Paris: Gauthier-Villars, 1869. Notes sur la thérie des équations, Note XIV (*Où l'on donne la résolution générale des équations à deux termes*).

LaMacchia, Samuel E. "Polynomials with Galois Group PSL(2,7)." *Communications in Algebra* 8, no. 10 (1980).

Lang, S. *Algebra.* 3rd ed. Addison Wesley, 1993.

Lidl, R., and Niederreiter, H. *Finite Fields (Encyclopedia of Mathematics and Its applications).* New York, Cambridge University Press, 2008.

Malle, G., and Matzat, B. H. *Inverse Galois Theory.* Berlin: Springer-Verlag, 1999.

MacKay, J. "On Computing Discriminants." *American Mathematical Monthly* 94, no. 6, (1987): 523–7. June–July 1987.

McCarthy, Paul J. *Algebraic Extension of Fields.* New York: Dover Publications Inc., 1991.

McKay, J. "Some Remarks on Computing Galois Groups." *SIAM Journal of Computing* 8, no. 3 (August 1979): 344–7.

Mead, D. "Newton's Identities." *The American Mathematical Monthly* 99, no. 8, (1992): 749–51.

Morandi, P. *Fields and Galois Theory,* New York: Springer-Verlag, 1996.

Niven, I. "The Transcendence of π." *The American Mathematical Monthly* 46 (1939): 469–71.

Osada, H. "The Galois Groups of the Polynomials $X^n + aX^l + b$, J." *Number Theory* 25, no. 2 (1987): 230–8.

Osofsky, B. "Nice Polynomials for Introductory Galois Theory." *Mathematics Magazine* 72 (1999): 218–22.

Passman, Donald S. *Permutation Groups.* Mineola, NY: Dover Publication, 1968.

Rotman, J. *Galois Theory.* New York: Springer-Verlag, 1990.

Safarevic, I. R. *Construction of fields of algebraic numbers with a given soluble Galois group.* Izv. Akad. Nauk SSSR 18 (1954), 525–78 (in Russian), English translation in *American Mathematical Society Translations* 4 (1956) 185–237.

Schur, I. *Gleichungen ohne Affeckt (1930). Gesammelte Abhandlungen,* Band III, No. 67, 191–7.

Snapper, E. "An Ingrained Error Concerning Resultants." *American Mathematical Monthly* 92, no. 2 (February 1985): 14.

Soicher, L. "The Computation of Galois Groups." Masters Thesis, Department of Mathematics, Concordia University, Montreal, Quebec, Canada, April 1981.

Soicher, L., and McKay, J. "Computing Galois Groups over Rationals." *Journal of Number Theory* 20 (1985): 273–81.

Spindler, K. *Abstract Algebra with Applications.* New York: Marcel Dekker Inc., 1994.

Stauduhar, R. "The Determination of Galois Groups." *Mathematics of Computation* 27, no. 124 (October 1973): 981–96.

Stevenhagen, P., and Lenstra, H. W., Jr., "Chebotarev and His Density Theorem." *The Mathematical Intelligencer* 18, no 2 (1996): 26–37.

Stewart, I. *Galois Theory.* New York: Chapman and Hall, 1989.

Swan, Richard G. "Factorization of Polynomials over Finite Fields." *Pacific Journal of Mathematics* 12, no. 3 (1962): 1099–106.

Tignol, Jean-Pierre. *Galois' Theory of Algebraic Equations.* Singapore: World Scientific, 2001.

Vandermonde, A. T. *Mémoire sur la résolution des équation.* Histoire de l'Acad. Royale des Sciences (avec les mémoires de Math. & Phys. pour la même année, tirés des registres de cette Acad.), Paris. 1771, 365–416.

Van Der Linden, F. J. "The Computation of Galois Groups." In *Computational Methods in Number Theory,* edited by H. W. Lenstra Jr., and R. Tijdeman, 199–211. Amsterdam: Mathematisch Centrum, 1982.

Van der Waerden, B. L. *Modern Algebra.* Vol. 2. New York: Ungar, 1953.

Weintraub, Steven H. *Galois Theory.* New York: Springer, 2005.

Wielandt, H. *Finite Permutation Groups.* New York: Academic Press, 1964.

Williams, Kenneth S., and Hardy, K. *The Red Book of Mathematical Problems.* Mineola, New York: Dover Publications Inc., 1996.

Yahaghi, Bamdad R. *Iranian Mathematics Competitions 1973–2007.* New Delhi, India: Hindustan Book Agency, 2010.

Index

www.ingramcontent.com/pod-product-compliance
Lightning Source LLC
LaVergne TN
LVHW081252011026
849162LV00006BA/1508